COCROFT

PHYLOGEOGRAPHY

JOHN C. AVISE

PHYLOGEOGRAPHY

THE HISTORY AND FORMATION OF SPECIES

HARVARD UNIVERSITY PRESS

Cambridge, Massachusetts   London, England   2000

Library of Congress Cataloging-in-Publication Data
Avise, John C.
   Phylogeography : the history and formation of species /
John C. Avise.
      p.   cm.
      Includes bibliographical references (p.   ) and index.
      ISBN 0-674-66638-0 (alk. paper)
      1. Phylogeography.   I. Title.
QH84.A95   2000
578'.09—dc21   99-19648

CONTENTS

PREFACE

Sometimes a novel word or phrase encapsulates a concept and becomes part of the working lexicon of a discipline. Such was the case for "biodiversity," coined in the title of an ecological treatise promoting the notion of biological variety as a cherished resource (Wilson, 1988). Other evocative yet utilitarian terms that helped to define and stimulate emerging fields of biological inquiry include DNA fingerprinting, genetic engineering, sociobiology, sperm competition, and island biogeography.

"Phylogeography" has served similar functions. The word itself was introduced little more than a decade ago (Avise et al., 1987a), more by need than design. In earlier molecular surveys of mitochondrial (mt) DNA lineages in natural populations, cumbersome phrases had been employed to summarize a straightforward observation: branches in intraspecific gene trees often displayed a striking geographic pattern. In other words, a genealogical component typified the spatial distributions of genotypes within and among related organisms. Following coinage of the term, various relationships between gene genealogies and geography could be referred to simply as phylogeographic patterns. Furthermore, christening the phenomenon prompted focused efforts to characterize phylogeographic outcomes, to relate them to population demography and coalescent theory, and to identify the place of phylogeography within the broader framework of biodiversity analysis.

Measured by the burgeoning scientific literature on phylogeography, what budded as a mere utilitarian word has blossomed in recent years into a vigorous adolescent discipline with rich connections to biology, paleontology, and historical geography. Phylogeographic perspectives have revolutionized conceptual as well as empirical interpretations of microevolutionary processes in nature. In this book, I aim to synthesize this emerging field in a manner that makes the material accessible to advanced undergraduates and graduate students, as well as to practicing ecologists, geneticists, ethologists, molecular biologists, population biologists, conservation biologists, and others who would like a simplified and mostly graphical (as opposed to mathematical) explanation of the subject matter.

More specifically, this book recounts the genesis and ontogeny of phylogeography, reviews relevant literature in some detail and summarizes the field's empirical and conceptual findings, and describes the intellectual richness of the discipline in a way that captures the innovative spirit that phylogeographic perspectives have brought to ecological and evolutionary studies. To date, most empirical efforts in phylogeography have concentrated on multicellular animals, so this book inevitably will reflect this bias. However, phylogeographic principles should apply to plants and microbial taxa also, as noted where relevant in the text.

Thanks go to Joan Avise, Andrew DeWoody, Anthony Fiumera, Mike Goodisman, Glenn Johns, Adam Jones, Joe Neigel, Bill Nelson, Guillermo Ortí, Svante Pääbo, Devon Pearse, DeEtte Walker, and Kurt Wollenberg for excellent comments on various drafts of the manuscript. I would also like to thank my editors—Michael Fisher and Kate Brick—for encouragement and assistance. I am grateful to the National Science Foundation and the University of Georgia for longstanding support of my laboratory, and to the Pew Fellowship Program for recent underwriting that enabled me to complete this book.

PHYLOGEOGRAPHY

HISTORY AND CONCEPTUAL BACKGROUND

This section introduces the meaning and purview of phylogeography, describes the history of the discipline on both the empirical and theoretical sides, and highlights novel perspectives that the field has spawned. It concentrates in particular on the tight connections between genealogy and population demography, and the nonequilibrium (i.e., historical) nature of population genetics.

THE HISTORY AND PURVIEW
OF PHYLOGEOGRAPHY

PHYLOGEOGRAPHY is a field of study concerned with the principles and processes governing the geographic distributions of genealogical lineages, especially those within and among closely related species. As the word implies, phylogeography deals with historical, phylogenetic components of the spatial distributions of gene lineages. In other words, time and space are the jointly considered axes of phylogeography onto which (ideally) are mapped particular gene genealogies of interest (Fig. 1.1). The analysis and interpretation of lineage distributions usually requires extensive input from molecular genetics, population genetics, ethology, demography, phylogenetic biology, paleontology, geology, and historical geography. Thus, phylogeography is an integrative endeavor that lies at an important crossroads of diverse microevolutionary and macroevolutionary disciplines (Fig. 1.2).

As a sub-discipline of biogeography, phylogeography serves to place into broader temporal context traditional ecogeographic perspectives that emphasize the role of contemporary ecological pressures in shaping the spatial distributions of organismal traits. A rich literature exists on selection in nature (Endler, 1986; Bell, 1997), including illustrations of how

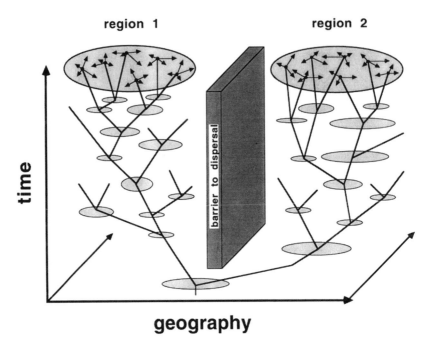

FIGURE 1.1   Hypothetical gene genealogy for a species displaying restricted gene flow within each of two physically separated regional populations (after Avise, 1996a). Shaded ovals represent geographic ranges of particular lineages, and arrow vectors in the extant populations (top) denote spatial magnitudes of contemporary dispersal of individuals from their natal sites.

environmental gradients can generate clinal patterns in organismal adaptations (Endler, 1977). Many of these selection-mediated responses are idiosyncratic to particular species, whereas others reflect trends sufficiently pronounced or recurrent as to constitute *ecogeographic rules* (Table 1.1). Yet, natural selection is not the only evolutionary force capable of generating geographic patterns in genetic attributes. For populations that have been sundered historically and have experienced little or no gene flow for long periods of time, evolutionary divergence proceeds rather inexorably in selectively neutral as well as non-neutral genes.

Phylogeographers seek to interpret the extent and mode by which historical processes in population demographics, including but not confined to those related to natural selection over extended periods of time, may

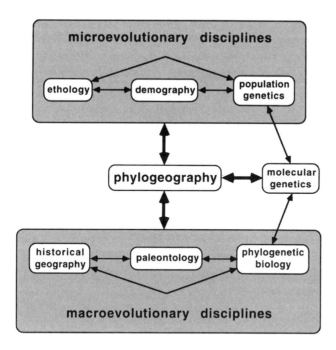

FIGURE 1.2   Phylogeography lies at a critical juncture among several well-established fields.

have left evolutionary footprints on the contemporary geographic distributions of gene-based organismal traits. Thus, phylogeography serves to expand and to balance the perspectives of ecogeography (Fig. 1.3a; Thorpe et al., 1995).

However, ecogeographic and phylogeographic hypotheses are not mutually exclusive (Vermeij, 1978; Brooks, 1985). Suppose, for example, that two geographic populations had been separated for hundreds of thousands of years. Appropriate phylogeographic analyses should uncover this historical partition even if the gene lineages analyzed were strictly fitness-neutral. Historically isolated populations are prime candidates for differences in genetic adaptations also, because they would have had long potential exposure to divergent selection pressures without the homogenizing influences of gene flow. Indeed, in applications such as biodiversity assessment, phylogeographic units identified by supposedly "neutral"

TABLE 1.1   Examples of ecogeographic rules: Reported trends within and among species traditionally interpreted as outcomes of geographical gradients in selection pressures.[a]

Allen's rule. Homeotherms tend to have shorter appendages in regions with cold climates.
   Hypothesized nature of selection: physiological advantages associated with conservation of body temperature.

Bergmann's rule. Homeotherms tend toward larger body sizes (lower surface/volume ratios) in regions with low temperature and humidity.
   Hypothesized nature of selection: physiological advantages associated with conservation of body temperature.[b]

Gloger's rule. Organisms tend toward darker pigmentation in geographic areas with high humidity.
   Hypothesized nature of selection: background-matching that reduces detectability by predators, prey, or competitors.

Clutch-size rule. Birds tend to have larger clutch sizes at higher latitudes.
   Hypothesized nature of selection: more hours of daylight for foraging; higher primary productivity in spring and summer, and associated flush of insects and other food items; proportionately fewer predators than in the tropics.

a. A nonexcluded possibility in most cases is that the traits monitored are flexible in development and respond more or less directly to variable environmental conditions. See Zink and Remsen (1986) for a critique of the evidence and interpretations of ecogeographic rules in birds. Of course, the broader field of ecogeography is concerned with how natural selection molds the geographic distributions of any genetic trait, regardless of whether repeatable trends across taxa emerge.

b. For this and other ecogeographic rules, alternative selection hypotheses also exist. For example, McNab (1971) interpreted Bergmann's rule as an evolutionary response to interspecific competition.

molecular markers are of special significance precisely because they also constitute likely sources of divergent organismal adaptations that conservationists might wish to preserve.

Phylogeography also serves as a useful conceptual umbrella covering alternative historical scenarios to account for the spatial arrangements of organisms and their features. For example, vicariance and dispersal (Fig. 1.4) are two oft-competing possibilities invoked to account for the origins of spatially disjunct taxa (Ronquist, 1997). Under vicariance interpretations, related populations or taxa became separated when the more-or-less

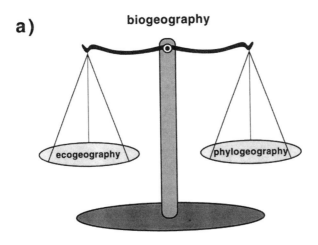

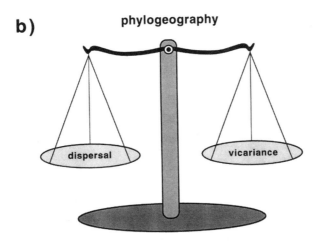

FIGURE 1.3   Balancing services provided by the field of phylogeography. *(a)* Within the broader discipline of biogeography, phylogeographic perspectives serve to balance traditional ecogeographic views that emphasize patterns produced by contemporary natural selection. *(b)* Within phylogeography, appropriate genealogical appraisals can help to weigh the influences of dispersal versus vicariance (as well as other historical evolutionary processes) in shaping the geographic distributions of genetic traits.

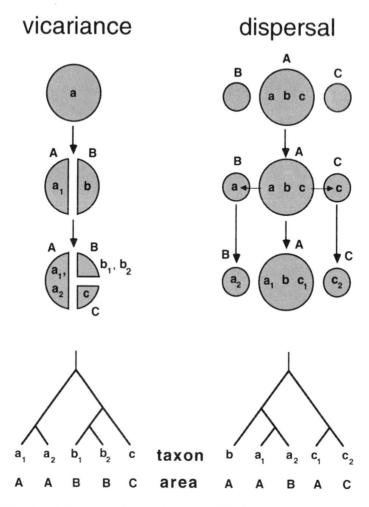

FIGURE 1.4   Phylogenetic relationships of spatially disjunct populations or species under vicariance and dispersal (after Futuyma, 1998). Lowercase letters represent taxa; uppercase letters, geographic areas. Under vicariance, the phylogeny of taxa may mirror the order of separation of areas, whereas areas and taxa can show more varied historical relationships to one another under dispersal. All outcomes are impacted by additional factors such as historical population demography, the evolutionary times of the vicariant or dispersal events, and the resolving power of the molecular or other assays.

continuous ranges of ancestral forms were sundered by environmental events (Croizat et al., 1974; Nelson and Platnick, 1981; Nelson and Rosen, 1981; Humphries and Parenti, 1986; Myers and Giller, 1988): e.g., the rise of a mountain range that separated lowland taxa, the breakup of a continental landmass that severed terrestrial populations, or the physical subdivision of a body of water that split aquatic lifeforms. Under dispersalist interpretations, a taxonomic assemblage came to occupy its present range through active or passive dispersal from one or more ancestral centers of origin (Briggs, 1974). Vicariance and dispersal are historical phenomena whose relative roles in particular instances can be weighed on the scales of phylogeographic analysis (Fig. 1.3b).

In purest form, phylogeographic appraisals deal with the spatial distributions of alleles whose phylogenetic relationships are known or can be estimated. However, empirical or theoretical treatments that consider phylogenetic aspects of the spatial distributions of any genetic trait (morphological, behavioral, molecular, or other) also can qualify as phylogeographic under a broader definition of the term. To retain a manageable framework, this book will address phylogeography primarily in the narrow sense of the spatial analysis of gene lineages. In this arena lie some of the most novel and imaginative of recent scientific findings.

EMPIRICAL ROOTS OF MITOCHONDRIAL DNA RESEARCH

Research publications employing the word phylogeography in the title or as an index term have roughly doubled in number during each successive two-year interval since 1987 (Fig. 1.5), and by early 1999 stood at more than 300 total. These articles represent only a small tip of the iceberg because many more studies have dealt with the topic implicitly though not by name. In what has been a mixed blessing to the field (Bermingham and Moritz, 1998), approximately 70 percent of phylogeographic studies conducted to date involved analyses of animal mitochondrial (mt) DNA either primarily or exclusively (Fig. 1.6). Any account of the history of empirical phylogeography must, therefore, trace the roots of interest in this

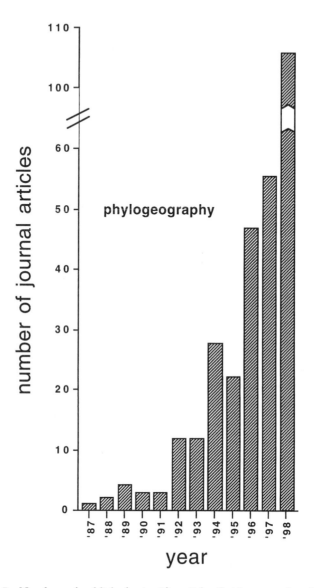

FIGURE 1.5   Numbers of published scientific articles that have employed the words phylogeography or phylogeographic in the title or as key terms. These represent only a small fraction of the papers that have dealt with phylogeographic issues.

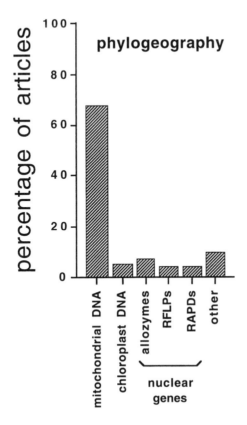

FIGURE 1.6   Breakdown of the phylogeographic articles from Fig. 1.5 according to the molecule or assay procedure employed.

cytoplasmically housed genome and its phylogenetically favorable properties of maternal transmission, extensive intraspecific variation, and
usual absence of intermolecular genetic recombination.

The broader history of molecular phylogenetics has been reviewed
elsewhere (Avise, 1994). By the late 1960s, streamlined protein-electrophoretic approaches as introduced by Harris (1966) and Lewontin and
Hubby (1966) had begun to revolutionize the field of population genetics
by providing the first ready access to an empirical wealth of genotypic information from any organism. By uncovering allele-based (Mendelian)
differences in the net charges of proteins, allozyme assays permitted the

rapid accumulation of genotypic data, typically from 20 or more nuclear loci per individual in large population samples.

Although allozyme data soon were recognized to have systematic value, particularly when viewed collectively in distance-based analyses (Avise, 1974), they had an inherent limitation for some phylogenetic purposes in that the historical relationships of alleles remained unknown. In other words, allozyme alleles at a locus are qualitative multistate traits whose phylogenetic relationships cannot be deduced safely from the observable output of the molecular assays: electromorph location on a gel. Many of the equations of traditional population genetic theory (Fisher, 1930; Wright, 1931; Haldane, 1932) could be represented in a manner that permitted descriptions of the temporal and spatial dynamics of the frequencies of historically unordered alleles under the microevolutionary forces of mutation, gene flow, mating pattern, genetic drift, and natural selection. Protein electrophoretic approaches produced data that could be interpreted in the conventional language and framework of population genetics, but one net effect was to channel thought away from more explicit phylogenetic perspectives that increasingly characterized other areas of biology, notably higher-level systematics.

Thus, the protein electrophoretic revolution did little to narrow the fundamental gap that always existed between the conceptual orientations and operational practices of microevolutionary biology (i.e., population genetics) on the one hand, and macroevolutionary biology (including systematics) on the other (A. C. Wilson et al., 1985; Avise, 1989b). From the perspective of subsequent gene-genealogical studies on mtDNA, the primary contribution of the protein-electrophoretic revolution was to alter the general intellectual climate of population biology in a way that would permit and eventually foster expanded contact with the formerly separate discipline of molecular biology.

Preludes to Phylogeographic Efforts

The immediate technical stage for phylogeographic efforts using mtDNA was set at about this same time through molecular research on several fronts. By the mid- to late 1970s, edited volumes summarized a growing knowledge on the genetic functions and biogenesis of extra-chromosomal DNA (Kroon and Saccone, 1974; Saccone and Kroon, 1976; Gillham, 1978; Cummings et al., 1979). In the late 1960s, restriction endonucleases were discovered (Linn and Arber, 1968; Meselson and Yuan, 1968). Known to cut duplex DNA at specific recognition sites, these enzymes soon were proved feasible for generating RFLPs (restriction fragment-length polymorphisms) and restriction site maps for animal mtDNA (e.g., Brown and Vinograd, 1974; Upholt and Dawid, 1977).

In 1977, Upholt reported the first statistical algorithm for estimating mtDNA sequence divergence from restriction digests, followed soon thereafter by mathematical extensions by Gotoh et al. (1979), Kaplan and Langley (1979), Nei and Li (1979), Engels (1981), Kaplan and Risko (1981), Li (1981), Nei and Tajima (1981), and Hudson (1982). Several early studies used RFLP markers to document at the molecular level what had been suspected from cytological and genetic investigations earlier in the century: that mtDNA is transmitted maternally in higher animals, such that offspring of either gender inherit most if not all mitochondria from the mother (Dawid and Blackler, 1972; Hutchinson et al., 1974; Hayashi et al., 1978; Avise et al., 1979a; Francisco et al., 1979; Giles et al., 1980).

In 1975, Brown and Wright published the first significant analysis of mtDNA variation in nature in a brief paper on parthenogenetic lizards. This paper pioneered a series of studies spanning two decades that documented the power of mtDNA analysis in deciphering the evolutionary origins and ages of numerous unisexual vertebrate taxa (e.g., Brown and Wright, 1979; Wright et al., 1983; Densmore et al., 1989; Echelle et al., 1989; Moritz, 1991; Moritz and Heideman, 1993; review in Avise et al., 1992a). In 1979, Wes Brown and colleagues published another influential article announcing an unexpected fast pace of mtDNA sequence evolution as gauged by interspecies comparisons of higher primates (Fig. 1.7). This led to a widely used "clock" calibration for animal mtDNA: about 2 percent sequence divergence between pairs of lineages per million years (or, 1 percent sequence evolution per lineage per 10^6 years).

The finding of a high evolutionary rate for animal mtDNA came as a complete surprise. At face value, it appeared to violate a fundamental

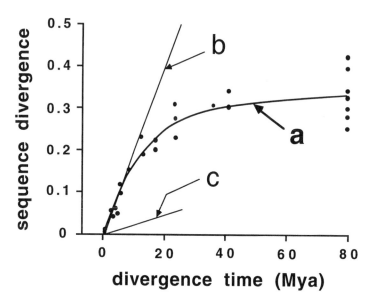

FIGURE 1.7   The evolutionary dynamics of mtDNA nucleotide sequence divergence as documented originally by Brown et al. (1979). Curve *a* plots mtDNA sequence divergence as a function of time for various primates and other mammalian species whose separation dates were estimated from independent fossil or biogeographic evidence. Note that beyond about 10–20 million years ago (Mya), the curve gradually approaches a plateau as substitutable positions in the molecule presumably approach saturation. Most empirical examples discussed in this book involve lineages that separated well within the past 15 My, and hence are not in a zone where severe nucleotide saturation effects would be anticipated. Line *b* describes the initial slope of the mtDNA curve and yields an estimated rate of sequence evolution of about 1 percent per lineage per My. Line *c* depicts the lower rate of sequence evolution characteristic of much of the nuclear genome.

principle of molecular evolution: that constraint on function implies constraint on macromolecular structure. Prior to 1979, most molecular evolutionists had assumed that mtDNA's small size and tight packaging of coding sequences (the genetic economy of the molecule—Attardi, 1985) would have made mtDNA one of the most *slowly* evolving of all the cell's nucleotide sequences.

Another surprise came with a later discovery that plant mtDNA, which also dwells in the cell cytoplasm and encodes similar suites of genes, does not abide by the evolutionary rules that apply to animal mtDNA. Instead, plant mtDNA evolves rapidly with respect to gene order yet slowly in nucleotide sequence (Palmer 1985, 1990; Palmer and Herbon, 1988). These features, plus other technical difficulties of assay, have conspired to limit the applications of plant mtDNA for intraspecific phylogeography (but see Palmer, 1992). Another cytoplasmic genome in plants, chloroplast (cp) DNA (Wolfe et al., 1987), also displays a leisurely pace of sequence evolution. Nonetheless, cpDNA variation is present and known to be structured geographically in several plant species (Hosaka and Hanneman, 1988; Doyle et al., 1990; Lavin et al., 1992; Soltis et al., 1992a; Dong and Wagner, 1994; McCauley, 1994; Demesure et al., 1996; El Mousadik and Petit, 1996; Dumolin-Lapegue et al., 1997; Fujii, 1997; Vandijk and Bakzschotman, 1997; King and Ferris, 1998).

Several hypotheses have been advanced to account for the rapid evolution of animal mtDNA (A. C. Wilson et al., 1985; Gillespie, 1986; Richter, 1992; Li, 1997; Nedbal and Flynn, 1998): (a) relaxation of functional constraint (because mtDNA does not code for proteins involved directly in its own replication or transcription, and because a molecule that produces only 13 kinds of polypeptides might tolerate less accuracy in translation); (b) a high mutation rate (due to an inefficiency of DNA repair mechanisms, high exposure to mutagenic free radicals in the oxidative mitochondrial environment, or fast replicative turnover within cell lineages); and (c) the fact that mtDNA is naked (not complexed with histone proteins that are evolutionarily conserved and might constrain rates of nuclear DNA evolution). Such possibilities are not mutually exclusive. Regardless of the cause, rapid sequence evolution is a prerequisite for mtDNA's utility as a microevolutionary phylogenetic marker.

Pioneering Phylogeographic Studies

The first publications to capitalize upon mtDNA polymorphisms in nature from what now would be considered an explicit phylogeographic perspective involved population surveys of *Peromyscus* mice (Avise et al., 1979a; see also Lansman et al., 1982, 1983a; Avise et al., 1983) and *Geomys* pocket gophers (Avise et al., 1979b). Close inspection of the latter study can serve here to introduce phylogeographic approaches as still practiced today (in expanded and refined forms).

The molecular survey of the southeastern pocket gopher, *Geomys pinetis*, was conducted at a time when many restriction enzymes were unavailable commercially but instead were purified and swapped routinely by the laboratories of individual investigators. Six restriction enzymes were employed to cleave mtDNA samples purified from each of 87 pocket gophers collected throughout the species' range. Digestion products were separated by molecular weight on agarose gels and stained with ethidium bromide to reveal RFLP patterns, one per specimen/enzyme combination. In general, a useful check on proper scoring of mtDNA digestion patterns is permitted by the closed-circular nature of the molecule. The number of fragments produced in a given digest is equal to the number of restriction sites in the mitochondrial genome, and, thus, observed fragment lengths should sum to the total length of mtDNA (typically about 16.5 kilobases).

Observed differences between some of the RFLP patterns in *Geomys* could be accounted for by the gain or loss of one restriction site (Fig. 1.8). For example, a *Bst*EII pattern "N" included two mtDNA fragments, ca. 10.0 and 1.1 kb long, that were missing in *Bst*EII pattern "M," which instead included a fragment 11.1 kb in length. Thus, if "M" was ancestral, the acquisition of a *Bst*EII restriction site (probably by a single nucleotide substitution at an appropriate position within the 11.1 kb sequence) would have produced the derived "N" pattern. Conversely, if "N" was ancestral, a nucleotide substitution at any base pair in this *Bst*EII recognition site would have resulted in the evolutionary loss of the restriction site and the appearance of a derived "M" digestion pattern. By similar lines of reasoning, other RFLP patterns in *Geomys* (e.g., "M" and "O" in Fig. 1.8) were deduced to differ by two or sometimes more changes in restriction sites. Data were accumulated across enzymes, coded as haplotypes, used to quantify sequence divergence among mtDNA clones, and (in the groundbreaking aspect of the study) employed to estimate an intraspecific matriarchal phylogeny for the species (Fig. 1.9a).

Despite the use of only six restriction enzymes, 23 different mtDNA haplotypes were uncovered in the *Geomys* survey. Phylogeographic facets

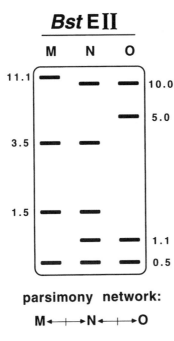

FIGURE 1.8   Diagrammatic representation of mtDNA fragments (and their approximate sizes in kilobase pairs) in the three *Bst*EII digestion profiles observed among 87 specimens of the southeastern pocket gopher (Avise et al., 1979b). Also shown is a parsimony network interconnecting these genotypes, where each slash across a network branch indicates a single deduced restriction-site change.

of the analysis crystallized when the matriarchal phylogeny, estimated strictly from the genetic data, was overlaid onto the geographic sources of the collections (Fig. 1.9b). This representation revealed striking localization of all mtDNA haplotypes and an evident historical component to the matrilineal relationships, including a relatively deep genetic separation (mean sequence divergence $p = 0.034$) between all specimens from eastern versus western portions of the species' range. This east-west subdivision in mtDNA genealogy also was concordant with major shifts in population allele frequencies at two nuclear allozyme loci (Avise et al., 1979b; Laerm et al., 1982).

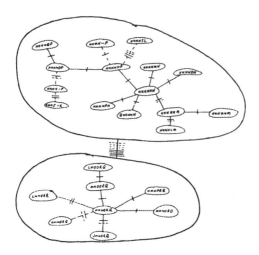

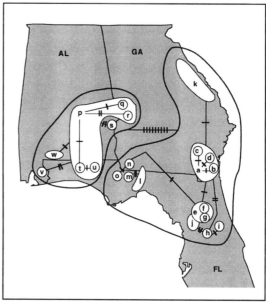

FIGURE 1.9 *Above:* Original hand-drawn phylogenetic network connecting 23 different mtDNA haplotypes in pocket gophers, as scribbled onto a piece of scratch paper from the RFLP data that eventually was to appear in Avise et al. (1979b). Uppercase letters provide coded summaries of the digestion profiles produced by six restriction enzymes. Slashes across network branches reflect the minimum numbers of inferred mutational steps along a pathway. Heavier lines encompass two distinctive mtDNA clades that differed by at least nine mutations. *Below:* Same network, now superimposed over the geographic sources of the samples in Alabama, Georgia, and Florida. Letters a–w indicate the 23 mtDNA haplotypes, and circles or extended ellipses encompass the geographic distributions observed for each haplotype.

These early population surveys of mtDNA were revolutionary in at least two regards beyond assay methodology *per se*. First, they introduced an unorthodox notion that individual organisms could be treated as OTUs (operational taxonomic units) in a population genetic analysis. This suggestion, though radical at the time, followed logically from the fact that each animal displayed a specifiable mtDNA haplotype inherited intact, without intermolecular genetic recombination, through maternal ancestors.

In most prior intraspecific analyses based on allozymes (except those involving immediate parentage), the Mendelian complications of diploidy, genetic segregation, and independent assortment typically necessitated that populations rather than individuals be treated as OTUs. This meant that populations had to be defined in some way (usually by geographic criteria) prior to the analysis of nuclear gene data. Yet once a population is so defined, it will appear to have some relationship to other such populations in any study based on allele frequencies, and thus will seem justified. Following introduction of mtDNA methods, this unwanted element of circularity could be removed (with respect to matrilines) from the interpretation of population patterns. No longer did population frequencies of nuclear alleles or genotypes constitute the sole basic data of population genetics. The matrilineal histories of individual organisms now could be estimated without complications stemming from *a priori* definitions of population units. Furthermore, *for the particular individuals considered,* matrilineal interpretations could be made without concern about sampling errors at the population level otherwise attendant in analyzing allelic or genotypic frequencies from finite collections.

A second revolutionary aspect of the early mtDNA studies was the introduction of explicit phylogenetic concepts to intraspecific evolution. Before these studies (and continuing today in many systematic circles; e.g., Goldstein, 1997), an entrenched notion was that phylogeny had no meaning at the intraspecific level because, for sexually reproducing organisms, conspecific lineages are anastomotic rather than hierarchically branched. Hennig (1966) characterized biological speciation as the demarcation between the realms of tokogenetic associations (genetic relationships among individuals, where phylogenetic concepts supposedly did not apply) and phylogenetic associations among species. Later authors have characterized speciation as a line of peril below which phylogeneticists must not tread. However, due to the asexual mode of mtDNA transmission, mtDNA gene trees *are* non-anastomose and hierarchically branched even within sexually reproducing species. Thus, the extended matrilineal component of organismal history (at the least) *can* be assessed using the algorithms and perspectives of phylogenetic biology.

Thus, the challenge before the field is not to deny a phylogenetic relevance for mtDNA (or nuclear) gene genealogies at the intraspecific level, but rather to interpret properly the status of gene trees as meaningful components of extended organismal pedigrees. The latter are ineluctable realities that define the historical transmission pathways available to alleles, and hence provide a logical and secure starting point for attempts to introduce genealogical and phylogenetic perspectives to intraspecific evolution.

As noted by the paleontologist George Gaylord Simpson more than a half-century ago (1945), "The stream of heredity makes phylogeny; in a sense it is phylogeny. Complete genetic analysis would provide the most priceless data for the mapping of this stream." These hereditary streams originate as rivulets in generation-to-generation transmission pathways. As evidenced throughout this book, phylogeography from its outset has been characterized by attempts to build rather than to burn empirical and conceptual bridges between the traditionally disengaged disciplines of phylogenetic biology and population genetics (Avise et al., 1987a; Hey, 1994).

Subsequent Mitochondrial Findings

In the early and mid-1980s, knowledge expanded rapidly on the nature of molecular evolution of the mitochondrial genome. Extensive mtDNA polymorphism was uncovered within numerous invertebrate and vertebrate animals ranging from horseshoe crabs (Saunders et al., 1986) and fruit flies (Shaw and Langley, 1979; Fauron and Wolstenholme, 1980b) to great apes and humans (Brown and Goodman, 1979; Brown, 1980; Ferris et al., 1981a,b; Brown et al., 1982; Cann et al., 1982, 1984; Greenberg et al.,

1983). Most of the mtDNA variation proved to involve nucleotide substitutions, but changes in genomic size ranging from a few base pairs to a kilobase or more were uncovered also (Crews et al., 1979; Fauron and Wolstenholme, 1980a; Brown, 1981; Aquadro and Greenberg, 1983).

Complete sequences of the mtDNA genome soon were published for the mouse (Bibb et al., 1981), human (Anderson et al., 1981), and cow (Anderson et al., 1982), and in recent years they have been determined for many additional species (see Staton et al., 1997). Typically, each mtDNA genome consists of 37 functionally distinct genes (Fig. 1.10) without large intergenic spacers (but see McKnight and Shaffer, 1997). These loci encode

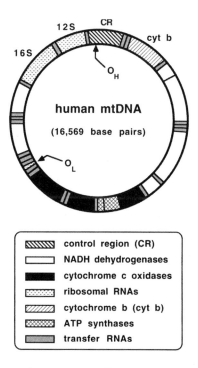

FIGURE 1.10   Structure and organization of genes in mammalian mtDNA. O_H and O_L are origins of replication of the two strands of the molecule. MtDNA molecules normally display identical gene orders within and among related animal species, but occasional rearrangements have occurred in evolution and can be exploited as markers to identify deep branches in macroevolutionary trees (Smith et al., 1993; Boore et al., 1995).

22 different transfer RNAs, two ribosomal RNAs, and 13 messenger RNAs specifying polypeptide subunits of proteins involved in electron transport and oxidative phosphorylation that take place on the mitochondrion's inner membrane. From a phylogenetic view, however, the entire mtDNA genome constitutes a single locus, because character states are linked genealogically by virtue of the molecule's asexual mode of transmission (Saville et al., 1998 describe an exception involving fungi).

INTRACELLULAR POPULATIONS OF MOLECULES

In the late 1970s and early 1980s, it became widely appreciated that entire populations of mtDNA molecules cohabit somatic and germ-cell lineages (Fig. 1.11). A typical somatic cell contains hundreds to thousands of mitochondria, each with several mtDNA copies. A mature oocyte is particularly rich in mtDNA, transmitting perhaps 10^5 molecules to an offspring. These situations differ dramatically from those for autosomal loci in diploid organisms, where each somatic cell carries two copies of a gene and each gamete contributes only one allele to a zygote. Thus, attention soon was directed toward a new level in the genetics hierarchy—populations of mtDNA molecules within cell lineages—that had no precedent in earlier studies of nuclear Mendelian genes (Birky and Skavaril, 1976; Bogenhagen and Clayton, 1977; Birky, 1978, 1983; Ohta, 1980; Thrailkill et al., 1980; Takahata and Maruyama, 1981; Birky et al., 1982, 1983; Chapman et al., 1982). One interesting way to think about this added layer of complexity is to appreciate that for nuclear loci, Mendelian rules of inheritance are made possible precisely because a bottleneck of one molecule per germ cell is entailed in the transmission of a gene across each animal generation. Quite apart from the immediate genealogical implications, these key differences between nuclear and cytoplasmic genomes have ramifications for evolutionary thought about the phenomena of aging, sexual reproduction, and DNA repair (Avise, 1993).

Heteroplasmy.   On the empirical side of the neophyte field of intracellular population genetics, sporadic cases soon were documented in which two

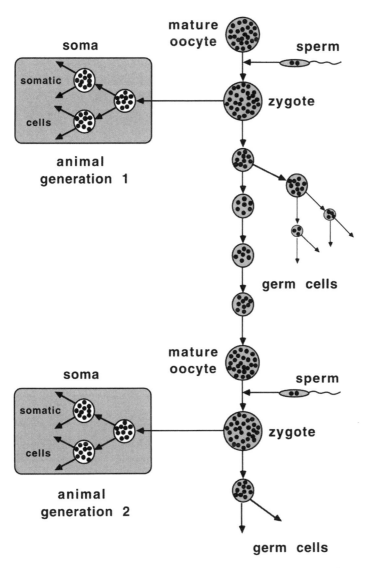

FIGURE 1.11   Mitochondria (black dots) exist as large populations of organelles within the cytoplasmic lineages of somatic cells and germ cells. In most animals, about 20–50 germ-cell generations intervene between organismal generations.

or more distinguishable mtDNA haplotypes occurred jointly within an in-dividual (Olivo et al., 1983; Solignac et al., 1983; Hauswirth et al., 1984; Monnerot et al., 1984; Densmore et al., 1985; Harrison et al., 1985; Berming-ham et al., 1986). Rare instances of extreme sequence heteroplasmy also were discovered (Petri et al., 1996). Although most of the variants produc-ing heteroplasmy probably are fitness neutral to the organism, research on humans revealed that some such mtDNA mutations can have serious health consequences when they reach high frequencies in the somatic cells of an individual (Wallace, 1986, 1992; Lightowlers et al., 1997).

Usually, the haplotypes within each heteroplasmic individual proved to be closely similar genetically, suggesting that the variants probably arose as mtDNA mutations within an isolated female lineage (or within the somatic cells of an individual) rather than from an exogenous source such as introduction via a sperm cell during zygote formation. (Notable exceptions to constrained heteroplasmy later were discovered in some bi-valve mollusks where highly divergent mtDNA lineages with distinct pa-ternal and maternal transmission routes cohabit individual animals [Hoeh et al., 1991; Zouros et al. 1992, 1994a,b; Skibinski et al., 1994; Liu et al., 1996]).

In several early studies, heteroplasmic animals were put to service in breeding programs designed to monitor or project the rate of genotypic sorting from the heteroplasmic condition to homoplasmic states through multiple generations of an organismal pedigree (Hauswirth and Laipis, 1982; Solignac et al., 1984; Rand and Harrison, 1986). Typically, mtDNA genotypic sorting proved to be rapid, usually taking place within a few to several hundred animal generations. These observations suggested that effective population sizes of mtDNA molecules in germ-cell lineages must be vastly smaller than otherwise might be supposed from the huge census sizes of mtDNAs in mature egg cells (Dawid and Blackler, 1972; Jenuth et al., 1996).

Why are the effective population sizes of mtDNA within germlines unexpectedly small, and might various forms of natural selection such as intermolecular competition be involved in the relatively rapid genotypic sorting within cell lineages? These topics are complex in theory (Birky et al., 1989) and remain poorly understood at the cellular level (Howell et al., 1992; de Stordeur, 1997; Lightowlers et al., 1997; Marchington et al., 1997; Turnbull and Lightowlers, 1998; Zouros and Rand, 1999). Furthermore, a homoplasmic condition once achieved is not absolute because low frequency variants continually arise within an individual and normally escape detection unless exceptionally stringent assay conditions are employed (Comas et al., 1995). Nonetheless, for practical purposes of genotypic assignment, there is usually a close approximation to mtDNA homoplasmy for somatic cells within most individuals. This is another fortuitous prerequisite for the utility of mtDNA as a phylogenetic marker.

Paternal Leakage.   Experimental approaches using unidirectional backcross strains soon appeared as the first critical tests of possible low-level "paternal leakage" of male-derived mtDNA into otherwise isolated female cytoplasmic lineages (Lansman et al., 1983b; Gyllensten et al., 1985; Avise and Vrijenhoek, 1987). These studies involved successive backcrosses of hybrid females to males of a strain or species carrying distinct mtDNA markers, and subsequent examination of later-generation backcross progeny for male-derived mtDNA. Early reports failed to detect spermmediated paternal leakage of mtDNA, but later studies based on more sensitive PCR assays finally did identify such incidences in several species (Satta et al., 1988; Kondo et al., 1990; Gyllensten et al., 1991a; Magoulas and Zouros, 1993). However, paternal leakage when it occurs normally appears to be a low-level and transient phenomenon (Meusel and Moritz, 1993; Anderson et al., 1995; Shitara et al., 1998), perhaps in part because an active mechanism exists in some species that degrades paternal mtDNA (Kaneda et al., 1995). In any event, as is true for other sources of heteroplasmy (*de novo* mutations, and rare instances of recombination between genetically different molecules—Lunt and Hyman, 1997), paternal leakage has not compromised mtDNA's utility as a historical chronicle of matrilines for most animal species.

Mitochondrial DNA Clocks

Many early attempts were made to calibrate mtDNA clocks in various taxa and to refine estimates for particular mtDNA gene regions. Several RFLP studies of the whole mtDNA genome suggested broad generality for the rapid sequence evolution originally discovered by Brown et al. (1979) for primate mtDNA. Similar evolutionary rates were reported for horses (George and Ryder, 1986), bears (Shields and Kocher, 1991), wolves (Lehman et al., 1991), and geese (Shields and Wilson, 1987). However, different rates of mtDNA evolution also were noted, including a several-fold faster pace in rodents (Auffray et al., 1990) and several-fold slower paces in some turtles (Lamb et al., 1989; Avise et al., 1992c; Bowen et al., 1992, 1993a; but see also Seddon et al., 1998; Walker and Avise, 1998), amphibians (Wallis and Arntzen, 1989; Caccone et al., 1997), sharks (Martin et al., 1992b), and some other fishes (Canatore et al., 1994). Occasionally, pronounced rate heterogeneity is reported even among closely related lineages, suggesting a need for considerable caution in estimating separation times from observed sequence differences (Zhang and Ryder, 1995).

Martin and Palumbi (1993) suggest that differences in metabolic rate between poikilothermic and homeothermic vertebrates generate mutational and selective pressures that contribute to a perceived trend toward slower rates of mtDNA substitution in many ectothermic taxa (see also Adachi et al., 1993; Rand, 1993, 1994; Martin, 1995; Mindell and Thacker, 1996). Several other biological factors including generation time (Wu and Li, 1985), DNA repair efficiency (Britten, 1986), and DNA replication interval (nucleotide generation time) also have been proposed as important influences on rates of nucleotide sequence evolution. A general complication in evaluating such hypotheses is that unrelated taxa often differ in more than one of these characteristics, and that several of the factors may be correlated.

Another complication is that mtDNA rate calibrations often are highly provisional because they rest upon uncertain biogeographic or fossil evidence concerning separation times of the relevant taxa. At least as much uncertainty attends mtDNA rate estimates across invertebrates. For example, conventional wisdom has been that mtDNA in insects may not evolve much faster than single-copy nuclear (scn) DNA, yet a recent study reported severalfold higher rates of synonymous substitutions in mitochondrial than in nuclear genes of *Drosophila* (Moriyama and Powell, 1997).

Far greater certainty applies to the conclusion that different sites and genes in the mtDNA genome evolve at widely varying rates within a lineage. For recently separated taxa, silent substitutions (concentrated in first-codon and especially third-codon positions of protein-coding genes) typically are severalfold more abundant than replacement substitutions (Irwin et al., 1991); and, rates of nonsynonymous substitution vary considerably among the mitochondrion's 13 protein-coding genes (Li, 1997).

Of special utility for phylogeographic analyses over extreme microevolutionary time scales (e.g., thousands or tens of thousands of years—Ward et al., 1993) is the mtDNA control region (CR; Fig. 1.12), which often

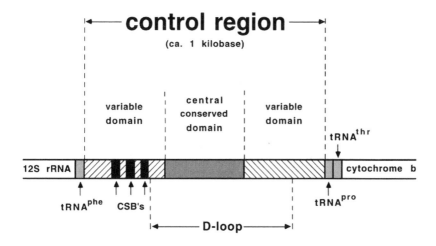

FIGURE 1.12   Schematic diagram of three major domains of the mammalian mtDNA control region (after Taberlet, 1996). Orientation from left to right is as in Fig. 1.10. Shown are two hypervariable sections of the control region sandwiching a more conserved central domain, plus three (sometimes two) typically conserved sequence blocks (CSBs) within one otherwise variable area. The "D-loop" is a well-known portion of the control region visible under the microscope as a displacement strand associated with mtDNA replication.

shows an exceptionally fast pace of nucleotide substitution and high level of intraspecific polymorphism (e.g., McMillan and Palumbi, 1997; Lunt et al., 1998). Suggested clock calibrations for the CR as a whole can be three to five times higher than for the remainder of the mtDNA genome (Aquadro and Greenberg, 1983; Cann et al., 1984; Vigilant et al., 1989). Some hypervariable CR sequences in vertebrates reportedly evolve faster still (i.e., 10 percent sequence divergence per lineage per My; Quinn, 1992; Stewart and Baker, 1994; Baker and Marshall, 1997). Even within the CR, however, some vertebrate species display little or no sequence variation (A. J. Baker et al., 1994; Walker et al., 1998b), and the paradigm of an exceptionally fast rate for CR evolution may be violated altogether by some insects (Zhang et al., 1995; Zhang and Hewitt, 1997). Calibrating mtDNA clocks remains an uncertain and contentious enterprise (Gibbons, 1998).

Influential Reviews on MtDNA Evolution

Early and midstream review articles that helped to direct the burgeoning interest in mtDNA polymorphism and evolution were provided by Lansman et al. (1981), Brown (1981, 1983), Avise and Lansman (1983), A. C. Wilson et al. (1985), Avise (1986), Avise et al. (1987a), Moritz et al. (1987), and Harrison (1989). Table 1.2 summarizes the major empirical findings on animal mtDNA and two of the unorthodox perspectives (Avise, 1991) they entail for phylogenetic appraisals at microevolutionary scales.

CONCEPTUAL ROOTS OF COALESCENT THEORY

In the late 1970s, excitement generated by the emerging mtDNA discoveries ran high as many intriguing questions vied for attention. Why do individual organisms typically display only a few mtDNA gel bands in a restriction digest, rather than a smear of fragments from the hundreds of billions of mtDNA molecules that must be included in each assay? As described above, it must be because most individual specimens carry one predominant sequence. Why then do different individuals within a local population typically display distinct RFLP patterns (such that the great ma-

TABLE 1.2   Molecular and transmission genetics of animal mtDNA: unanticipated
             discoveries and unorthodox conceptual orientations for
             microevolutionary analysis.

Observations:

1) Animal mtDNA displays extensive intraspecific polymorphism and often evolves
   faster than typical single-copy nuclear DNA.

2) Most mtDNA variants involve nucleotide substitutions or small length changes;
   gene order is highly stable over short evolutionary time.

3) Populations of mtDNA molecules inhabit somatic-cell and germ-cell lineages.

4) Most individuals typically are nearly homoplasmic for a single prevalent mtDNA
   sequence; genetic sorting from heteroplasmy is relatively rapid.

5) MtDNA inheritance is asexual, maternal (almost exclusively), and normally
   apparently without intermolecular genetic recombination.

6) Plant mtDNA violates nearly all the rules for animal mtDNA (see text).

Immediate phylogeographic ramifications:

1) Individual animals can be viewed as OTUs in phylogenetic appraisals.

2) MtDNA genotypes record matrilineal relationships within and among species.

jority of population variation in mtDNA was distributed among rather
than within individuals)? With hindsight, this must be because mtDNA
mutations are common and sometimes reach fixation rapidly (within a
small number of organismal generations) within the population of mtDNA
molecules in a germ cell lineage from which the assayed soma were derived. Why do mtDNA genotypes in organismal populations appear con-
nectable to one another in phylogenetically intelligible ways? Because (as
noted above), in the absence of biparental inheritance and the genetic
scrambling effects of intermolecular recombination, the matrilineal histories of sequential mutations are recorded in relatively ungarbled fashion.
As phrased by Dawkins (1995), "Mitochondrial DNA is blessedly celibate."

Another puzzling question was as follows: Why do members of sexually reproducing populations and species usually group together in

mtDNA genotypes when the evolutionary agents of mating and genetic recombination that supposedly hold a species' gene pool together seem not to apply to these uniparentally transmitted genomes? The answer now appears obvious (but was seen only dimly by some of us working on mtDNA at the time): Under normal population demographic conditions, coalescent processes ensure phylogenetic connections among genotypes within a species via vertical pathways of descent, even in the absence of interlineage genetic exchange mediated by mating events and gene flow.

For many molecular researchers originally trained in the allozyme era, the transition to genealogical thought at the intraspecific level initially was tortuous. A useful device now used widely for introducing mtDNA genealogical concepts involves drawing parallels with surname evolution in many human societies (Avise, 1989c). Just as sons and daughters "inherit" their father's nontransfigured surname (before recent rule changes in some societies), so too do progeny receive nonrecombined mtDNA from their mothers. Furthermore, just as "mutations" arise occasionally in surnames (my own name was a nineteenth-century misspelling of Avis), point mutations arise and cumulatively differentiate related mtDNA genotypes. Thus, looking backward in time, mtDNA haplotypes (and those of any nuclear gene) eventually coalesce to common ancestors. MtDNA molecules record the histories of matrilines much as surnames record the histories of patrilines. A major difference is that nature's matrilineal records extend back much further in time—surnames were invented only within the last several centuries.

These insights about coalescent processes were hardly new to the field. Indeed, much of classical population genetic theory originally had been framed in terms of transition equations describing how mutation, migration, natural selection, and genetic drift affect the dynamics of allelic lineages that trace to identical-by-descent conditions (Crow and Kimura, 1970). As mentioned before, this theory could be (and was) modified to accommodate the identical-by-state alleles provisionally identified as protein electromorphs during the allozyme era, but one net effect had been to channel attention away from more explicit genealogical aspects of intra-specific evolution.

Even more germane to the lineage issues raised by mtDNA were theoretical studies first conducted early in the century by statistical demographers interested in the dynamics of surname turnover in human populations (Lotka, 1931a,b). These researchers developed models of lineage branching processes that now could be applied with little modification to gene lineages such as those provided by mtDNA (Schaffer, 1970). Such models stimulated modern efforts to examine the theoretical ties between population demography and phylogeographic patterns within populations (Chapman et al., 1982; Kingman, 1982a,b; Avise et al., 1984a, 1988; Watterson, 1984; Donnelly and Tavaré, 1986, 1995) and among populations and species (Hudson, 1983; Tajima, 1983; Neigel and Avise, 1986). They also stimulated attempts to address these expectations in a series of empirical mtDNA studies on a wide variety of organisms in nature. An important legacy of these efforts has been to engender within the field a far deeper appreciation of the fundamental distinction between gene trees and organismal trees (Avise, 1989b; Baum and Shaw, 1995; Maddison, 1995, 1996; Doyle, 1997) (Fig. 1.13).

Coalescent theory is the name now applied to formal mathematical and statistical treatments of gene genealogies within and among related species (Felsenstein, 1971; Griffiths, 1980; Tavaré, 1984; Hudson, 1990, 1998). As discussed by Harding (1997), modern coalescent theory is merely the latest incarnation of long-standing concepts in population genetics dealing with branching processes and aspects of lineage sorting that had found earlier expression in the study of family surnames and in a host of other topics such as the lineage fate of rare alleles (Fisher, 1930). Although coalescent theory and phylogenetic systematics both are concerned with hierarchical branching structures in evolutionary trees, they developed as independent scientific disciplines with separate historical roots in micro- and macroevolutionary analysis, respectively. However, as this book will attest, these two major branches of evolutionary biology offer kindred philosophical perspectives that are central to nearly all molecular phylogeographic interpretations.

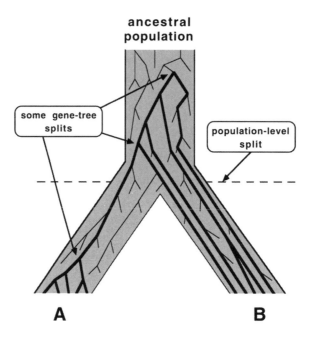

FIGURE 1.13 Fundamental distinction between a gene tree and a population tree or species tree. Note that branching events in a gene tree leading to extant individuals can either postdate *(A)* or predate *(B)* a population-level split.

OTHER RECENT DEVELOPMENTS

Through the late 1980s and the subsequent decade, no further empirical or conceptual revelations of impact comparable to those listed in Table 1.2 have appeared concerning modes of transmission genetics and evolution of animal mtDNA. Rather, the period was characterized by refinements of thought, accretion of information on molecular aspects of mtDNA variation in many species, and burgeoning numbers of mtDNA applications to natural animal populations in a phylogeographic context. Some of the major historical landmarks in the field of phylogeography itself are summarized in Table 1.3. Progress in several related fields, not the least of which are molecular and computer technologies, also contributed greatly to the scientific climate that permitted the flowering of phylogeographic studies.

TABLE 1.3   Brief chronology of some of the important developments in the history of phylogeography.[a]

1974	Brown & Vinograd demonstrate how to generate restriction site maps for animal mtDNAs.
1975	Watterson describes basic properties of gene genealogies, marking the beginnings of modern coalescent theory.
	Brown & Wright introduce mtDNA analysis to the study of the origins and evolution of parthenogenetic taxa.
1977	Upholt develops the first statistical method to estimate mtDNA sequence divergence from restriction digest data.
1979	Brown, George, & Wilson document rapid mtDNA evolution.
	Avise, Lansman, & colleagues present the first substantive reports of mtDNA phylogeographic variation in nature.
1980	Brown provides an initial report on human mtDNA variation.
1983	Tajima and also Hudson initiate statistical treatments of the distinction between a gene tree and a population tree.
1986	Bermingham & Avise initiate comparative phylogeographic appraisals of mtDNA for multiple co-distributed species.
1987	Avise & colleagues coin the word phylogeography, define the field, and introduce several phylogeographic hypotheses.
	Cann & colleagues describe global variation in human mtDNA.
1989	Slatkin & Maddison introduce a method for estimating interpopulation gene flow from the phylogenies of alleles.
1990	Avise & Ball introduce principles of genealogical concordance as a component of phylogeographic assessment.
1992	Avise summarizes the first extensive compilation, involving multiple species and genetic assays, of phylogeographic patterns for a regional fauna.
1994	Moritz emphasizes a conceptual distinction between shallow and deep intraspecific phylogenies by promoting the recognition of management units and evolutionarily significant units (see also Ryder 1986; Avise, 1987; Waples, 1991; Dizon et al., 1992; Riddle, 1996).
1996	Volumes edited by Avise & Hamrick and by Smith & Wayne summarize the many roles for molecular phylogeographic analysis in conservation biology.
1998	A special issue of the journal *Molecular Ecology* is devoted to phylogeography.

a. The text of this book describes many additional historical contributions. Table is after Avise, 1998a.

Improvements in the efficiency of laboratory assays (reviews in Ferraris and Palumbi, 1996; Hillis et al., 1996) have greatly facilitated phylogeographic efforts. Nucleotide sequencing procedures that were introduced more than 20 years ago (Maxam and Gilbert, 1977; Sanger et al., 1977) have been streamlined and automated such that many phylogeographic appraisals now involve direct analyses of particular DNA sequences rather than restriction sites. Sequencing studies can be geared to the level of resolution required. For example, the rapidly evolving CR often is sequenced in fine-scale studies of microevolutionary divergence, whereas the leisurely evolving cytochrome *b* gene is employed routinely for broader phylogeographic patterns and species' relationships.

In the mid-1980s, the polymerase chain reaction (PCR) was introduced as a means of *in vitro* amplification of DNA from even minuscule amounts of starting tissue (Mullis et al., 1986; Mullis and Faloona, 1987). Elaborations of PCR methods have opened a world of novel opportunities for molecular assay (Erlich, 1989). Another noteworthy development was the publication of "universal" PCR primers that could amplify mtDNA segments from many species (Kocher et al., 1989). The availability of these primers helped to spur comparative analyses of the strengths and weaknesses of particular mitochondrial sequences such as cytochrome *b*, ribosomal genes, or the CR for phylogenetic assessments at varying taxonomic levels (Esposti et al., 1993; Meyer, 1994; Simon et al., 1994; Taberlet, 1996).

Recent years also have seen great improvements in phylogenetic algorithms for extracting historical information from molecular data (Page and Holmes, 1998). Following Hennig (1966), qualitative phylogenetic approaches, including the search algorithms of maximum parsimony, have been improved and made available through user-friendly computer programs (e.g., Maddison and Maddison, 1992; Swofford, 1996). At the microevolutionary scales of most phylogeographic studies, such qualitative approaches to gene-genealogical analysis are particularly germane (in addition to distance-based methods) because they permit direct interpretations of historical interrelationships among molecular character states, and in some cases permit provisional identification of allelic clades whose members share derived genotypic features. These molecular clades are to be interpreted carefully: for mtDNA, they pertain strictly to the matrilineal component of an organismal pedigree.

In recent years, coalescent theory too has been refined and extended to populations of varying demographies and structures (Hudson, 1998). Several related statistical and phylogenetic methods have been developed expressly for analysis of gene genealogies in a phylogeographic context. Most of these methods in principle can be applied to nonrecombining sequences within the nuclear as well as the mitochondrial genome. Thus, the past few years have seen the start of empirical efforts to obtain genealogical data from nuclear genes (or their rapidly evolving introns; Villablanca et al., 1998) in a fashion analogous to that for mtDNA (Lessa, 1992; Slade et al., 1993; Palumbi and Baker, 1994, 1996; Palumbi, 1996a; Friesen et al., 1997; Prychitko and Moore, 1997; Hare and Avise, 1998).

Phylogeography is a young field that displays the exuberance and growing pains that might be expected of a vigorous adolescent discipline.

SUMMARY

1. Phylogeography is a field of study concerned with the principles and processes governing the geographic distributions of genealogical lineages. The historical roots of the science are intertwined with empirical studies of animal mitochondrial DNA. Research in the 1970s and early 1980s uncovered the major molecular and transmission-genetic features of mtDNA that instill this molecule with special value as a microevolutionary phylogenetic marker: maternal inheritance, typically without intermolecular genetic recombination; rapid evolution at the nucleotide sequence level; and extensive intraspecific polymorphism the great majority of which is apportioned among rather than within individuals.

2. These special properties of animal mtDNA promoted the emergence of several unorthodox perspectives on microevolution: that matrilineal histories of conspecific organisms could be recovered;

that particular gene trees as exemplified by mtDNA constitute hierarchical and branched phylogenetic components of otherwise anastomose organismal pedigrees; that individual organisms accordingly could be treated as "operational taxonomic units" in some data analyses; and in general that phylogenetic outlooks could be incorporated justifiably into discussions of intraspecific evolution.

3. With its focus on historical processes, phylogeography contextualizes and balances the perspectives of ecogeography that tend to emphasize natural selection's role in microevolution. Within historical biogeography, phylogeography offers a useful conceptual framework for weighing the relative merits of oft-competing vicariance and dispersalist scenarios to account for the geographic distributions of genetic traits.

4. Phylogeography lies at a critical juncture between several otherwise disengaged disciplines in micro- and macroevolution. In particular, phylogeography provides an empirical and conceptual bridge between the traditionally separate disciplines of population genetics and phylogenetic biology.

If we count backwards this multiplication of individuals in each species, in the same way as we have multiplied forward, the series ends up in one single parent.

—Carolus Linnaeus, 1758

Any method that depends on phylogenies of genes requires a shift in perspective on within-species variation.

—Montgomery Slatkin
  and Wayne Maddison, 1989

2

DEMOGRAPHY-PHYLOGENY CONNECTIONS

DEMOGRAPHIC considerations seldom arise in discussions of higher-level phylogenetics (though perhaps they should—see Chapter 6). However, the historical demographies of populations are of profound relevance to phylogeographic patterns over microevolutionary time scales by virtue of their inevitable impact on the structures of gene genealogies. Historical demographies and pedigree-based genealogies of conspecific populations are meshed inextricably. Branching-process theory and coalescent theory are conceptual disciplines that address these genealogy-demography connections. These subjects can be highly mathematical (Griffiths and Tavaré, 1997; Herbots, 1997; Taib, 1997). The intent of this chapter is to introduce these theoretical frameworks for intraspecific gene lineages mostly in simplified graphical formats, and to emphasize conclusions that are particularly germane to empirical phylogeographic interpretations.

INTRA-POPULATION MATRILINES

The concept of a hierarchical gene tree within a sexual population can be introduced by considering the branching structure of matrilines through an organismal pedigree (Fig. 2.1). Most of what follows will refer to the theoretical properties of gene trees as distinguished from imperfect estimates of gene trees, which might come from molecular surveys of particular mtDNA sequences.

If each female replaced herself with exactly one daughter in each generation, there would be no lineage sorting, no hierarchical branching process for matrilines, and no coalescent. However, females in all real populations show variance in contributions to the progeny pool. As a consequence, an intraspecific matrilineal gene tree continually grows and self-prunes as successful branches proliferate and others die off. For matrilines, the means and variances in reproductive success are measured in the currency of daughters produced by mothers. Frequency distributions of family sizes underlie the probability models that describe the lineage sorting process.

Branching Processes and Coalescent Theory

Imagine that females produce daughters according to a Poisson distribution with mean (and hence variance) equal to 1.0. Under this model, the expectation that a female contributes zero daughters to the next generation (or, the expected population frequency of daughterless mothers) is $e^{-1} = 0.368$ (e is the base of natural logarithms), and her probabilities of producing n offspring ($n \geq 1$) are given by $e^{-1}(1/n!)$. Thus, the chances of 1, 2, 3, 4, and 5 or more daughters are 0.368, 0.184, 0.061, 0.015, and 0.004, respectively. Any Poisson distribution assumes that sequences of events (successive daughters from a mother in this case) are independent and random, i.e., neither positively nor negatively correlated. This might be a reasonable null expectation in the unlikely circumstance that reproduction is neither concentrated in particular mothers nor unduly uniform across females for environmental or genetic reasons.

These probabilities apply across a single organismal generation. To calculate recursively the probabilities of matrilineal extinction across multiple (nonoverlapping) generations, mathematical generating functions can be employed (Li, 1955; Crow and Kimura, 1970; Spiess, 1977). These

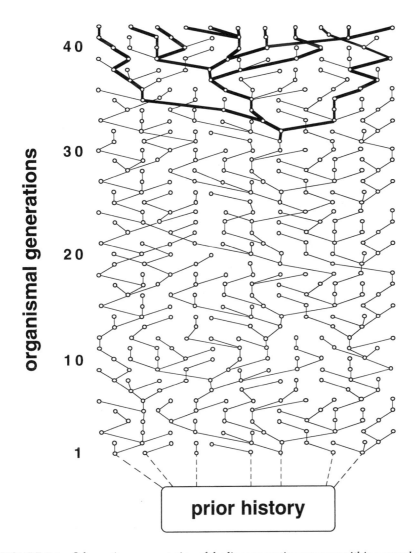

FIGURE 2.1   Schematic representation of the lineage sorting process within a population. Shown are matrilineal pathways connecting mothers and daughters across more
than 40 generations. Lines that cross do not imply any genetic event such as recombination. Note that all seven founding lineages except one have gone extinct during this interval, and that all extant lineages *(top)* coalesce to a common female parent 10
generations earlier. Note also that these matrilineal pathways collectively describe a
hierarchical and branched gene tree reflecting the hereditary transmission routes traversed by mtDNA molecules.

functions are available for several parametric frequency distributions of family sizes (Table 2.1). Application of the Poisson generating function, for example, yields cumulative probabilities of lineage extinction that increase asymptotically from $p = 0.368$ across one generation to $p = 0.981$ by generation 100 (Table 2.2). As exemplified in Table 2.2, additional calculations for populations of any specified size can yield other branching-process statistics: the cumulative probability (q) that varying numbers of matrilines survive across G generations (Fig. 2.2); the probability (p^N) of population extinction from exhaustive matrilineal losses; and the survival expectation (γ) through time for two or more founding matrilines.

The latter statistic is of special interest in a phylogenetic context. If γ is close to 1.0, a population founded G generations earlier likely retains lineages descended from at least two original founders. Conversely, a value of γ near zero indicates a high probability that all lineages separated less than G generations prior, and, thus, that extant matrilines trace to a single ancestor within that time. Such lineages then constitute a monophyletic matrilineal group or clade from that point forward in the gene tree. Such expectations also can be conditioned ($^c\gamma$ values) on the probability that

TABLE 2.1   Examples of mathematical generating functions for various distributions of family size.[a]

Family-size Distribution	Mean	Variance	Generating Function $[p_G(x)]$
Poisson	μ	μ	$e^{\mu(x-1)}$
binomial	np	npq	$(q + px)^n$
geometric	q/p	q/p^2	$p/(1-qx)$
negative binomial	rq/p	rq/p^2	$[p/(1-qx)]^r$

a. A generating function is used to calculate recursively the cumulative probability of loss of a lineage by each generation. Let x equal the probability of loss for the previous generation ($G-1$). Then $p_G(x)$ is the probability of loss by generation G. Other parameters are as follows: n, number of lineages; p and q, probabilities of extinction and survival, respectively, of a single lineage computed recursively as in Table 2.2; r, a parameter in the negative binomial distribution that can be thought of as r extinctions in the outcomes of n random lineages. Table is after Avise et al., 1984a.

TABLE 2.2   Probabilities of extinction (p) and survival (q) of single lineages through G generations under a Poisson distribution of family sizes (daughters per mother) with mean $\mu = 1.0$.[a]

			Binomial Expansion (population of $N = 4$ females)			
G	p	q	p^4	$4p^3q$	Sum of Other Terms (γ)	$^c\gamma^b$
1	0.3679	0.6321	0.0183	0.1259	0.8558	0.8718
2	0.5315	0.4685	0.0798	0.2814	0.6338	0.6925
5	0.7319	0.2681	0.2869	0.4205	0.2926	0.4103
10	0.8417	0.1583	0.5019	0.3775	0.1206	0.2421
20	0.9125	0.0875	0.6933	0.2659	0.0408	0.1330
100	0.9807	0.0193	0.9250	0.0728	0.0022	0.0293
general	$p_G = e^{[p(G-1)-1]}$	$1-p$	p^N	$np^{N-1}q$	$1-p^N- np^{N-1}q$	$1-[(Np^{N-1}q) /(1-p^N)]$
			pop. extinct	pop. extant		

a. After Avise et al., 1984a. Also shown, for a particular small population of $N = 4$ females, are additional lineage sorting statistics. The first and second terms of the binomial expansion, for example, are the probabilities that zero or one lineage, respectively, survive to generation G (see text).

b. The probability of survival of two or more lineages given that the population remains extant.

populations survived (did not go extinct) and thereby remained available for contemporary observation (Table 2.2).

When viewed from the present looking back, a small value for γ or $^c\gamma$ indicates a high likelihood that all extant lineages coalesced to a single common ancestor. Expectations for coalescent times are related to population size (Fig. 2.3). In a population of 10,000 females, for example, γ remains near one for 10^4 generations and thereafter declines quickly toward zero by about 10^5 generations. In a population of 10 females, γ values are high only when $G < 10$, and lineage coalescence is almost certain to occur within 100 organismal generations. Similar relations between population

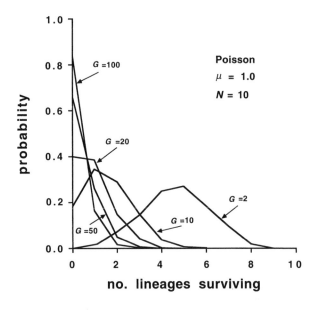

FIGURE 2.2   Frequency distributions of the probabilities of survival of founding lineages through G generations in populations initiated with 10 unrelated females producing daughters according to a Poisson distribution with mean $\mu = 1.0$ (after Avise et al., 1984a). These distributions were generated by the approach illustrated in Table 2.2.

size and coalescent time hold for other parametric distributions of family size such as the negative binomial (Fig. 2.4, left).

   In the traditional literature of population genetics, related concepts often were couched in somewhat different language, such as expected times to loss or fixation of newly arisen mutations (Kimura and Ohta, 1969; Burrows and Cockerham, 1974) or separation times for alleles that were identical by descent (Malecot, 1948). An inevitable relationship between such traditional views and the concepts of branching-process theory and coalescent theory can be seen by noting that whenever a new mutation reaches fixation in a population or species, all alleles at that locus trace to (are identical by descent from, or coalesce to) the single copy of the allele in which that mutation arose.

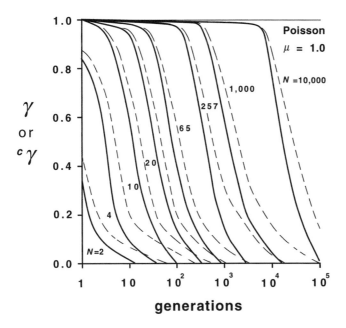

generations

FIGURE 2.3   Probabilities of survival of two or more founding lineages through G generations in populations initiated with N (values ranging from 2 to 10,000) unrelated females producing daughters according to a Poisson distribution with mean $\mu = 1.0$ (after Avise et al., 1984a). Solid lines, γ; dashed lines, $^c\gamma$ (the probability of survival of two or more lineages conditioned on the population having remained extant and, thus, available for observation).

Coalescence to a single ancestor in a matrilineal genealogy does not imply that only one female was alive in the coalescent generation. It merely indicates that any other females alive at that time were not survived by matrilineal descendants. Nor does coalescence in a matrilineal tree imply that other females in the ancestral population failed to contribute genetically to later generations. Nuclear genes from other females likely are represented in descendants by alleles that percolated through the multitude of nonmatrilineal pathways within an organismal pedigree.

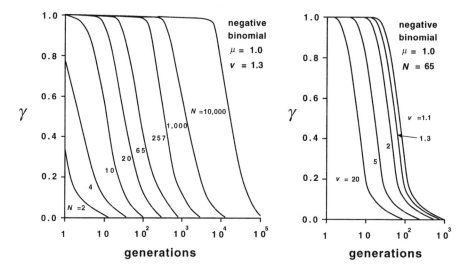

FIGURE 2.4 Probabilities of survival of two or more founding lineages in populations characterized by negative binomial distributions of family size with mean of $\mu = 1.0$ daughters per female (see
legend to Fig. 2.3). *Left:* Populations with differing numbers N of founding females but the same variance ($v = 1.3$) in progeny numbers across females. *Right:* Populations with differing variances v in
progeny numbers across females but the same number ($N = 65$) of founders.

These calculations assume that survivals (and extinctions) of different
lineages are independent events. This assumption is violated routinely in
nature when, for example: (a) environmental factors such as climatic
changes affected multiple lineages jointly, in which case overall population size might increase or decrease dramatically for deterministic reasons;
or (b) density-dependent factors contributed to the regulation of population size. All such demographic factors impact family sizes, which in turn
influence the dynamics of branching processes and coalescence for gene
lineages.

Even minor changes in family size can alter population demography
and coalescent probabilities profoundly. For example, the left graph in Fig.
2.5 shows probabilities through time of matrilineal survival in a growing
population (mean of $\mu = 1.1$ daughters per female) under a Poisson distri-

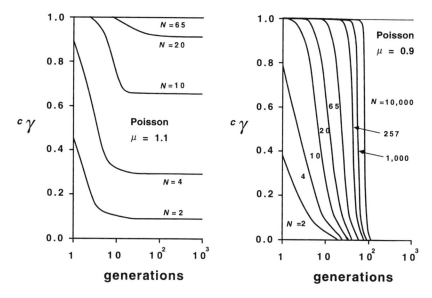

FIGURE 2.5   Conditional probabilities of survival of two or more founding lineages in populations of various size N. *Left:* Growing populations (mean of $\mu = 1.1$ daughters per female). *Right:* Declining populations ($\mu = 0.9$) (see Fig. 2.3 caption).

bution of family sizes. Population growth inhibits lineage extinction and thereby raises dramatically the probability of lineage survival over a specified time (compare to Fig. 2.3). Conversely, lineage survival normally is reduced dramatically in declining populations (right graph in Fig. 2.5).

Another influential factor in the matrilineal branching process, even in stable-sized populations, is the variance in reproductive success among females. This can be illustrated by appealing to non-Poisson distributions of family size where the effects of changes in the mean and variance can be examined separately. For example, in a population with a negative binomial distribution of family size and mean of $\mu = 1.0$ daughter per female, a larger variance in offspring numbers accelerates the lineage sorting process greatly (Fig. 2.4, right). All else being equal, larger variances in family size produce more rapid coalescence because surviving lineages in effect are channeled through fewer parents in each generation.

Many species have huge interfamily variances in reproductive success. In the marine realm, for example, broadcast spawners such as oysters are astoundingly fecund as individuals but face exceptional challenges in matching reproductive activity to stringent oceanographic conditions conducive to successful gamete production, fertilization, larval development, larval settlement, and recruitment into the adult population. Hedgecock and colleagues (1982, 1992; Hedgecock and Sly, 1990; Li and Hedgecock, 1998) argue that "sweepstakes" variances in reproductive success are attendant with this life history pattern. Empirical genetic signatures of sweepstakes reproduction may be registered in two otherwise unanticipated molecular observations for several marine shellfish species: (a) chaotic patchiness—pronounced but seemingly stochastic variation in genotypic frequencies over small spatial and temporal scales; and (b) low levels of genetic diversity (implying recent coalescent times in gene genealogies) relative to what might have been anticipated given these species' often huge present-day populations.

Related Perspectives from Inbreeding Theory: Generational Times to Coancestry

The coalescent is basically a model of lineage sorting and genetic drift run backward in time to common ancestors (Harding, 1996). Another way to consider the coalescent process is to derive from inbreeding theory the expected frequency distributions of times to common ancestry for extant matrilines within a population. The following is based on Tajima's (1983) formulation for nuclear gene trees under selective neutrality, as modified by Avise et al. (1988) for matrilineal pathways.

Imagine an idealized population, stable in size with non-overlapping generations, in which females contribute to a large pool of potential offspring from which N_F daughters are drawn at random to form the female population of the next generation. The probability that two randomly chosen daughters share a mother is $1/N_F$. This is also the probability that their time to common ancestry was exactly one generation ago ($G = 1$). The probability of coancestry two generations ago ($G = 2$) is derived as follows. First note the probability that two randomly chosen females in the current generation had different mothers: $1 - 1/N_F$. The probability that they had different mothers but the same grandmother is, therefore, $(1 - 1/N_F)(1/N_F)$. By extension, the probability that two randomly chosen females in the current generation stem from a shared female ancestor G generations ago is:

$$f(G) = (1/N_F)(1 - 1/N_F)^{G-1} \cong (1/N_F)e^{-(G-1)/N_F} \qquad (1)$$

This equation gives the probability distribution of times to common ancestry, measured in organismal generations, for extant females. These distributions (Fig. 2.6) are geometric, with mean approximately N_F and

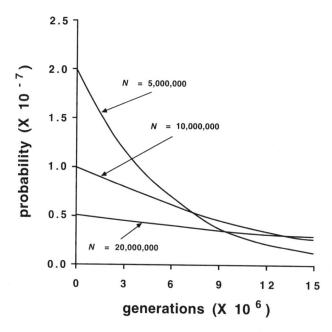

FIGURE 2.6 Probability distributions (plotted per generation) of times to shared matrilineal ancestry for female populations with indicated evolutionary effective sizes (after Avise et al., 1988). The lines cross one another because the areas under each curve sum to 1.0.

variance $N_F(N_F - 1)$. In other words, for random pairs of females in a population of constant size N_F, the expected mean number of generations back to a shared female ancestor is $G \cong N_F$. An important point is to distinguish between the expected mean coancestry times for lineage pairs versus the expectation for the coalescent of an entire suite of lineages. When the number of lineages is large, the former value (under neutrality theory) is about one-half as great as the latter (Nei, 1987; Ayala, 1995a) or, for matrilines, approximately N_F versus $2N_F$ generations.

Such expectations must be interpreted cautiously because the model assumes an idealized population in which a constant number of breeding females replaces itself each generation and females attempt production of many daughters but succeed with small probability. These assumptions in conjunction with large N_F imply a distribution of family sizes approximated by the Poisson with a mean (and variance) of 1.0 daughter per female. Few populations in nature display these features exactly. Variances in progeny numbers may be large and temporally variable, population sizes often fluctuate through time, and generations may overlap. However, the impact of these and related demographic factors on the coalescent process are approximated via their influence on effective population size N_e (Wright, 1931; Nei, 1987): the number of breeding individuals yielding an idealized population that would have genetic properties (expected times to common ancestry in this case) similar to those observed in a real population. Thus, if the long-term effective population size is known or hypothesized, expected times to common maternal ancestry can be approximated by substituting $N_{F(e)}$ for N_F in the formulations above.

Another reason for caution in applying these predictions to real species stems from the stochastic nature of historical genealogies. The theory is "global" in the sense that times to common ancestry for female pairs are assumed to be independent. The outcomes are mean expectations for large numbers of replicate pedigree trials, but for any one pedigree, matrilines are historically connected and genealogically nonindependent (Ball et al., 1990). In particular, the earliest matrilineal split leading to surviving lineages in a pedigree exerts a disproportionate influence on separation times among pairs of extant females (Felsenstein, 1992a). Indeed, bimodal or even multi-peaked ("ragged") distributions of pairwise times to co-ancestry sometimes can appear in the gene trees of stable-sized populations (Slatkin and Hudson, 1991). Thus, the history of lineage branching within any pedigree imposes a correlation on pairwise times to common ancestry that compromises precise predictions for populations considered individually (Hudson, 1990; Slatkin and Hudson, 1991; Kuhner et al., 1995). Nevertheless, empirical data *can* be compared against theory to address order-of-magnitude issues regarding long-term historical population sizes, as described next.

EVOLUTIONARY EFFECTIVE POPULATION SIZE

From restriction-site or sequence surveys, mtDNA genetic distances between individuals can be obtained, and, based on assumptions of evolutionary rate, converted to provisional estimates of absolute time to matrilineal coancestry. Such estimates accumulated for many individuals yield pairwise frequency distributions of coancestry times ("mismatch distributions," for short) that can be compared to expectations from the inbreeding theory described in the previous section (Rogers and Harpending, 1992; see also Felsenstein, 1992a,b; Fu 1994a,b). One advantage of this approach is that in some cases (e.g., under rapid population expansion) a gene tree might be nearly devoid of information on clade structure because many nodes are temporally close, yet a frequency histogram of coancestry times could be edifying about historical population demography (Rogers, 1997).

In principle, data sets appropriate for these comparisons might involve mtDNA genetic distances among individuals within an isolated local population of effective size $N_{F(e)}$ females. In practice, however, such comparisons are mostly uninformative for at least two reasons. First, short-term effective sizes of local populations usually are too low (and coalescent times for maternal lineages within them therefore too recent) to have permitted an accumulation of many *de novo* mutations detectable in conventional mtDNA assays. Second, when high mtDNA polymorphism within a small local deme is observed, a significant fraction of sequence variants quite likely originated exogenously: e.g., as ancestral lineages retained from larger source populations, and/or as recent lineage immigrants. For these reasons, mtDNA surveys seldom permit realistic appraisals of $N_{F(e)}$ from the expected frequency distributions of times to coancestry within small local populations considered individually.

Data sets more appropriate for such inferences involve large, high-gene-flow species that have not been sundered historically. Then an entire species can be considered a single population from an extended temporal perspective, and the expected coalescent times refer to the evolutionary effective numbers of females in this broader demographic unit. For such species, mtDNA polymorphisms provide ballpark gauges of empirical coalescent times that permit meaningful comparison against theoretical expectations for species of varying evolutionary $N_{F(e)}$. To avoid complications arising from spatial population structure, species that qualify ideally must have had historically high levels of interpopulation gene flow: i.e., $N'_{F(e)}m \gg 1.0$, where m is the female migration rate per generation between geographic populations each of size $N'_{F(e)}$ (Maruyama and Kimura, 1974; Slatkin, 1985a).

Consider as a relatively clean example the American eel (*Anguilla rostrata*), a fish that spends most of its life in freshwater but spawns in the ocean (a catadromous life cycle). Juveniles inhabit freshwater and coastal habitats of North America until sexual maturity, which for females in temperate waters is reached at about 10 years of age. Eels then migrate seaward to spawn in the western tropical mid-Atlantic Ocean (Sargasso Sea) from whence larvae disperse in water currents back to continental waters. Thus, collections of eels from any continental locale probably are samples of the same breeding population (Williams and Koehn, 1984). Available genetic data from allozymes (Williams et al., 1973; Koehn and Williams, 1978) and from mtDNA (Avise et al., 1986) are consistent with this scenario: shared alleles are distributed widely in collections of eels from Maine to Louisiana.

To generate a mismatch distribution of matrilineal coancestry times for eels, empirical estimates of mtDNA sequence divergence (p) were con-

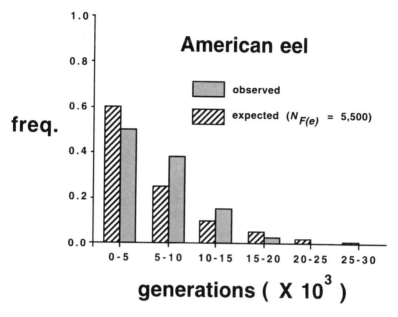

FIGURE 2.7 Mismatch frequency distributions of times to shared maternal ancestry, measured in generations, in pairwise comparisons of 109 American eels (after Avise et al., 1986, 1988). Solid bars represent empirical times estimated from mtDNA restriction site data for collections throughout a major portion of the species' range. Lined bars are expected times based on a model of neutral lineage sorting (from inbreeding theory for a stable-sized population, see text) given an evolutionary effective population size of 5,500 females.

verted to absolute time (t) by the equation $t = (0.5 \times 10^8)(p)$, which assumes a conventional clock calibration of 1 percent sequence evolution per lineage per million years. These absolute times then were converted to generational counts based on an eel generation span of 10 years. For eels sampled randomly across the species' range, the mismatch distribution proved remarkably close in magnitude and pattern to theoretical expectations for a single population of evolutionary effective size $N_{F(e)}=5,500$ (Fig. 2.7). This value is much smaller than the current adult census size of many millions of breeding individuals (Williams and Koehn, 1984).

Similar appraisals of long-term $N_{F(e)}$ have been conducted on several other high-gene-flow species that presumably have been unsundered historically, such as the hardhead catfish, *Arius felis,* along coastlines in the southeastern United States (Avise et al., 1987b), and the red-winged black-bird, *Agelaius phoeniceus,* and downy woodpecker, *Picoides pubescens,* across North America (Ball et al., 1988; Ball and Avise, 1992). Millions of breeding individuals now comprise these species, and as gauged by mtDNA surveys each species lacks substantial geographic population structure across a huge area. As for the American eel, the mtDNA data were used to estimate times to matrilineal coancestry for conspecific individuals, and the mismatch distributions were compared to theoretical expectations for idealized populations of varying $N_{F(e)}$. In each case, inferred depths of the mtDNA lineage separations indicate evolutionary effective population sizes that are orders of magnitude below present-day breeding census sizes for these species (Fig. 2.8).

Similar conclusions extend to other high-gene-flow species (Fig. 2.9). Especially noteworthy was a molecular survey of two of the most abundant eukaryotic species on Earth—the planktonic marine copepods *Calanus finmarchicus* and *Nannocalanus minor.* From observed population densities and from area distributions of these species in the world's oceans, census estimates of female population size are in excess of 10^{15} individuals. Yet, analyses of mtDNA sequence data by the coalescent procedures described above yield estimates of evolutionary effective population size of $N_{F(e)} \cong 10^5$ for each species, a 10^{10}-fold difference (Bucklin and Wiebe, 1998). The authors speculate that historical fluctuations in population size account for the relative paucity of genetic variation and for a truncation of the coalescent to far more recent dates than might have been anticipated for these superabundant species.

Similar conclusions have been reached for sardines *(Sardina, Sardinops)* and anchovies *(Engraulis).* Regional populations of these fishes are huge, with commercial catches in the thousands of metric tons. Yet, coalescent depths in mtDNA genealogies remain strikingly shallow (Grant and Bowen, 1998; Grant et al., 1999).

Thus, a general rule has emerged from the superabundant and high-gene-flow species analyzed to date. Despite the usual presence of extensive

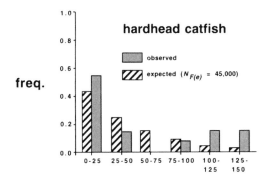

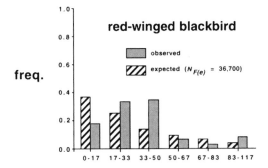

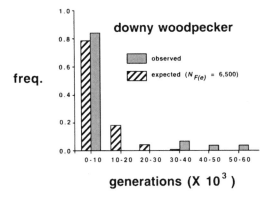

generations (X 10³)

FIGURE 2.8 Mismatch frequency distributions of times to shared maternal ancestry, measured in generations, in pairwise comparisons of conspecific individuals within one species of marine fish and two species of birds that are abundant yet exhibit little or no geographic population structure in mtDNA (after Avise et al., 1988; Ball and Avise, 1992). Solid bars represent empirical times estimated from mtDNA restriction sites for collections throughout major portions of the respective species' ranges. Lined bars represent expected times based on a model of neutral lineage sorting given the evolutionary effective population sizes of females indicated.

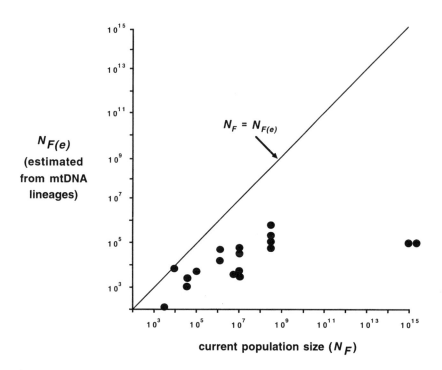

FIGURE 2.9   Current female population size (N_F) versus evolutionary effective population size ($N_{F(e)}$) as estimated from empirical mtDNA data within each of several marine species that exhibit minimal geographic population structure in mtDNA genotypes (updated from Avise, 1992). Note that most points fall orders of magnitude below a line in which $N_F = N_{F(e)}$.

mtDNA variation within species, the temporal depths estimated for matrilineal separations can be brought into alignment with coalescent theory for neutral haplotypes only when it is assumed that long-term evolutionary effective population sizes of females are relatively small.

The low $N_{F(e)}$ values might be the result of historical population demographic factors. Most populations probably fluctuate dramatically over time scales ranging from decades to millennia as they experience periodic population bottlenecks due to disease outbreaks (O'Brien and Evermann,

1988), climatic changes, or other challenges in the biotic or physical environment. Such demographic fluctuations would depress long-term $N_{F(e)}$ to levels dramatically below population standing crops at most points in time (Fig. 2.10). If their periodicity falls outside the usual time frame of field observations, severe demographic contractions often might escape notice in contemporary populations, but they do leave signatures on extant genomes.

Another potential explanation for the low $N_{F(e)}$ values involves cytoplasmic "selective sweeps." When selectively favored mtDNA variants (or other cytoplasm-hitchhiking elements such as maternally inherited microbes [Hoffmann et al., 1986; Turelli and Hoffmann, 1991, 1995; Johnstone and Hurst, 1996]) arise occasionally in nature, positive directional selection may cause them to sweep through a high-gene-flow species, thereby temporarily eliminating all preexisting mtDNA polymorphism. In effect, coalescent times would be abbreviated because surviving matrilines were channeled through a single female in which the *de novo* mtDNA mutation or microbial variant arose.

Regardless of the evolutionary explanation in particular instances, the usual truncation of coalescent times stemming from low intraspecific $N_{F(e)}$ values is of great pragmatic importance to higher-level phylogenetic systematics. A relatively rapid sorting of mtDNA lineages within most high-gene-flow species should tend to generate intraspecific genealogical monophyly over time frames often shorter than the internodes between speciation events. This provides a justification (that otherwise would be lacking) for the traditional use of small species' samples as provisional OTUs in interspecific phylogenetic exercises. It also helps to explain a phenomenon usually taken for granted but that otherwise would present a profound enigma: why branched and hierarchical species' phylogenies usually appear to be recoverable with some reliability from gene-tree data. Nevertheless, low intraspecific variation in relation to between-species divergence should not be assumed *a priori,* but instead should be evaluated empirically for any taxonomic group under examination (Smouse et al., 1991; Hoelzer, 1997).

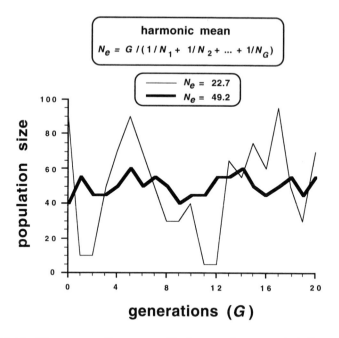

FIGURE 2.10   The concept of long-term effective population size (N_e). Shown are two populations with the same arithmetic mean number of individuals across generations but different effective population sizes (harmonic means of adult census sizes). A harmonic mean is closer to the smaller rather than to the larger of a series of numbers such that N_e for an unstable population is lower than that for a stable one. For a focused discussion of population bottlenecks and their effects on genetic variation, see Nei et al. (1975).

DYNAMICS OF POPULATION SIZE

Effective population size predicts general genealogical depth in a fluctuating species, but alternative temporal demographies (even for populations with identical N_e) can leave different signatures on the distributions of coalescent events in matrilineal trees. For example, the slowed pace of lineage extinction characteristic of a population that grew rapidly in the past could produce a temporal clustering of older nodes in a gene tree. Thus, the structure of a gene tree as estimated from extant lineages can be influenced greatly by the date of the expansion and by the population's prior and subsequent demographic histories (Tajima, 1989; Harpending et al., 1993; Eller and Harpending, 1996).

Many species (humans included) probably have undergone sustained increases in population size when new ecological opportunities arose and/or during recovery from population bottlenecks. In an exponentially growing population starting from a small source stock, a "star phylogeny" can be anticipated with many genealogical lineages tracing to a restricted span of times near the initial population expansion (Slatkin and Hudson, 1991). With respect to mismatch distributions of coancestry times, gene trees in growing populations typically display "waves" (Rogers and Harpending, 1992) whose crests fall to the right of the modes in the corresponding geometric curves for constant-sized populations.

An example involving red-winged blackbirds (an abundant avian species across North America) is provided in the middle panel of Fig. 2.8. This species was sampled across the continent for mtDNA genotypes, and the mismatch distribution displayed an evident wave pattern that departed from the geometric distribution expected for a demographically stable species of the same effective population size. Quite likely, blackbird numbers increased explosively as the species expanded its range into areas that were glaciated during much of the Pleistocene.

As noted above, mismatch distributions of lineage divergence times have drawbacks as a sole source of inference about population history because the plots are influenced heavily by early bifurcations in a gene tree. In particular, a disproportionate number of pairwise comparisons come from lineages on opposite branches tracing to the first split in the gene-tree root. Nee et al. (1996a) suggested a way to circumvent this difficulty by focusing on gene-tree structure itself, and in particular on temporal nodal placements (which should be statistically independent of one another).

The basic idea involves plotting the number of lineages in a gene tree from its origin to the present, as estimated for example from empirical data on extant mtDNA lineages. If this "lineages-through-time" plot (with log-transformed lineage numbers) displays a concave pattern ("A" in Fig. 2.11), then the rate of gene-tree branching has increased as the present

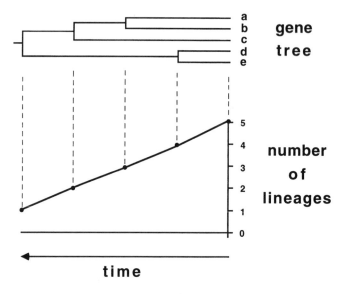

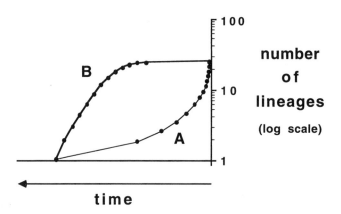

FIGURE 2.11 Explanation of one gene-tree approach to assess patterns of fluctuation in historical population size (after Nee et al., 1996a). *Above:* From a gene tree whose structure has been estimated (in this case, for five lineages a–e), a temporal plot is made of the number of gene lineages in existence up to the present (arrow points toward the past). *Below:* Expected patterns for such plots (graphed semi-logarithmically) under scenarios of constant population size *(A)* and exponentially increasing population size *(B)* across the periods of time covered by the coalescences in a gene tree (see text).

is approached. Such a pattern implies that the lineages could have been drawn from a population perhaps of constant size, because a gene tree in such a population is expected to grow hyperbolically through time. By contrast, a convex pattern in a lineages-through-time plot ("B" in Fig. 2.11) would evidence a slowed rate of gene-tree branching in recent history. Such an outcome would be consistent with rapid exponential growth in the population from which the gene tree was drawn.

Nee et al. (1996a) applied their method to a mtDNA data set for world-wide collections of humpback whales (Baker et al., 1993). The structure of the mtDNA gene tree was consistent with the idea that the humpback population remained roughly constant in size over the period of time covered by the coalescents. Nonetheless, it may be premature to draw firm conclusions from such lineages-through-time analyses in their current state of development. First, the limited time scales over which populations could grow exponentially may not match well the empirical resolution afforded by current mtDNA assays. Second, the model assumes an absence of spatial population structure, yet any such structure might inhibit lineage extinction in such a way as to falsely mimic a lineages-through-time plot expected for an exponentially growing population.

Another approach to inferring the general demographic history of a population from gene-tree data involves examination of two different measures of haplotype variation. Haplotype diversity ($h = 1 - \Sigma f_i^2$, where f_i is the frequency of the ith haplotype) condenses information on the numbers and frequencies of different alleles at a locus, regardless of their sequence relationships. Nucleotide diversity ($p = \Sigma f_i f_j p_{ij}$, where p_{ij} is the sequence divergence between the ith and jth haplotypes) is a weighted sequence divergence between individuals in a population, regardless of the number of different haplotypes. Intuitively, a population with low h and low p may have experienced a prolonged or severe demographic bottleneck (or, perhaps, a selective sweep) in recent times. Conversely, high values for h and p are an expected signature for a stable population with large long-term N_e; or, they also might be observed in an admixed sample of individuals from historically sundered populations. High h and low p

suggests rapid population growth from an ancestral population with small N_e, provided the time was sufficient for recovery of haplotype variation via mutation yet too short for an accumulation of large sequence differences. Conversely, low h and high p could result from a transient bottleneck in a large ancestral population, because an extremely short crash can eliminate many haplotypes without necessarily impacting p severely (Nei et al., 1975). Low h and high p also might reflect an admixture of samples from small, geographically subdivided populations. Grant and Bowen (1998) elaborate the rationales for these possibilities and apply empirical comparisons between h and p to draw inferences on the demographic histories of several marine fishes.

EXTENSIONS TO SPATIALLY STRUCTURED POPULATIONS

Most species consist of geographically structured populations, some of which may experience little or no genetic contact for long periods of time. Other species may be characterized by recent range expansions such that their populations are tightly connected genealogically (Ibrahim et al., 1996). Historical and contemporary demography can affect the spatial structure of conspecific populations and thereby influence intraspecific matriarchal genealogies in many ways. The empirical challenge of mtDNA phylogeographic surveys is to decipher the past and present demographic factors that are likely to have produced an observed spatial arrangement of matrilines in particular instances. Here I introduce some of the theoretical expectations for gene genealogies among spatially structured populations.

Isolated Populations

The quantitative conclusions of coancestry theory presented earlier apply to a species that in effect is comprised of a single evolutionary population and, thus, falls at the high end of a continuum of historical connectedness regimes. At the opposite end of this continuum are species composed of long-isolated populations.

RETENTION OF ANCIENT LINEAGES

One genealogical consequence of the long-term isolation of viable populations is an inevitable retention within species of anciently separated matrilines. In effect, the evolutionary continuance of isolated populations buffers against lineage extinction within a species, and thereby extends coalescent times well beyond those expected for a single nonsundered population of the same total size (Nei and Takahata, 1993). For any species whose extant populations have been isolated completely for G generations, the coalescent in an intraspecific gene tree can have occurred no less than G generations ago.

Historical population sundering can produce frequency distributions of times to lineage coancestry that depart dramatically from the unimodal geometric distributions expected (with the caveats mentioned above) for single, nonsundered populations. Molecular surveys of mtDNA commonly identify phylogeographic patterns that presumably register the effects of long-standing population structure. For example, in species-wide surveys of both the sharp-tailed sparrow, *Ammodramus caudacutus* (Rising and Avise, 1993), and the seaside sparrow, *A. maritimus* (Avise and Nelson, 1989), frequency histograms of interindividual mtDNA genetic distances were bimodal (Fig. 2.12), reflecting large sequence differences among conspecific individuals from two different geographic regions. In the sharp-tailed sparrow, the mtDNA clades were confined to northern versus southern portions of the species' range in Canada and the United States, and they mark populations (later named separate species [AOU, 1995]) also distinguishable by multivariate morphology, song, and flight displays (Montagna, 1942; Greenlaw, 1993; Rising and Avise, 1993). In the seaside sparrow of the southeastern United States, the two mtDNA clades agree with population units (from coastlines of the Atlantic versus Gulf of Mexico) suspected from traditional biogeographic evidence to have been sundered historically (Funderburg and Quay, 1983).

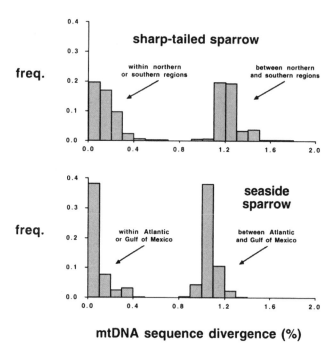

FIGURE 2.12   Bimodal mismatch distributions of mtDNA genetic distance estimates within each of two avian species sampled across major portions of their respective ranges (after Avise et al., 1992b; Avise, 1996c).

Actually, the matrilineal coalescent for a sundered species is likely to be older than the times of population isolation to an extent influenced by the effective size and preexisting lineage structure in the ancestral population(s). From a traditional population genetic perspective, one way to correct for ancestral mtDNA polymorphism in estimating the genetic distance between two extant populations is to assume that mean levels of variation in the latter are representative of variation that had been present in the common ancestor (Edwards, 1997). The corrected or net genetic distance between two isolated populations A and B is thus

$$p_{\text{net}} = p_{AB} - 0.5(p_A + p_B) \tag{2}$$

where p_{AB} is the mean genetic distance in pairwise comparisons of individuals from A versus B, and p_A and p_B are mean genetic distances among individuals (nucleotide diversities) within these populations. To the extent that relationships between genetic divergence and evolutionary time are understood, corrected distances imply population separation times. For example, in the sharp-tailed sparrow, net sequence divergence between the two major matrilineal clades was $p_{net} = 0.012$, which under a conventional rate calibration for avian mtDNA implies a date of population separation about 600,000 years ago.

PHYLOGENETIC CATEGORIES OF RELATIONSHIP

Another approach is more explicitly genealogical and involves an extension of lineage sorting theory to daughter populations (A and B) descended from a common ancestor. In terms of maternal genealogy, three phylogenetic patterns are possible (Fig. 2.13): (I) *reciprocal monophyly*, in which case all extant matrilines within each daughter population are

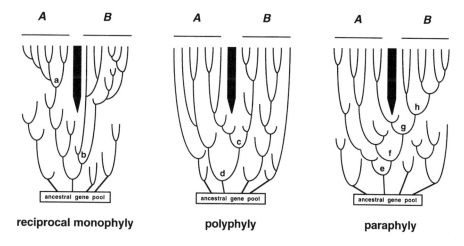

FIGURE 2.13   Schematic representation of three categories of phylogenetic relationship possible for extant matrilines of two daughter populations (A and B) stemming from a common ancestral gene pool (after Avise et al., 1983). Solid bars indicate firm barriers to gene flow, and letters identify key nodes that help to define the phylogenetic categories (see Table 2.3 and text).

closer genealogically to one another than to any matrilines in the second population; (II) *polyphyly*, where some but not all extant matrilines in *A* join with some but not all extant matrilines in *B* to form a clade; and (III) *paraphyly*, wherein all matrilines within one population form a monophyletic group nested within the broader matrilineal history of the second daughter population. These categories are defined formally by the inequalities presented in Table 2.3. Categories II and III illustrate how the topology of a gene tree can differ in fundamental branching pattern from the topology of a population tree. The discordances arise because some gene-tree separations normally predate population separations.

These three categories of phylogenetic relationship frequently characterize the same pair of daughter populations at various time depths following their separation at the population level. For example, close inspection of the extended matrilines pictured in Fig. 2.14 reveals that populations *A* and *B* have been reciprocally monophyletic in the matriarchal tree for only about the last 10 generations. Immediately prior, some matrilines in population *A* were closer genealogically to some matrilines in

TABLE 2.3   Formal definitions of the phylogenetic status of two daughter populations with respect to a gene genealogy.[a]

Phylogenetic Category	Phylogenetic Status	Distance Relationship
I	*A* and *B* monophyletic	$\max t_{AA} < \min t_{AB}$ and $\max t_{BB} < \min t_{AB}$
II	*A* and *B* polyphyletic	$\max t_{AA} > \min t_{AB}$ and $\max t_{BB} > \min t_{AB}$
IIIa	*A* paraphyletic with respect to *B*	$\max t_{AA} > \min t_{AB}$ and $\max t_{BB} < \min t_{AB}$
IIIb	*B* paraphyletic with respect to *A*	$\max t_{AA} < \min t_{AB}$ and $\max t_{BB} > \min t_{AB}$

a. After Neigel and Avise, 1986. Deciding criteria are the maximum coancestry times (t) for all pairs of lineages within the two populations ($\max t_{AA}$ or $\max t_{BB}$) versus minimum coancestry times for pairs of lineages between the daughter populations ($\min t_{AB}$) (see Fig. 2.13 and text).

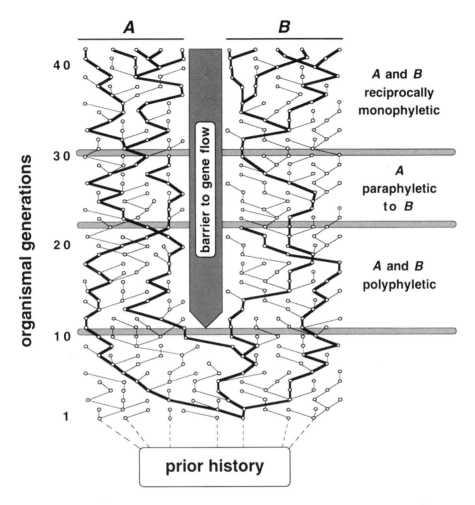

FIGURE 2.14   Schematic representation of the matrilineal sorting process in two populations separated by a firm, long-standing barrier to gene flow. Following the sundering of the ancestral population at generation 11, lineage sorting was such that isolated daughter populations *A* and *B* first appear polyphyletic in matriarchal ancestry, then paraphyletic (of *A* with respect to *B*), and eventually reciprocally monophyletic. Heavier lines highlight key lineage coalescences that help to define the phylogenetic categories.

population B than to others in A, such that A was paraphyletic to B. Before that time, populations A and B were polyphyletic in matriarchal ancestry.

Neigel and Avise (1986) used computer simulations to monitor the phylogenetic status of daughter populations as functions of their mode of founding, time since separation, and population size. If the founders of isolated daughter populations A and B are drawn at random from a single well-mixed ancestral gene pool, the usual chronology of events is that these populations proceed successively through stages of polyphyly and paraphyly in a gene tree before eventually achieving a status of reciprocal monophyly. Under neutrality, the expected post-separation times (in organismal generations, G) spent in these transitional states are functions of effective population sizes of the daughter populations (Fig. 2.15). Probabilities of genealogical polyphyly are high when $G < N_{F(e)}$, whereas reciprocal monophyly for A and B is likely to have been achieved through continued lineage sorting when $G > 4N_{F(e)}$. At intermediate times of population separation, daughter populations could be polyphyletic, paraphyletic, or reciprocally monophyletic in matrilineal phylogeny.

These expectations apply to random draws of founders from a well-mixed ancestral population. More often, an ancestral gene pool is phylogeographically structured, and foundresses of the daughter populations may represent a limited subset of this spatial genealogical heterogeneity. The number of demographic permutations in the founder process is unbounded, but some general expectations concerning the phylogenetic status of gene trees in daughter populations are relatively easy to intuit.

For example, a firm but recent barrier to dispersal between contemporary daughter populations might be coincident with prior genealogical structure within the species, thereby producing reciprocal monophyly for A and B at the outset (Fig. 2.16, top). Or, the contemporary dispersal barrier might be discordant with prior genealogical history within the species, in which case A and B initially would be polyphyletic in matrilineal ancestry (Fig. 2.16, middle). In this case, discordance between the structure of the population tree and the structure of the gene tree could be frozen into

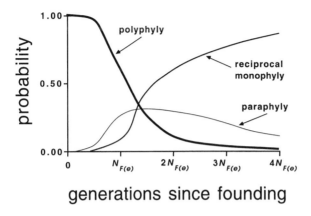

FIGURE 2.15   General form of the probability curves for reciprocal monophyly, poly-phyly, and paraphyly for two daughter populations following their separation from a well-mixed ancestral gene pool (after Neigel and Avise, 1986). These probabilities are functions of effective population sizes in the daughter populations.

place, or converted eventually to paraphyly and reciprocal monophyly, depending on which lineages in *A* and *B* survive into the future. (The topic of permanent discordance between a gene tree and a species tree will be deferred to Chapter 6.) Finally, the dispersal barrier might demarcate a peripheral population, in which case the main geographic body of the parent population would be paraphyletic with respect to the isolated derivative (Fig. 2.16, bottom).

The latter outcome is biologically realistic in many instances. Island populations, for example, often are small and isolated from their mainland counterparts. Likewise, populations on the distributional periphery of a species may be founded by small numbers of individuals and remain separated from the primary range by ecological obstacles to gene flow. Under these and related demographic circumstances, paraphyletic outcomes in gene genealogies are virtually inevitable (Patton and Smith, 1989; Harrison, 1991). In such cases, the polyphyletic phase of the phylogenetic chronology simply is bypassed.

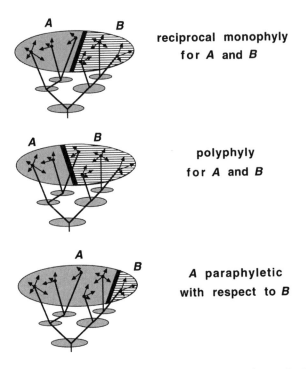

FIGURE 2.16 Demographic effects at founding on the initial genealogical status of two geographically isolated daughter populations, *A* and *B*. Shown are the phylogenetic histories of matrilines before a contemporary sundering event (black bar) separates the extant daughter populations. *Top:* The sundering event is concordant with earlier phylogeographic subdivisions of the species; *A* and *B* are reciprocally monophyletic from the outset. *Middle:* The sundering event is discordant with earlier phylogeographic subdivisions; *A* and *B* are polyphyletic at the outset. *Bottom:* The sundering event is at the periphery of the species' range; *A* is paraphyletic with respect to *B* at the outset.

A polyphyletic or paraphyletic status in the matrilineal genealogy of two daughter populations also might result from post-separation (secondary) gene flow between them. It can be difficult to distinguish between the competing hypotheses of incomplete lineage sorting and secondary gene flow to account for particular instances of gene-tree polyphyly or paraphyly among geographic populations. However, under some scenarios these two possibilities leave distinct genealogical signatures (Fig. 2.17). If the gene flow between otherwise long-isolated populations was recent only, heterotypic lineages producing a polyphyletic or paraphyletic appearance in the gene tree normally will be identical or nearly identical to some extant lineages in the second daughter population. On the other hand, if incomplete lineage sorting alone is responsible, heterotypic lineages should be highly divergent from homotypic lineages because they must coalesce to times no more recent than the age at which the populations were sundered. In essence, the two possibilities pictured in Fig. 2.17 can be thought of as "gene-flow-early" versus "gene-flow-late" alternatives, or as genealogical footprints stemming from ancient vicariance versus recent dispersal across a gene flow barrier.

An empirical illustration of these concepts involves menhaden fishes in the southeastern United States. Two morphologically distinct forms (often classified as *Brevoortia tyrannus* and *B. patronus*) inhabit temperate coastlines of the Atlantic and the Gulf of Mexico, respectively, but for the most part remain disjunct because the peninsula of southern Florida protrudes into subtropical and tropical waters unsuitable for these fishes. A molecular survey of the menhaden complex revealed two major branches (α and β) within the mtDNA gene tree (Bowen and Avise, 1990; see also Avise et al., 1989), but this genealogical partition was discordant with the anticipated distinction between Atlantic and Gulf populations. Menhaden samples in the Gulf were fixed for the α lineage, whereas Atlantic populations displayed both the α and β mtDNA lineages. Thus, Atlantic menhaden appeared paraphyletic in matrilineal genealogy with respect to Gulf menhaden (Fig. 2.18). Closer inspection revealed that several α-lineage mtDNA haplotypes shared by the Atlantic and Gulf fishes were virtually identical at restriction sites for all 18 endonucleases employed in the survey, despite the fact that the α and β mtDNA lineages appeared highly divergent in nucleotide sequence (about 5 percent on average) in these same assays. Thus, although Atlantic and Gulf menhaden populations may have been sundered historically, recent interregional movements

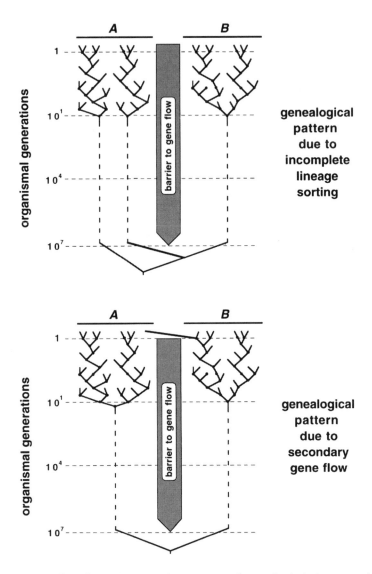

FIGURE 2.17   Two alternative scenarios to account for a polyphyletic or paraphyletic status for extant population *A* with respect to *B* in a gene tree. *Above,* incomplete sorting of lineage polymorphisms from a relatively ancient common ancestor; *below,* recent inter-population gene flow.

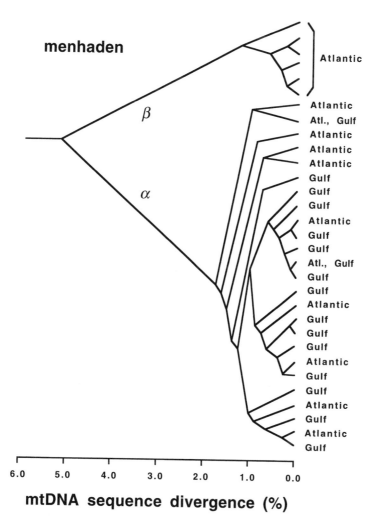

FIGURE 2.18 Empirical gene tree for menhaden fishes showing the coastal region for 31 different mtDNA haplotypes in two major lineages, α and β (after Bowen and Avise, 1990).

(perhaps of α-lineage mtDNA haplotypes from the Gulf into the Atlantic) probably have occurred as well.

SHORT-PERIOD ISOLATIONS

Long-standing population isolation tends to extend intraspecific coalescent times far beyond those expected for a nonsundered species of comparable total size, but other forms of historical population structure, ironically, can produce the reverse effect. Consider a species subdivided into numerous local populations characterized by a high rate of extinction and recolonization. Isolation in effect is short-term, and all populations may be connected tightly in a historical sense even if interdemic gene flow in most generations is low or nonexistent. For example, under a source-sink ecological model (Pullium, 1988), a species may consist of many temporary local populations (sinks) that occasionally are restocked from one or more stable viable populations (sources) located elsewhere.

Analogous evolutionary models can be envisioned wherein ephemeral local populations are restocked or recolonized periodically from relatively stable centers of origin (Fig. 2.19). In such cases, the coalescent process for a species is governed primarily by demographic processes in the source population(s) and by dynamics of the extinction-recolonization process. Under some conditions, local populations could evolve more or less in concert genetically, notwithstanding perhaps complete isolation at most horizons in time. Thus, the mean coalescent for a subdivided species could be much more recent than a face-value appraisal of the species' composite size and current geographic range otherwise might suggest.

Conditions under which population extinction and recolonization become important cohesive forces in retarding (as opposed to promoting) genetic differentiation depend on population demographic details of the process (Slatkin, 1977, 1985a; Wade and McCauley, 1984; Whitlock and McCauley, 1990; McCauley, 1991; Nei and Takahata, 1993; Neuhauser et al., 1997; Whitlock and Barton, 1997). Under various theoretical models of metapopulation structure (Hanski and Gilpin, 1996) examined by Slatkin (1985a, 1987), a general rule emerged: If the average time in generations to extinction of local populations is less than the effective number of locally

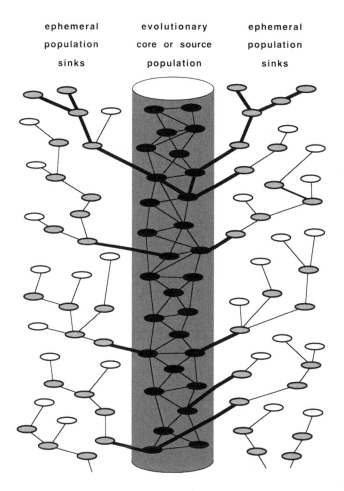

FIGURE 2.19   Lineage coalescence under an evolutionary source-sink scenario of population extinction and recolonization. Shown are many local populations (small ovals) which, with high probability in peripheral areas, tend to go extinct (open ovals) yet are recolonized (heavy lines) by propagules from a source or core population.

breeding adults, then extinction and recolonization inhibit genetic differentiation of local populations by random drift. Wade and McCauley (1988) showed that outcomes also depend on the numbers of individuals colonizing new populations relative to the numbers of individuals per generation that move between populations. Such metapopulation models introduce a realistic notion that genealogical ties among geographic populations can involve various mixes of historical connection and contemporary gene flow.

Intermediate Situations

Many species are composed of large numbers of local geographic populations among which gene flow and population connectedness have been sporadic in time, variable in space, or otherwise idiosyncratic in historical details. A philosophical strength of phylogeographic analysis is the field's attempt to augment the kinds of equilibrium expectations often invoked in traditional population genetic models with an added element of realism that only an explicit recognition of historical contingency can provide. This is not to imply that adequate recovery of the genealogical history of a species is a simple task. This section reviews briefly the major classical methods for estimating levels of genetic exchange among populations, and then describes some recent phylogeographic approaches that capitalize upon historical information from genealogical data.

TRADITIONAL APPROACHES TO GENE-FLOW ESTIMATION

Conventional population genetic methods for estimating gene flow typically deal with the spatial frequencies of phylogenetically unordered alleles such as those revealed in allozyme assays (reviews in Felsenstein, 1982; Slatkin, 1985a, 1987; Slatkin and Barton, 1989; Neigel, 1997). Most of these approaches are based on equilibrium expectations derived from theoretical models of population structure under neutrality theory. For example, an island model (Wright, 1931) supposes that a species is subdivided into populations of equal size N, all of which exchange alleles with equal probability, whereas a stepping-stone model (Kimura, 1953) supposes that any genetic exchange occurs between adjacent populations only. From observed geographic variation in population allele frequencies, a combination parameter Nm typically is estimated, where m is the migration rate (the proportion of alleles in a population that were migrants each generation). This Nm parameter is interpreted as a mean per-generation estimate of the absolute number of migrants exchanged among populations. Outcomes of Nm greater than about 1–4 indicate that the homogenizing influences of gene flow over time have overridden the diversifying effects of genetic drift, whereas values of $Nm < 1$ suggest the converse (Birky et al., 1983).

Several methods to estimate Nm are available. For example, Wright (1951) showed that for an island model of population structure, Nm at equilibrium between genetic drift and gene flow is related to F_{ST}, the interpopulational component of genetic variation (or the standardized variance in allele frequencies across populations), according to

$$Nm \cong (1 - F_{ST}) / 4F_{ST} \tag{3}$$

This theoretical relationship is plotted in Fig. 2.20 (top). Note the inflection of the curve at $Nm \cong 1$ ($F_{ST} \cong 0.20$), a value conventionally interpreted as the approximate demarcation point between "high-gene-flow" species (where at equilibrium the homogenizing influence of gene flow predominates) and "low-gene-flow" species (where the diversifying effect of genetic drift assumes prominence). The F_{ST} values themselves can be calculated from empirical allele frequency data by procedures described in Weir and Cockerham (1984) and Cockerham and Weir (1993). Analogous statistics have been proposed to accommodate nucleotide sequence data involving many alleles (Takahata and Palumbi, 1985; Lynch and Crease, 1990; Slatkin, 1991).

A second approach to Nm calculation involves a focus on private alleles—those found in only one population. The rationale is that private alleles are likely to have attained high frequency within a population only when Nm is low. Slatkin (1985b) showed by computer simulations that for a variety of theoretical population models, the natural logarithm of Nm is related to the average frequency of private alleles [$p(1)$] according to:

$$\ln(Nm) = \frac{(\ln[p(1)] + 2.44)}{-0.505}. \tag{4}$$

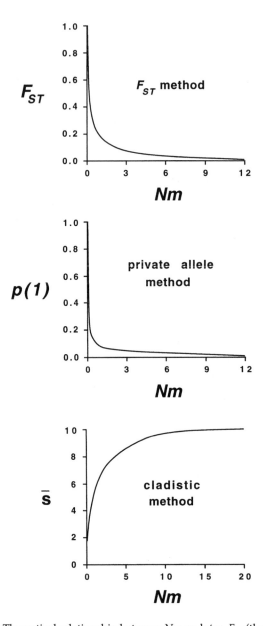

FIGURE 2.20 Theoretical relationship between Nm and: *top*, F_{ST} (the interpopulation component of genetic variation; Wright, 1951); *middle*, *p(1)* (the frequency of private alleles; Slatkin, 1985b); and *bottom*, *s* (the minimum number of historical migration events consistent with an allelic phylogeny; Slatkin and Maddison, 1989). See text and the original papers for particular demographic assumptions of the models underlying these curves.

This formula describes an expected curvilinear relationship between Nm and $p(1)$ (Fig. 2.20, middle).

A third approach to Nm calculation (Slatkin, 1989; Slatkin and Maddison, 1989) is closer to the true spirit of a phylogeographic approach because it considers the geographic distributions of branches in a gene tree. Given a correct gene-tree topology and knowledge of the geographic populations where particular allelic clades occur, a parsimony criterion is applied to estimate the minimum number of historical migration events (s) consistent with the tree. By computer simulations, Slatkin and Maddison (1989) showed that the distribution of s is a function of Nm, which therefore can be estimated from empirical data by comparison to results tabulated from their simulated populations. An example of the nature of the theoretical relationship between Nm and s for a particular set of simulated parameter values is shown in Fig. 2.20 (bottom).

Notwithstanding their considerable popularity and widespread use, these and related approaches to Nm calculation have serious limitations when used as a basis for interpreting contemporary levels of gene flow from molecular data (Bossart and Prowell, 1998). First, the theoretical expectations are bound (to varying degrees) by the parameters of the models, and by the common presumption that an equilibrium or quasi-equilibrium (Barton and Slatkin, 1986) has been achieved within a species between genetic drift and interpopulation gene flow acting on neutral alleles. However, the approximate number of generations required to approach equilibrium [$G = 1/(2m + 1/(2N_e))$; Crow and Aoki, 1984] can be large; on the order of the effective population size when migration rates are low. A related concern about Nm values is that classical estimates typically represent averages for a species and thus, as usually applied, fail to dissect potential differences in levels of genetic contact among particular populations. However, most species in nature probably have idiosyncratic histories and eccentric legacies that make nonequilibrium outcomes a sine qua non of intraspecific evolution.

Second, the curvilinear relationships between Nm and F_{ST}, $p(1)$, or s (Fig. 2.20) have the unfortunate consequence of concealing potentially important quantitative differences in the magnitudes of interpopulational genetic exchange (Templeton, 1998). Consider, for example, the theoretical relationship between Nm and F_{ST}. Two major portions of the plotted curve are almost flat such that small differences in one variable translate to huge differences in the other. Thus, any values of F_{ST} less than about 0.1 suggest high gene flow, but whether this entails the exchange of many thousands of individuals or only a few per generation cannot realistically be distinguished. Conversely, any values of F_{ST} greater than about 0.2 yield low but indistinguishably different values of Nm. Similar conclusions apply to the unfavorable behavior of Nm with respect to $p(1)$ and s. These difficulties are compounded by the fact that gene-flow theory concerns the relationship of Nm to a parametric F_{ST}, rather than to an empirical statistic F_{ST}. The latter contains, in addition to the actual variance in allele frequencies among populations, sampling variance due to finite numbers of individuals and subpopulations assayed (Nei and Chesser, 1983; Weir and Cockerham, 1984).

Even a secure knowledge on general magnitude of Nm may be insufficient for many practical purposes. An Nm estimate of 10.0 ($F_{ST} \cong 0.025$), for example, might be interpreted to document high gene flow, yet even if literally correct, would entail vastly different *demographic* impacts on large as opposed to small populations. Thus, two huge fish populations exchanging $Nm = 10$ individuals per generation would be almost completely independent in a demographic sense notwithstanding approximate genetic uniformity at equilibrium. As a consequence, Nm values could be misconstrued in certain kinds of wildlife management programs such as the identification of commercial fisheries stocks and the design of harvest allocations (Waples, 1998). In such applications, pragmatic decisions might be served better by more direct information on interpopulational movements of individuals.

A final but related reason for reservations about classical gene flow estimates is that they fail to distinguish ongoing genetic exchange from the effects of historical associations among populations. A large Nm value, for example, could mean high contemporary gene flow among a collection of populations at equilibrium, or recent historical association with zero gene flow at present, or some unspecified mix of these two possibilities (Slatkin and Maddison, 1989; Templeton and Georgiadis, 1996).

Several other statistical approaches from population genetics exist that can be applied to molecular data to address the effects of population history and geographic structure. These include spatial autocorrelation analysis (Sokal et al., 1989a,b; Slatkin and Arter, 1991; Epperson, 1993), principal component analysis (Cavalli-Sforza et al., 1994; Bertorelle and Barbujani, 1995; Cavalli-Sforza, 1997), and multidimensional scaling (Lessa, 1990). These methods typically are applied to genetic identity or distance summaries of allele frequency data, often from multiple loci, and are appropriate particularly for situations in which alleles have not been ordered into a genealogical tree.

PHYLOGEOGRAPHIC STATISTICS

With the advent of molecular methods (such as mtDNA assays) that permit estimates of genealogical relationships in addition to frequencies of alleles, new opportunities exist to consider historical gene flow and population fragmentation in a more explicit phylogenetic framework. Although the classical population genetic approaches mentioned above can be applied to gene-tree data, unless appropriately modified they fail to make full use of the historical information these data provide (Excoffier et al., 1992; Excoffier and Smouse, 1994; Barton and Wilson, 1996; Kuhner et al., 1998).

Intraspecific mtDNA trees are non-anastomose and hierarchical. Thus, methods of phylogeny reconstruction traditionally used to reconstruct species trees in macroevolutionary studies can be employed to estimate matrilineal trees for conspecific haplotypes. These include analyses that begin with quantitative genetic distances between OTUs, as well as character state methods including Hennigian cladistics, maximum parsimony, and maximum likelihood that examine matrices of qualitative trait data such as nucleotide sequences or restriction sites (Swofford et al., 1996).

Maximum parsimony methods should be well suited for intraspecific studies because the relatively short branch lengths usually entailed mostly obviate the potential difficulties of extreme rate heterogeneity and long-branch attraction that can compromise parsimony approaches in higher level systematics (DeBry, 1992; Huelsenbeck and Hillis, 1993; Kuhner and Felsenstein, 1994). Distance-based analyses likewise have an advantage at these levels because the relatively small genetic distances typifying intraspecific comparisons mean that DNA sequences are far from saturated with superimposed nucleotide substitutions. On the other hand, modest genetic distances also can compromise assignments of statistical significance to many of the putative clades in intraspecific gene trees (Crandall et al., 1994; Smouse, 1998).

The many philosophies, optimality criteria, and algorithmic methods of phylogenetic inference are reviewed elsewhere (Felsenstein, 1988, 1993; Maddison and Maddison, 1992; Hillis et al., 1993; Hillis and Huelsenbeck, 1995; Nei, 1996; Swofford et al., 1996; Weir, 1996). I intend to describe briefly and in simple terms some of the recent statistical methods that incorporate demographic-phylogenetic methods into analyses of spatial variation in intraspecific gene lineages. However, dedicated phylogeographic statistics and algorithms are still in rather preliminary stages of development and deployment (Takahata, 1988, 1991; Takahata and Slatkin, 1990; Kaplan et al., 1991; Felsenstein, 1992a; Hudson et al., 1992; Barton and Wilson, 1995, 1996; Nee et al., 1995, 1996a; Kuhner et al., 1997; Hoelzer et al., 1998).

The Neigel Approach. In rapidly evolving molecules such as animal mtDNA, mutations that delineate particular lineages may not be dispersed at rates sufficient to attain an equilibrium between genetic drift and gene flow. The Neigel approach (Neigel et al., 1991; Neigel and Avise, 1993) considers nonequilibrium distributions of gene lineages under a population scenario of isolation by distance (Wright, 1943, 1946). The method capitalizes upon a reasonable expectation that restricted gene flow in a continuously distributed species leads to a positive correlation between the spatial extent and the age of a lineage. In other words, older lineages should be more widespread than younger lineages when dispersal is limited.

In this sense, the philosophy underlying the Neigel approach bears some resemblance to the aforementioned private-allele method of Slatkin (1985b), where alleles confined to a single location are assumed to be of relatively recent origin and restrictions on interpopulation gene flow are inferred accordingly. However, the Neigel method makes no *a priori* assumptions about possible relationships between the frequency of an allele and its origination time, nor does it confine attention to rare alleles. Instead, the method examines the geographic distributions of haplotypes and clades at all temporal depths in a gene tree.

The Neigel approach envisions a spatial continuum in which mtDNA lineages are dispersed by a multigeneration random walk process. In any generation, the absolute dispersal distance of daughters from their mothers has a variance σ_F^2, the square root of which can be interpreted as a measure of the single-generation standard dispersal distance, σ_F. If two members of the same lineage coalesce G generations in the past, the probability distribution for the cumulative spatial distance between them will thus have a variance $\sigma_G^2 = 2G\sigma_F^2$. So, from an appropriate sample of pairs of mtDNA sequences of known spatial orientation, and with estimates of G based on a molecular clock as applied to the data set at hand, standard single-generation dispersal distances can be estimated as $\sigma_F = (\sigma_G^2 / 2G)^{1/2}$. Such estimates can be made for mtDNA clades grouped either by location or by age. Another attractive feature of this model is that it provides estimates of absolute dispersal distances (rather than dispersal rates) that in turn can be compared against known or suspected vagilities in the species under investigation.

The first empirical application of the Neigel approach involved a reanalysis of the phylogeographic distributions of mtDNA lineages from a continent-wide survey of the deer mouse, *Peromyscus maniculatus*. This small rodent is distributed more or less continuously throughout North America (including vast areas formerly glaciated), and shows extensive variation and geographic population structure in mtDNA restriction sites (Fig. 2.21). Based on the mtDNA phylogeographic data, Neigel et al. (1991) estimated the standard dispersal distance for the species to be about 200

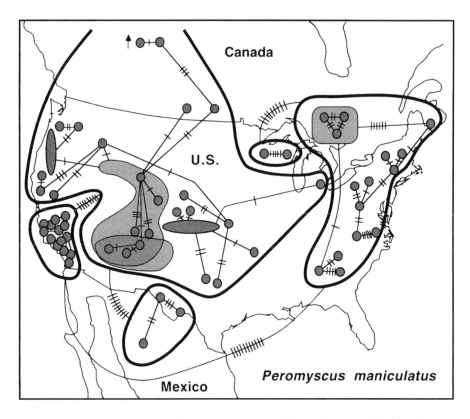

FIGURE 2.21 Geographic arrangement of a parsimony network for 61 distinct mtDNA haplotypes (shaded areas) observed in a restriction site survey of the deer mouse (after Lansman et al., 1983a). Slashes across network branches indicate inferred numbers of restriction site changes. Heavy lines encompass phylogeographic assemblages that differed by unusually large numbers of restriction sites. Due to space limitations, the parsimony network for the 14 genotypes in the Southern California assemblage is not displayed.

meters per generation for all lineage ages (Fig. 2.22). Mark-and-recapture experiments in the field indicate that individuals typically move about 250 meters between birthplaces and breeding sites (Blair, 1940; Dice and Howard, 1951). Thus the genetic estimate of single-generation historical dispersal distances in deer mice, based on the long-term cumulative dis-

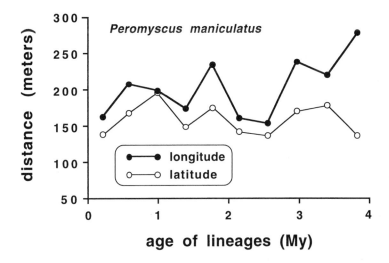

FIGURE 2.22 Standard single-generation dispersal distances (in both longitude and latitude) versus lineage ages as estimated from the phylogeographic distributions of mtDNA lineages in the deer mouse (after Neigel et al., 1991). Calculations were based on a standard molecular clock for mammalian mtDNA and a generation time of 0.2 years for deer mice.

persion of mtDNA haplotypes and clades, was consistent with a direct estimate of contemporary dispersal based on field observations.

Nonetheless, as with any indirect genetic method for estimating gene flow and dispersal, several reservations attend Neigel's approach as applied to real data sets (Barton and Wilson, 1995). First, the theory entails that independent codescendant lineages have been sampled at random with respect to geographic location. However, overlapping sets of conspecific lineages are not independent in ancestry, and most genetic surveys involve nonrandom sampling due to a concentration of collecting efforts at particular sites. Second, implementation of the model requires information on lineage coalescent times. This information typically comes from (perhaps questionable) applications of a molecular clock. Third, because the analysis yields composite estimates of dispersal distance for lineages of particular age, it fails to identify (or at least to focus explicitly upon) particular historical genealogical units that might be present within a species. Fourth, the model assumes an unconstrained random-walk dispersion of lineages, yet most species exhibit local population density regulation and barriers to dispersal. Range limits could impose an upper limit on σ_G^2 and thereby weaken the expected correlation of lineage age with geographic distance.

However, this latter concern can be turned to an advantage by noting that particular patterns of departure from the model's predictions can themselves be informative about the population history of a species. Computer simulations by Neigel and Avise (1993) showed that for a low-movement species whose total range is constrained by ecological or geographic limits, three historical stages of lineage dispersal can be recognized (Fig. 2.23):

> *Stage 1, young lineages.* Geographic constraints have not yet been encountered; σ_G^2 increases linearly with lineage age, and σ_F (standard single-generation dispersal distance) is independent of lineage age.
>
> *Stage 2, intermediate-age lineages.* Dispersal continues but is limited by geographic constraints; σ_G^2 shows a positive but nonlinear correlation with lineage age, and σ_F declines with older lineages.
>
> *Stage 3, old lineages.* An equilibrium between genetic drift and migration has been achieved; σ_G^2 shows no correlation with lineage age, and σ_F declines with lineage age.

The absolute time scales over which these stages are reached depend on the geographic range of the species relative to the magnitudes of the dispersal distances in the supposed random walks.

As judged by the independence of σ_F with mtDNA lineage age in the deer mouse (Fig. 2.22) and by the increase of σ_G^2 with lineage age (not shown), this species would appear to be in a nonequilibrium phase (Stage 1) of phylogeographic differentiation. Similar conclusions were reached for two other rodent species with limited dispersal, the pocket gopher, *Geomys pinetis,* and the old-field mouse, *Peromyscus polionotus*

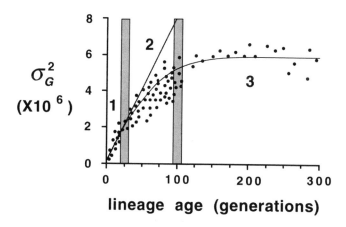

lineage age (generations)

FIGURE 2.23 General form of the theoretical relationship between σ_G^2 (the variance in the spatial distance between lineages) and lineage age measured in generations, under an isolation by distance model with constraints on the geographic range of the species. The points are from computer simulations with demographic conditions specified in Neigel and Avise (1993). Note that the variances are truncated for older lineages because the available space becomes saturated. Boundaries (shaded bars) between stages 1, 2, and 3 are somewhat arbitrary because the transitions are gradual.

(Neigel and Avise, 1993). By these same criteria, two avian species (the red-winged blackbird, *Agelaius phoeniceus,* and common grackle, *Quiscalus quiscala*) and two marine fish species (the American eel, *Anguilla rostrata,* and hardhead catfish, *Arius felis*) fell into the equilibrium Stage 3 of phylogeographic divergence, as might have been predicted based on their high vagilities. Furthermore, estimates by the Neigel approach of the standard dispersal distances for these vagile marine fishes and avian species (3–11 km per generation) were more than an order of magnitude greater than those obtained for the relatively sedentary rodents (< 0.2 km per generation).

The Templeton Approach.   This method (Templeton, 1993, 1994, 1998; Crandall and Templeton, 1993) involves an overlay of geography on an estimated gene tree in a rigorous statistical framework designed to measure the strength of any geography/phylogeny associations and to interpret the evolutionary processes responsible. The approach begins by estimating an unrooted cladogram (i.e., from character-state data on mtDNA haplotypes) by statistical parsimony procedures detailed in Templeton et al. (1987, 1992), Templeton and Sing (1993), and Crandall et al. (1994). The initial result is a series of nested clades in a gene tree, each successive level in the hierarchy necessarily older than any clade contained within it.

Then geography is overlaid on the nested clades and two types of distance measures calculated (Fig. 2.24): a clade distance, D_c, defined as the mean spatial distance of members of a clade from the geographical center of the clade; and a nested-clade distance, D_n, defined as the mean spatial distance of members of a nested clade from its geographical center. These distances summarize spatial dispersion patterns. Depending on the biology of the species, the spatial distances could be straight-line distances, or they might be, for example, river distances for a riparian species. To assess departures from the null hypothesis of no geographical association, permutation tests are conducted (Templeton and Sing, 1993) that permit ascertainment of which clade and nested-clade distances are statistically small or large.

Also calculated are the mean differences between the clade and nested-clade distances for tip clades versus interior clades (Fig. 2.24). Tip clades are those connected to the rest of the cladogram by only one mutational pathway, whereas interior clades by definition do so by two or more mutational paths. Although the gene-tree networks are unrooted, tip clades tend strongly to be younger than interior clades (Castelloe and Templeton, 1994). Thus, these distance measures contrast geographic dispersion patterns for younger versus older clades. Random permutation tests again are conducted to assess statistical significance.

One advantage of the formal statistical design of the Templeton approach is that it permits rigorous tests of the null hypothesis of no association between geography and the inferred structure of a gene tree. Failure to reject the null hypothesis could be due to biological factors (namely, high contemporary gene flow or recent historical associations among populations) or to insufficient power of the statistical tests due to small sample

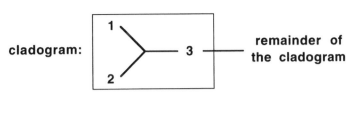

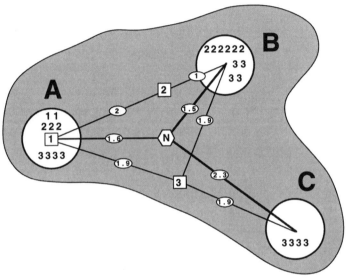

FIGURE 2.24   A hypothetical example illustrating the calculation of geographic clade distances (D_c) and nested clade distances (D_n) in Templeton's approach to statistical phylogeographic inference (after Templeton and Georgiadis, 1996). Observed representations of three haplotypes (1, 2, and 3) in each of three geographic sampling areas (A, B, and C) are used as a basis for calculating the D_c and D_n values and their differences for the interior clade (3) versus the two tip clades (1 and 2). The geographic center of a particular haplotype is indicated by a square containing the number of the haplotype, the geographic center of the entire nested clade N is indicated by the hexagon, and numbers within ovals indicate straight-line distances between these geographic centers and the sampling locations. Calculations are as follows:

$D_c(1) = 0$, $D_c(2) = (\frac{1}{3})(2) + (\frac{2}{3})(1) = 1.33$, and $D_c(3) = 1.9$;

$D_n(1) = 1.6$, $D_n(2) = (\frac{1}{3})(1.6) + (\frac{2}{3})(1.5) = 1.53$, and $D_n(3)$
$\quad = (1.6 + 1.5 + 2.3)/3 = 1.8$;

D_c (interior) $- D_c$ (tip) $= 1.9 - (0 + 1.33)/2 = 1.23$;

and D_n (interior) $- D_n$ (tip) $= 1.8 - (1.6 + 1.53)/2 = 0.23$

size or inadequate geographic sampling. Furthermore, when the null hypothesis is rejected, the particular behaviors of the distance statistics can be informative about underlying biological causes for the associations between phylogeny and geography (Templeton, 1993, 1994; Templeton et al., 1995).

For example, restricted gene flow among a set of geographic populations should produce significantly small D_c values particularly for tip clades, whereas a recent large range expansion should be registered as significantly large D_c values for tip clades. Long-distance colonization by only a few propagules in an otherwise low-gene-flow species should result in heterogeneous D_c estimates for tip clades, with a few values being significantly large and the remainder significantly small. Under historical allopatric fragmentation of an otherwise high-gene-flow species, one expects significantly large D_c values at low clade levels, significantly small D_c values at higher clades, D_n values that suddenly increase rapidly at a crucial clade level, and larger than average numbers of mutation steps connecting the geographically sundered clades to the remainder of the cladogram. Many additional predictions about the behavior of clade-distance statistics as functions of historical population demography are presented in Table 1 of Templeton et al. (1995). The same paper also presents an 18-step dichotomous key for progression through the chains of inference possible from alternative outcomes in the clade-distance measures.

The nested-clade design underlying an application of the Templeton method to a mtDNA data set for the tiger salamander (*Ambystoma tigrinum*) is illustrated in Fig. 2.25. Statistical analyses of the clade-distance values derived from this network formally supported the following molecular-based inferences, all of which are consistent with independent evidence about the dispersal behavior and probable historical population subdivisions within this species: (a) range expansions account for the distributions of mtDNA haplotypes nested within clades 1-1 and 2-2; (b) restricted gene flow via isolation by distance can account for haplotype distributions within several of the nested clades; and (c) allopatric fragmentation accounts for the genetic distinction between the two highest

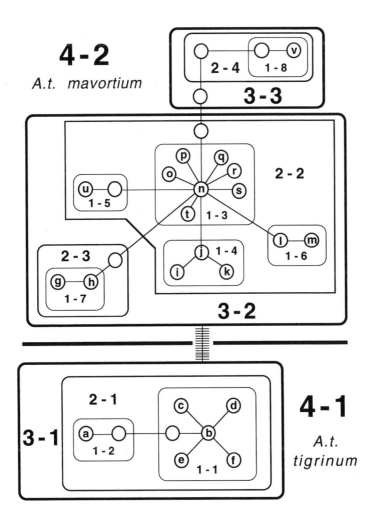

FIGURE 2.25   Nested clades in an mtDNA phylogenetic network for tiger salamanders collected in the central United States (after Templeton et al., 1995). Circles encom-
passing letters a–v indicate observed haplotypes, whereas open circles indicate
hypothetical intermediate haplotypes not observed. Branches connect haplotypes
differing by a single mutational step, except for the branch connecting clades 4-1
(*A.t. tigrinum*) and 4-2 (*A.t. mavortium*), which involved a minimum of 14 mutations.
Haplotypes are grouped into more-inclusive nested clades as indicated by increasingly
thick boxes and by increasing fonts in the numerical designations.

level clades, 4-1 and 4-2. These latter clades in the mtDNA gene tree correspond precisely to two named subspecies (*A.t. tigrinum* in Missouri and *A.t. mavortium* in Kansas, Nebraska, and Colorado) that also display pronounced morphological and life-history differences.

As noted by Templeton et al. (1995), all of these conclusions from quantitative analyses of the mtDNA data likely would have been reached also from a simple pictorial overlay of mtDNA network upon geography. The advantage of the formal statistical analyses (complicated though they are) is that they make the evolutionary inferences explicit and less subjective. Additional applications of this statistical method to mtDNA data sets for other species can be found in Crandall and Templeton (1996) and Templeton and Georgiadis (1996).

EXTENSIONS TO NUCLEAR GENEALOGIES

The introduction to lineage sorting concepts and coalescent theory presented thus far has been framed primarily as expectations for the nonreticulate, matrilineal component of an organismal pedigree (through which cytoplasmic genomes normally are transmitted). This section demonstrates that the same kinds of genealogical principles apply, at least in theory, to nuclear alleles whose multi-generation transmission routes involve both genders. It concludes that a coalescent process can be envisioned for each nuclear gene that is analogous to the coalescent description for a matrilineal gene tree. For autosomal genes in diploid sexual taxa, the primary theoretical difference is the fourfold adjustment required to account for the larger effective population sizes of alleles (all else being equal). However, additional complications arise in attempts to estimate nuclear gene trees empirically.

The Concept of Nuclear Gene Trees

The concept of hierarchical, nonreticulate gene trees within the nuclear genomes of sexually reproducing organisms can be introduced by reference to the branching structure of patrilines through an extended organismal pedigree (Fig. 2.26). Across multiple generations, a patrilineal

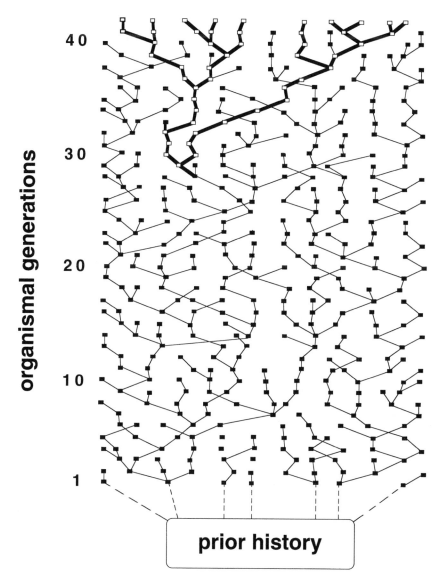

FIGURE 2.26 The patrilineal history of a population, analogous to but distinct from the matrilineal schematic presented in Fig. 2.1.

genealogy (traversed by the mammalian Y-chromosome, or by surnames in many human societies) is analogous to a matrilineal genealogy traversed by mtDNA except that males (M) rather than females (F) have been the transmitting gender. Thus, all of the theoretical expectations of lineage sorting and coalescent theory for matrilines can be applied to patrilines by merely substituting M for F (e.g., as in N_M for N_F) in any of the equations presented earlier.

The comparative coalescent patterns of patrilines versus matrilines are influenced by the demographic histories of the two sexes, which may differ dramatically. For example, breeding sex ratios in some species depart from 1:1, with evident lineage-sorting consequences arising from different effective population sizes for the two genders. In many species, members of one sex (often males) have larger variances in reproductive success such that (in that case) patrilineal sorting within a population would proceed faster and tend to cause the coalescent to be reached more recently than for matrilines. Differences between the sexes in fitness variance often are associated with the mating system. Males in highly polygynous species, for example, often have higher variances in offspring number than do males in monogamous taxa. Also, in many mammals and other species, males tend to be the more dispersive gender, whereas in birds the reverse often is true (Greenwood, 1980; Greenwood and Harvey, 1982). Such differences in philopatric tendencies between the sexes can produce distinct phylogeographic patterns for genes inherited along matrilineal versus patrilineal routes (Melnick and Hoelzer, 1992).

The matrilineal and patrilineal components of an organismal pedigree can be notated as the $F{\rightarrow}F{\rightarrow}F. . .{\rightarrow}F$ and the $M{\rightarrow}M{\rightarrow}M. . .{\rightarrow}M$ extended transmission pathways, respectively. Within any species, the entire collection of each of these ancestral-descendant paths describes a non-anastomose tree. Within the same organismal pedigree, these genealogical trees display different topologies uniquely influenced by the demographic histories of females and males. Another important point is that the patrilineal and matrilineal gene trees fall far short of describing the total genealogical information in an organismal pedigree. Huge numbers of other

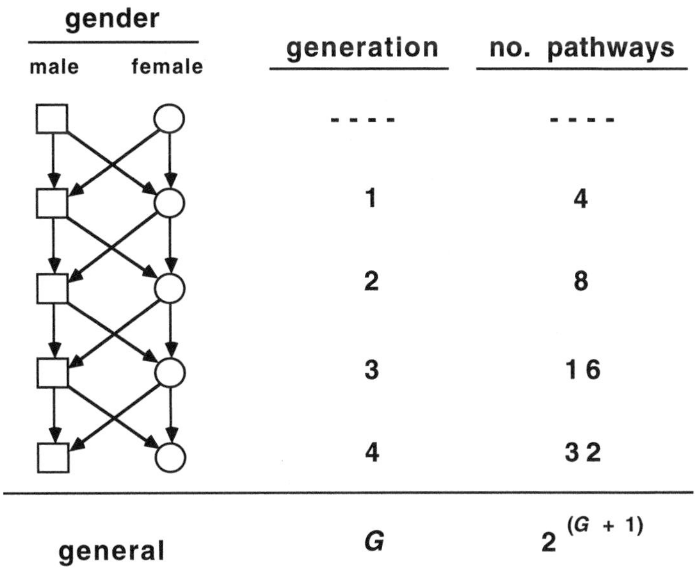

FIGURE 2.27   Total numbers of gender-defined transmission routes potentially available to alleles at autosomal loci within extended organismal pedigrees spanning G generations (after Avise, 1995).

extended transmission routes will have been available to nuclear alleles in each generation of sexual reproduction (Fig. 2.27).

The coalescent for multiple alleles at an autosomal locus will have been achieved via multifarious, gender-unspecified transmission routes through a pedigree. All else being equal, this entails for real autosomal genes a fourfold suppression in rates of lineage sorting and a fourfold increase in mean coalescent times in comparison to expectations for the gender-defined allelic genealogies such as those through which mtDNA or the Y-chromosome have been transmitted.

This baseline fourfold difference can be demonstrated in several ways. First, it can be seen to stem from a twofold effect due to diploidy for autosomal genes, coupled with a twofold effect due to transmission through two sexes rather than one. Second, the difference can be noted by reference to any generation of transmission as pictured in Fig. 2.27. For an auto-somal locus, four transmission routes connect one generation to the next, compared to the single route available to mtDNA or to the Y-chromosome. A third way to understand the fourfold difference can be illustrated by interdigitating the matrilineal (e.g., Fig. 2.1) and patrilineal (Fig. 2.26) components of an organismal pedigree (a twofold effect), and then inter-weaving these gender defined pathways into a single composite represen-tation that summarizes the complete history of parent pairs and their offspring (Fig. 2.28). To weave this completed pedigree, additional lines of transmission (another two-fold effect) had to be drawn each generation: one connecting each daughter to her father, and another connecting each son to his mother.

However, the fourfold effect is only a baseline expectation. Coalescent times for mtDNA actually can exceed those for nuclear genes under some special combinations of demographic parameters related to the mating system, relative variances in reproductive success between the sexes, and magnitudes of population structure (Birky et al., 1989; Moore, 1995, 1997; Hoelzer, 1997).

The coalescent highlighted in Fig. 2.28 refers to alleles from a ran-domly chosen nuclear locus. Unlike the histories of matrilineal or patri-lineal trees, two autosomal alleles at a nuclear gene may trace to a single individual and then reemerge before finally coalescing to a single allele in some earlier ancestor. Noncoalescence within an individual occurs (with 50 percent chance in each case) because of diploidy and random Mendelian segregation. In any event, all extant alleles at each autosomal locus eventually will have coalesced to a single copy housed within an an-cestor who is expected to have lived about fourfold generations earlier (with caveats mentioned above) than the respective ancestors from which extant patrilines or matrilines descended. For the autosomal gene moni-tored in Fig. 2.28, the coalescent occurred in the fourth individual from the left in generation 1. Alleles at other unlinked loci likely would have coa-lesced elsewhere in the pedigree.

These quantitative expectations regarding lineage coalescence per-tain to neutral alleles. Various forms of natural selection can alter these

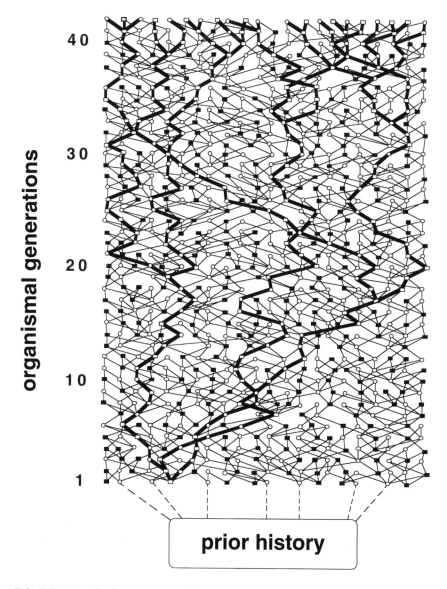

FIGURE 2.28 Coalescent process (heavy lines) for an autosomal gene within an organismal pedigree spanning more than 40 generations. This pedigree is an amalgamation of its matrilineal and patrilineal components as presented separately in Figs. 2.1 and 2.26, with additional pathways connecting each daughter to her father and each son to his mother.

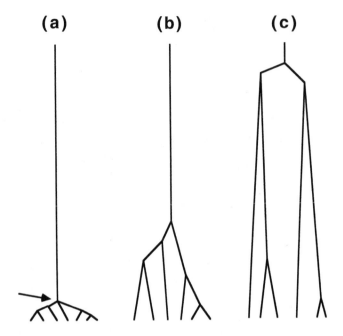

FIGURE 2.29   The expected shapes of gene genealogies within a population under different selection models (modified from Aguadé and Langley, 1994). *(a)* Star phylogeny under a recent selective sweep or hitch-hiking event (beginning at the arrow) mediated by positive selection on a favored mutant; *(b)* typical genealogy under a neutral model; *(c)* deeply branched genealogy under long-term balancing selection.

expectations (Fu and Li, 1993; Golding, 1997). As described earlier for mtDNA, positive directional selection on a nuclear allele may promote a selective sweep and thereby shorten the coalescent time for the gene and adjacent linked sequences. Background selection against deleterious mutations also eliminates neutral variation at linked sites, and, like positive selection, tends to reduce effective population size and coalescent time in a given gene region (Charlesworth et al., 1993; Hudson and Kaplan, 1996).

   On the other hand, balancing selection can inhibit allelic extinction at a locus and in adjoining sites, sometimes extending coalescent times for surviving alleles greatly (Fig. 2.29). Well-studied examples involve genes of the major histocompatibility complex in mammals (Takahata, 1990; Takahata and Nei, 1990; Nei and Hughes, 1991) and a self-incompatibility gene in plants (Ioerger et al., 1990; Clark, 1993). In both systems, balancing selection appears to have maintained some haplotype lineages over time scales far longer than expected under neutrality theory: e.g., tens of millions of years, and across multiple speciation events (Klein, 1986; Figueroa et al., 1988; Klein et al., 1993; Clark, 1997).

Empirical Complications

As a logical extension of Mendelian genetic principles and population demography, intraspecific nuclear gene trees exist in theory. However, their empirical recovery is less than straightforward. To date, technical and biological hurdles have conspired to hinder the retrieval of nuclear gene genealogies within most species.

TECHNICAL HURDLES

For diploid organisms, the major technical challenge is to isolate haplotypes one at a time from particular nuclear genes (a task nature routinely provides for mtDNA by producing individuals that are effectively homoplasmic). Traditional methods of nuclear DNA isolation normally are inadequate. In the polymerase chain reaction (PCR), for example, both alleles in a heterozygous individual usually are amplified from a target locus, such that subsequent assays fail to distinguish between alternative genetic configurations possible for the two haplotypes when they differ at multiple nucleotide positions (Fig. 2.30). Further complications can arise if heterogeneous sequencing products are present as a result of the target locus belonging to a family of genes with similar PCR priming sequences, or if (as can happen) the PCR reaction itself generates recombinant DNA products from two amplifying alleles of a single gene in a heterozygous individual (Bradley and Hillis, 1997).

Various approaches might circumvent these difficulties (Avise, 1994). Some proposed methods would take advantage of nature's own haplotype purification systems. For example, haplotypes from single-copy genes on

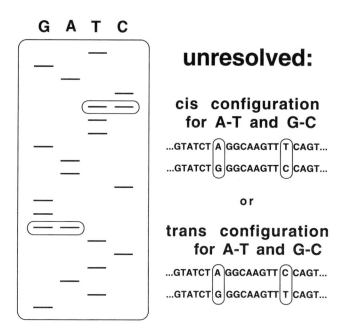

FIGURE 2.30 The nature of ambiguity in haplotype assignment from conventional assays of nuclear DNA isolated from a diploid individual who is heterozygous at multiple sites within a gene. *Left:* Portion of a sequencing gel indicating heterozygosity at two positions (ovals). *Right:* Two alternative but unresolved possibilities for the haplotype sequences carried by that specimen. Such ambiguities in haplotype assignment become more severe as the number of heterozygous sites increases.

sex chromosomes could be characterized directly in DNA isolated from the heterogametic sex. Or, especially in species with a prominent haploid phase of the life cycle, haplotypes at nuclear loci could be examined directly in DNA preparations from haploid tissue.

Other approaches rely on laboratory manipulations to isolate haplotypes. One method with broad taxonomic applicability involves cloning of PCR products through a biological vector (Scharf et al., 1986). However, cloning occurs through a single molecule, so the possibility of nucleotide misincorporation by *Taq* polymerase in the preceding PCR step is of some

concern (Keohavong and Thilly, 1989). In practice, possible misincorpora-
tions either are ignored in phylogenetic analysis (Palumbi and Baker, 1994;
Vogler and DeSalle, 1994a), or (in an expensive and laborious safeguard),
multiple separate clones from each individual are sequenced and compared to distinguish true allelic variants from PCR-cloning artifacts
(Bernardi et al., 1993). Another approach in haplotype isolation is applica-
ble to species that can be bred in the laboratory. In *Drosophila*, controlled
crosses are used routinely to "extract" individual haplotypes from nuclear
genes (Aquadro et al., 1986; Hey and Klinman, 1993).

In recent years, molecular procedures for DNA-screening have been
introduced (Lessa and Applebaum, 1993; Potts, 1996; Zhang and Hewitt,
1996) that generalize the possibilities for direct physical isolation of individual haplotypes from diploid tissue sources (Ortí et al., 1997). Two such
procedures are abbreviated SSCP (for single-strand conformational polymorphism; Orita et al., 1989a,b; Hongyo et al., 1993) and DGGE (denaturing gradient gel electrophoresis; Myers et al., 1986, 1989a,b). These
techniques begin with double-stranded DNA amplified via gene-specific
PCR primers, and both methods permit a physical separation in appropriate gels of the two homologous DNA molecules from a diploid hetero-
zygote. Then, the separated bands (haplotypes) can be cut from the gel and
further assayed, for example to determine nucleotide sequence.

Despite the recent availability of physical isolation procedures for individual haplotypes from diploid nuclear genes, few attempts as yet have
been made to capitalize upon these approaches as a starting point for estimating intraspecific nuclear gene trees (Hare and Avise, 1998). More often,
these methods have been used to screen for mutations and assess heterozygosity, without further genealogical characterization of the separated haplotypes.

BIOLOGICAL HURDLES

Even if the technical hurdle of isolating nuclear haplotypes could be overcome routinely, at least two inherent biological hindrances often remain to
complicate the recovery of intraspecific nuclear gene trees. First, a nuclear

locus may have evolved too slowly to provide a sufficient number of phylogenetically informative sites for the problem at hand. Second, nuclear DNA sequences may have experienced recent intragenic recombination. If so, the phylogenetic histories of mutational events may have been garbled by recombinational swaps between different branches in the intraspecific gene tree.

Molecular and cytological factors apparently can influence the frequency of recombination within a nuclear gene. In *Drosophila*, rates of recombination between adjacent nucleotides differ among chromosomal regions by more than 100-fold, with low rates typically characteristic of telomeric and centromeric sequences and the entire fourth chromosome. Surveyed genes in low recombination regions tend to be characterized by reduced natural selection (Klinman and Hey, 1993) and decreased levels of nucleotide sequence diversity, likely because of genetic "hitch-hiking" (Aquadro and Begun, 1993). When an occasional advantageous mutation sweeps to fixation within a population or species, adjacent linked sequences in effect are cleansed of preexisting variation, and such effects in theory and apparently in practice are more dramatic in nuclear regions characterized by low recombination (Aquadro, 1992; Begun and Aquadro, 1992; Aguadé and Langley, 1994; Aquadro et al., 1994). This finding, if general, poses a genealogical Catch-22 for microevolutionary appraisals of nuclear loci: Gene trees in principle are best recovered from genomic regions with low recombination, but such regions often may carry too few polymorphic markers for fine-scale phylogenetic resolution.

The nature of population subdivision also can influence effective rates of recombination. For example, in a low-gene-flow species with longstanding spatial structure, the field for genetic recombination is primarily within each population rather than among them (Baum and Shaw, 1995). In *Drosophila melanogaster*, only a few assayed nuclear loci (notably ADH; Fig. 2.31) have displayed adequate polymorphism *and* linkage disequilibrium (nonrandom associations of sequence variants) to permit robust estimates of nuclear gene trees (Aquadro, 1993). However, *D. melanogaster* experienced a global range expansion recently and in general is characterized by high gene flow, so this species may be an inappropriate model

ADH haplotype tree
Drosophila melanogaster

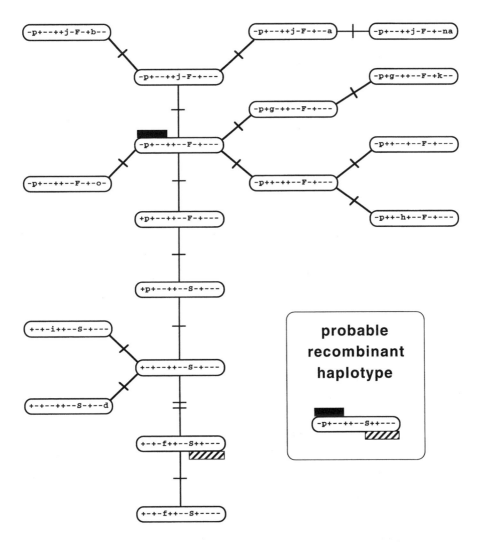

FIGURE 2.31 Classic example of an intraspecific gene tree (alcohol dehydrogenase gene, *Drosophila melanogaster*) exemplifying how interallelic recombination at a nuclear locus sometimes can be deduced from phylogenetic analysis of haplotype data (after a broader genealogy presented in Aquadro et al., 1986). Shown is a parsimony network for 18 haplotypes as identified by 15 restriction sites and other DNA sequence characters (coded from left to right in 5′→3′ direction for each circled haplotype). Note from the solid and hatched bars that the 5′ end of the presumed recombinant haplotype appears to stem from one portion of the network and the 3′ end from another.

from which to draw broad pessimistic lessons about the feasibility of recovering informative nuclear gene genealogies at the intraspecific level.
Although assessments of nuclear gene trees within other species have
barely begun, in one recent study of a strongly structured taxon (the pathogenic fungus *Coccidioides immitis*), several independent gene trees were
recovered successfully and proved to be highly informative about the phylogeographic history of the complex (Koufopanou et al., 1997).

Documenting the historical incidence of intragenic recombination
within a species is another difficult challenge. Conditions under which
particular instances of interallelic recombination can be identified from
haplotype data are restrictive. Typically required are two or more arrays of
phylogenetically divergent haplotypes whose defining sequence characters are in strong linkage disequilibrium. Then, occasional putative recom-
binants may be identified as clustered combinations of these diagnostic
molecular characters (Fig. 2.31). Such special circumstances are likely to be
rare, and in their stead several quantitative statistical, graphical, and
phylogenetic methods have been proposed to estimate recombination
rates from populational data on nuclear haplotype sequences (Stephens,
1985; Hudson, 1987; Sawyer, 1989; Hein, 1990, 1993; Maynard Smith, 1992;
Templeton and Sing, 1993; Crandall et al., 1994; McGuire et al., 1997; Maynard Smith and Smith, 1998; Weiller, 1998). Many of these methods treat
recombination as an evolutionary process that produces the appearance of
homoplasy (i.e., evolutionary convergence, parallelism, or reversals) in a
gene tree. For example, historical recombination might be inferred if two
or more otherwise separate homoplasies in a gene tree could be resolved
by invoking a single recombination event. However, most statistical approaches place only a conservative lower bound on the frequency of
interallelic recombination (Hudson and Kaplan, 1985). Furthermore, if recombination was frequent in the history of DNA sequences under examination, all attempts to recover a detailed genealogy of haplotypes are
compromised severely. Indeed, with respect to intraspecific gene genealogies, recombination violates a fundamental assumption of phylogenetic
analysis: that tree branches are nonreticulate.

MATRILINES: A SPECIAL CONNECTION
BETWEEN PHYLOGENY AND DEMOGRAPHY

Before closing this chapter, a final point should be emphasized concerning a special theoretical link between matriarchal phylogeny and population demography. For any gene tree, the transmission history of mtDNA represents only a minuscule fraction of the potential hereditary pathways within a population pedigree. This realization has prompted justifiable concern about drawing population-level inferences from such a small fraction of the genome (Cronin, 1993; Degnan, 1993), and has emphasized the importance of acquiring information from multiple nuclear loci before firm phylogeographic conclusions are drawn for any species (Avise and Ball, 1990). Yet, in one important sense (especially germane to conservation efforts; Milligan et al., 1994), a species' matrilineal history, by itself, can be uniquely informative with regard to population demography (Avise, 1995).

There are three elements to the argument for why mtDNA registers more than "just another" gene genealogy within an organismal pedigree. First, dispersal and gene flow are highly asymmetric by gender in many species, with females commonly philopatric to natal site. Second, in many animals (notable exceptions include marine species that disperse eggs widely into the environment), females and their young usually are proximate when offspring begin independent life. If female progeny remain philopatric to natal site or social group, either by active choice or passively because of limited dispersal capabilities, a species inevitably becomes spatially structured along matrilines. Third, a strong matrilineal population structure (as might be registered by geographic variation in mtDNA haplotypes) implies considerable demographic autonomy for local populations, at least over short (ecological) time.

This argument holds even for otherwise panmictic species. Imagine, for example, a broadly distributed species in which females are fidelic to natal site but males disperse widely and mate randomly with females from any locale. Frequencies of nuclear alleles (autosomal as well as sex-linked) quickly would homogenize across locales, yet the demographic fate of all local populations would be nearly independent. Because recruitment of young is contingent upon female reproductive success, any local population that was compromised or extirpated by humans or natural causes would be unlikely to recover or re-establish in the short term via recruitment of nonindigenous females. Thus, a critically important demographic feature of the species would be missed entirely in face-value geographic appraisals of nuclear genes. For this same species, however, the distribution of mtDNA lineages likely would reveal a dramatic spatial structure pointing correctly to the demographic autonomy of populations in different areas.

Fig. 2.32 summarizes theoretical relationships between population demography and interpopulational gene flow, the latter categorized with respect to genetic markers with alternative gender-based transmission pathways. When both sexes disperse widely from natal sites to reproduce (lower right quadrant), little population genetic structure is anticipated in any class of neutral genetic markers, the inference perhaps also being that the populations are connected demographically. Conversely, when both sexes are sedentary (upper left quadrant), strong population genetic structure should be registered in any suitable cytoplasmic or nuclear marker, implying a considerable autonomy in population demography.

When female dispersal is high and male dispersal low (upper right quadrant of Fig. 2.32), a Y-linked gene may or may not show strong population structure, depending on whether the vagabond females carry zygotes or unfertilized eggs. Suppose that females who move between populations carry haploid gametes only, which then are fertilized by local males. Only then might strong differentiation in Y-linked allele frequencies be anticipated, and these populations also would tend to be demographically independent from one another with respect to male reproduction (e.g., if all males at a location died, the population would go extinct due to lack of recruitment options). Otherwise, vagile females who dispersed zygotes (or juveniles) between populations could maintain demographic ties (as well as avenues for exchange of Y-linked alleles) between the geographic locales to which males were strictly fidelic.

female dispersal and gene flow

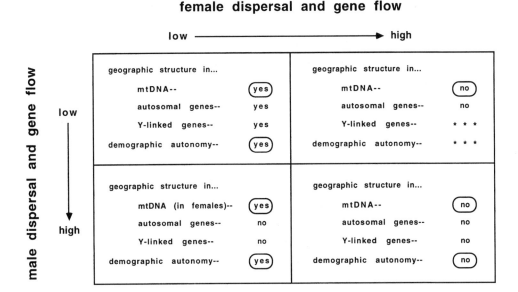

FIGURE 2.32   Relationships between population genetic structure and gender-specific dispersal and gene-flow regimes (after Avise, 1995). Interpopulation gene movement is categorized here with respect to neutral genetic markers with alternative gender-based transmission pathways. See text for details and qualifications.

The most intriguing connection between population demography and matrilineal structure occurs when female dispersal is extremely low and male dispersal high (lower left quadrant in Fig. 2.32). Then populations could be independent demographically even in the absence of significant spatial structure as registered in any nuclear genes. In that case, gene-flow estimates based solely on nuclear loci could provide a grossly misleading base for management decisions requiring a demographic perspective, such as how many population stocks exist, how they might respond to harvest, or how habitat corridors between refugia might couple otherwise separated populations.

However, there are conditions under which inferred patterns of female dispersal and matrilineal gene flow also could be misleading in demography-based population management. For example, if a strong geographic matrilineal structure resulted solely from density-dependent restrictions to female movement, overexploited populations nonetheless might recover quickly by foreign recruitment because of a relaxation of density-based impediments to female dispersal. This situation might apply with special force to high-fecundity species such as many marine fishes and invertebrates, where immigration of even a few gravid females might quickly replenish a local population. Another situation could involve species that conform to a source-sink demographic model in which most geographic populations would persist via continued recruitment from reproductively favorable areas. Matrilines would exhibit little geographic structure, suggesting at face value that local populations could recolonize quickly, yet extirpation of critical source population(s) could doom entire regional assemblages.

Several additional caveats apply to the exclusive reliance on mtDNA as a phylogeographic marker: the molecular assays may fail to capture some true matrilineal population structure; the empirical patterns might be attributable to evolutionary forces other than historical gene flow and genetic drift, such as habitat specific natural selection; and the historical gene flow might have been high enough to homogenize the frequencies of matrilines across locales ($N_F m >> 1$), yet too low to imply population unity in a demographic sense (especially likely to be true when N is large). Thus, in assessing whether populations exhibit sufficient demographic autonomy to qualify as quasi-independent entities in contemporary time, direct field observations and experiments on female dispersal will remain extremely important as adjuncts to phylogeographic information revealed in mtDNA.

SUMMARY

1. The concepts of gene phylogeny and population demography are related intimately in microevolutionary studies of conspecific populations and related species. Coalescent theory is a discipline that deals with the mathematical and statistical expectations for the his-

torical relationships among gene lineages as a function of population demographic parameters such as the means and variances in the distributions of family size, and magnitudes of interpopulational gene flow. Demographic variables are deciding factors that govern the depth, shape, and phylogeographic patterns of microevolutionary gene trees.

2. Within a large unsundered population or high-gene-flow species, the coalescent time for neutral genes is on the order of the evolutionary effective population size. A general rule has emerged from mitochondrial gene trees empirically evaluated against coalescent expectations: long-term evolutionary effective population sizes appear to be orders of magnitude smaller than present-day breeding census sizes for most currently abundant species. Such disparities are due to historical demographic factors (including, perhaps, selective sweeps) that have channeled surviving matrilines through fewer female ancestors.

3. Lineage sorting theories can be extended to spatially structured species. Collectively, viable populations that have remained isolated for long periods of time can display far greater depths in gene genealogies than nonsundered populations of the same total size. However, historical demographic details of metapopulation structure impact expected phylogeographic patterns and categories of relationship. Several dedicated methods of phylogeographic data analysis offer historical perspectives on population demography and gene flow that differ considerably from those employed in the more traditional theories of equilibrium population genetics.

4. With only minor modification, the expectations of branching process theory and lineage coalescence can be extended to nuclear genes. However, several complications commonly arise in empirical attempts to recover microevolutionary allelic genealogies from autosomal loci: (a) the identification of loci with suitably rapid evolution; (b) the technical problem of isolating haplotypes one at a time from diploid tissues; and (c) the possibility of intragenic

recombination. Recently developed laboratory methods (e.g., PCR-SSCP and DGGE) promise to alleviate the second difficulty. The other two complications are biological and can be avoided only if rapidly evolving genes are targeted that have been mostly free of interallelic recombination.

5. Any gene tree represents only a minuscule fraction of the genetic history (the pedigree) of a species. Thus, caution must be exercised in drawing population-level conclusions from gene-tree data at few loci. On the other hand, gene genealogies contain historical information beyond what can be retrieved from allelic frequencies alone. Furthermore, because of the usual proximity of females to their offspring, a species' matrilineal pedigree (as might be recovered from mtDNA) often yields demography-relevant conclusions about historical population structure that in principle are unobtainable from any nuclear gene.

EMPIRICAL INTRASPECIFIC PHYLOGEOGRAPHY

Theories of lineage sorting and the coalescent indicate that a great diversity of intraspecific phylogeographic patterns might be anticipated as functions of species' varied demographic histories. Empirical findings (primarily from mtDNA assays) appear to have borne this out. This section illustrates the wide assortment of idiosyncratic phylogeographic patterns documented to date. It asks whether general trends nonetheless have emerged that carry predictive power in relating comparative phylogeographic patterns across taxa to particular categories of natural history or environmental circumstance.

3

LESSONS FROM HUMAN ANALYSES

NOT SURPRISINGLY, no species has received greater attention in gene genealogical analyses than *Homo sapiens* (reviews in Boyce and Mascie-Taylor, 1996; Tashian and Lasker, 1996; Donnelly and Tavaré, 1997). A brief summary of this body of research can serve here to introduce the strengths as well as limitations of molecular phylogeographic appraisals. Several early population-level surveys examined mtDNA genealogy in humans (e.g., Brown and Goodman, 1979; Crews et al., 1979; Denaro et al., 1981; Aquadro and Greenberg, 1983; Johnson et al., 1983; Cann et al., 1984; Greenberg et al., 1986; Whittam et al., 1986). However, two studies stand out as having had exceptional historical and conceptual impact.

First, using restriction site assays, Brown (1980) observed only limited sequence divergence (p) in mtDNA (mean $p = 0.0036$) among 21 humans of diverse racial and geographic origin. Using mtDNA clock calibrations (1 percent or 2 percent sequence divergence between lineages per My), Brown (1980) concluded that this magnitude of sequence heterogeneity "could have been generated from a single mating pair that existed 180–360 $\times 10^3$ years ago, suggesting the possibility that present-day humans evolved from a small mitochondrially monomorphic population that existed at that time." The popular press soon dubbed this coalescent foundress "Eve," and the population bottleneck hypothesis as the "Garden of Eden" or the "Noah's Ark" scenario.

A second influential report extended the survey to 147 individuals and confirmed the limited mtDNA sequence divergence among humans worldwide (Cann et al., 1987). This study also promoted the notion that modern *Homo sapiens* colonized the globe recently from African stock, probably within the last 200,000 years. This "out-of-Africa" hypothesis was based on two lines of evidence beyond the relatively low level of se-quence divergence *per se:* higher mtDNA sequence variation in African than in non-African populations; and an inferred African root for the mtDNA gene tree, thus making African populations paraphyletic to those on other continents with respect to matrilineal genealogy (Fig. 3.1).

The major empirical findings of both studies—low intraspecific mtDNA divergence and limited geographic structure in humans relative to many other vertebrate species—were upheld in subsequent molecular surveys. These often involved direct sequence analyses of particular mito-chondrial genes (Horai et al., 1987, 1995; Vigilant et al., 1989, 1991; Horai and Hayasaka, 1990; Di Rienzo and Wilson, 1991; Hasegawa and Horai, 1991; Kocher and Wilson, 1991; Merriweather et al., 1991; Pesole et al., 1992; Hasegawa et al., 1993; Ruvolo et al., 1993, 1994). However, inferences about ancient human population demography drawn from the mtDNA studies (and those of other genes) have been subjects of vigorous discus-sion (reviews in Rogers and Jorde, 1995; Stoneking, 1997; Harpending et al., 1998; Jorde et al., 1998).

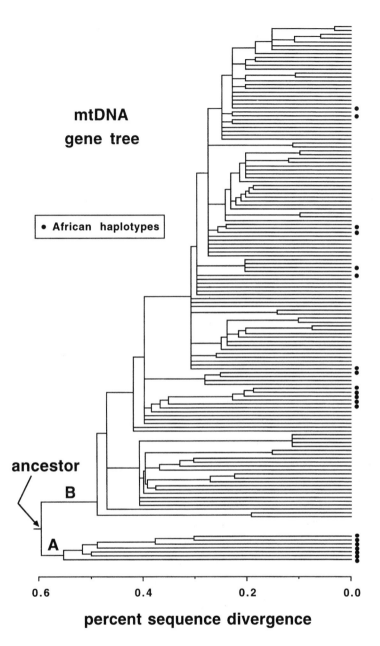

FIGURE 3.1 Redrawn parsimony network for human mtDNA haplotypes as reported in the classic study by Cann et al. (1987). Black dots indicate genealogical positions of haplotypes observed in native Africans. Haplotypes of native Asians, Australians, Europeans, and New Guineans generally are scattered throughout branch B of the gene tree.

REFINEMENTS OF DEMOGRAPHIC INTERPRETATIONS

Historical Population Size

The first matter of contention was the size of the female population in the "Garden of Eden."

MTDNA ASSESSMENTS

Early applications of branching process models to the problem pointed out that the coalescent foundress for human mtDNA lineages might have belonged, in principle, to a rather large population of unrelated females who (apart from Eve) left no matrilineal descendants to the present day due solely to stochastic lineage extinction attendant with reproduction (Avise et al., 1984a; Latorre et al., 1986). Assume, for example, that $N_F = 15,000$ females produce daughters according to a Poisson distribution with mean 1.0. Then, within approximately 15,000 generations (300,000 human years), random lineage turnover leads with reasonably high probability to the expected loss of all founder matrilines except one. The same can be said for a population of $N_F = 45,000$ unrelated foundresses reproducing according to a negative binomial distribution of daughters per mother with mean 1.0 and variance 3.0 (Avise et al., 1984a). In other words, because of the (counterintuitively) rapid pace of lineage sorting under reasonable variances in female reproductive success, a coalescent outcome in the human mtDNA gene tree within the last few hundred thousand years does not necessarily imply an extreme bottleneck in absolute population size in any generation.

From reanalyses of the mtDNA data, A. C. Wilson et al. (1985) suggested that the evolutionary effective population size of human females may have been $N_{F(e)} \cong 6,000$. Takahata (1993) summarized evidence that the effective size was $N_{F(e)} \cong 10,000$ over the past one million years, and that the population "has never dropped to a few individuals, even in a single generation." Ayala (1995a) refers to the earlier confusion of gene-tree coalescence with an extreme population bottleneck as "the myth of Eve."

Subsequent applications of coalescent theory to human origins sought to add refinement and realism to demographic interpretations of the data (Takahata, 1995). Di Rienzo and Wilson (1991) noted that the mismatch distribution of genetic distances between extant mtDNA haplotypes departs dramatically from equilibrium expectations under constant population size. Many nodes in the mtDNA phylogeny (Fig. 3.1) fall within a narrow range of genetic distances suggestive of a rapid population expansion perhaps 60,000 years ago (see also Sherry et al., 1994). Rogers and Harpending (1992) obtained an excellent fit to the empirical mismatch distribution (Fig. 3.2) in coalescent models that assumed a rapid population expansion in the late Pleistocene (some 60,000–120,000 years ago) from about 1,000 females to between 137,000 and 274,000 females. (The astounding growth of the human population in the last few centuries is largely irrelevant to the data interpretations, however, because this merely will have "frozen" for current observation most of the preexisting mtDNA lineages.)

One potential concern about the approach exemplified in Fig. 3.2 is that geographic structure might confound the expectations for a single well-mixed population (Harpending et al., 1993; Marjoram and Donnelly, 1994; but see also Rogers, 1997). A second reservation (which applies to all genealogical analyses) is that any gene tree is only one realization or observation from a statistical distribution of outcomes that is highly variable in theory.

This latter concept is illustrated in Fig. 3.3 in the context of the estimated age of coalescent Eve, our most recent common ancestor (MRCA) in a matrilineal tree. Marjoram and Donnelly (1997) used computer simulations of the coalescent to estimate the time to MRCA under various demographic assumptions. For example, under a constant-size population of 130,000 breeding females prior to 50,000 years ago, and exponential growth to a current population of $N = 5 \times 10^8$ breeding females, the modal expectation for date of the MRCA is about 400,000 years ago. However, the skewed theoretical distribution shows a long tail such that MRCA dates in excess of 800,000 years bp cannot be eliminated with great certainty (see also Wills, 1995). Marjoram and Donnelly (1997) also show that demographic details of population growth play a rather minor role on expected times to MRCA compared to prior population size and geographic structure (which for humans remain highly conjectural).

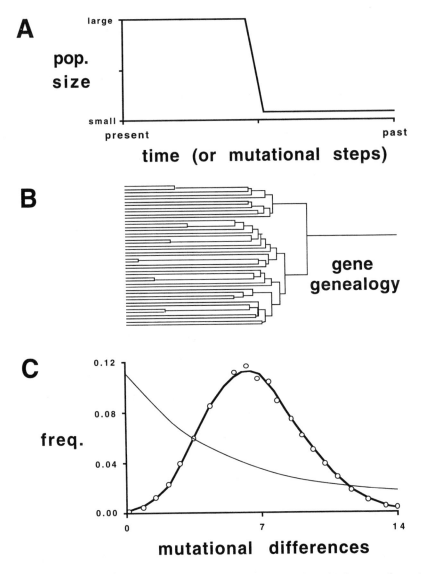

FIGURE 3.2 Conceptual rationale underlying the mtDNA-based inference of a rapid historical expansion of the human population (after Rogers and Harpending, 1992; Rogers, 1997). *(A)* and *(B)*: The hypothetical effects of a sudden population explosion on the structure of a mtDNA (or other) gene genealogy. Note the clustering of nodes near the time of the population expansion. *(C)*: Translation of such a mtDNA gene tree into a mismatch frequency distribution of pairwise genetic distances among extant individuals. The thick line is a theoretical curve fit by Rogers and Harpending (1992) to the empirical mtDNA data (open circles) of Cann et al. (1987). This form of frequency histogram is consistent with the demographic parameters described in the text for a Late Pleistocene population. The narrow line is the expected mismatch distribution of pairwise genetic distances under a stable human population with same mean size.

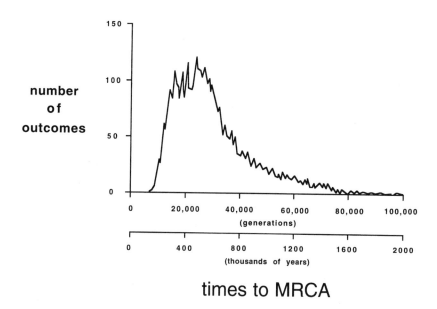

FIGURE 3.3   Simulated distribution of the time to most recent common ancestor (MRCA) for human mtDNA (after Marjoram and Donnelly, 1997). The coalescent model involved a current population of $N = 5 \times 10^8$ breeding females and assumed historical population parameter values described in the text.

A third reservation about the genetic patterns summarized in Fig. 3.2 is that they fail to distinguish between the effect of a population explosion *per se* and the effect of a historical selective sweep mediated by a rapid increase in frequency of an advantageous mtDNA mutation (Excoffier, 1990; Rogers, 1997). The genetic footprint of a selective sweep to fixation in a constant-size population can be similar to that for a rapid size expansion from a small founder population under neutrality (Donnelly and Tavaré, 1995). Thus, a concern in this or other approaches to demographic inference is whether mtDNA variants in the gene tree are selectively neutral. For all these reasons, historical human demographers and geneticists have sought genealogical data from nuclear loci to complement that from mtDNA.

NUCLEAR GENEALOGICAL ASSESSMENTS

A 729-bp intron from a sex-influencing locus (*ZFY*) on the Y-chromosome was sequenced in 38 men of diverse geographic and ethnic origins (Dorit et al., 1995). No variation was detected despite the presence of modest sequence divergence in this gene region between humans and the great apes. Coalescent calculations involving assumptions on mutation rates and branching times yielded an estimated date for the male MRCA at 270,000 years ago (95 percent confidence limits: 0–800,000 years bp). The findings were interpreted by Ayala (1995a) to indicate an effective male population size of $N_{M(e)} \cong 7,000$ (upper 95 percent confidence limit: $N_{M(e)} \cong 80,000$).

Reanalyses of the data identified uncertainties in assumptions underlying these calculations (Donnelly et al., 1996; Fu and Li, 1996; Rogers et al., 1996; Weiss and von Haeseler, 1996). Nonetheless, evidence from the Y-chromosome remains consistent with that from mtDNA in suggesting a rather shallow evolutionary depth for lineage separations among extant humans. Whether this results from low $N_{M(e)}$ or from purifying selection on portions of the Y-chromosome is uncertain (Dorit et al., 1995; Whitfield et al., 1995; Burrows and Ryder, 1997). However, patterns of nucleotide sequence diversity in another Y-linked region were interpreted as evidence against a recent selective sweep on the human Y-chromosome (Hammer, 1995). This latter study estimated 188,000 years as the date for the male MRCA (95 percent confidence limits: 51,000–411,000 years bp), and 10,000 individuals as the long-term effective population size.

Huang et al. (1998) examined nucleotide sequences in a *ZFX* gene on the X-chromosome. Only one polymorphic site was found in a worldwide sample of 29 individuals, and it was present in similar frequencies in Asian, European, and African samples. The authors calculate a modal time of 306,000 years ago (confidence limits: 162,000–952,000 years bp) for the MRCA of the sampled sequences. Interestingly, application of their calculation method to the sequence data of Dorit et al. (1995) yielded a mean time of 116,000 years ago for the MRCA (confidence limits: 61,000–416,000 years bp) for the *ZFY* gene on the Y-chromosome. Because the effective population size (and hence mean coalescent time) of an X-linked gene is expected to be about three times greater than that of a Y-linked gene, all else being equal, the somewhat older date inferred from *ZFX* as compared to *ZFY* is not unexpected.

An autosomal nuclear locus that shows exceptional sequence variation in humans is *DRB1*, one of about 100 genes that make up the HLA (human leukocyte antigen) complex. These genes encode cell surface proteins that bind and present processed antigens to immune cells. Many haplotypes in *DRB1* (and other) genes of the HLA region appear ancient as judged by large sequence differences and by the fact that they often form genealogical clades not with one another but with homologous sequences from other primates (Figueroa et al., 1988; Lawlor et al., 1988; Gyllensten et al., 1991b; Ayala et al., 1994; Ayala, 1995a). Apparently, HLA lineages have been retained for long periods of time (tens of millions of years in some cases) by balancing selection that confers a fitness advantage on heterozygotes.

By incorporating effects of overdominant selection into the coalescent calculations (Takahata, 1990, 1993), Ayala (1995a, 1996) concluded that the level of *DRB1* polymorphism in humans was consistent with the notion that no bottleneck smaller than about 4,000 individuals had occurred in any generation. Reanalyses of these and other HLA data led Erlich et al. (1996; see also Hickson and Cann, 1997) to propose an evolutionary effective population size for ancestral humans of $N_e \cong 10,000$ individuals. A similar estimate came from coalescent assessments of variable intron sequences *within* particular lineage classes of HLA alleles that otherwise have been maintained for long periods of time by balancing selection on exon regions (Bergström et al., 1998). Likewise, values of $N_e \cong 10,000$ also emerged from recent genealogical appraisals of nuclear β-globin haplotypes (Harding et al., 1997a) and some other nuclear DNA sequences (Takahata et al., 1995). Thus, all of these estimates of N_e based on coalescent approaches resemble those reported from mtDNA and the Y-chromosome. They also agree in general magnitude with earlier estimates based on heterozygosities at allozyme loci (Nei and Graur, 1984; Nei and Roychoudhury, 1982).

The Question of African Origins

A second matter of contention concerns the time of the supposed global expansion of *Homo sapiens'* ancestors out of Africa. First, some broader phylogenetic background is in order (this account follows Ayala, 1995b).

About 5–7 million years ago (Mya), the hominid lineage diverged from the lineage leading to chimpanzees. The hominid tree branched exclusively in Africa until about 2 Mya, producing on that continent a diversity of taxonomic forms: *Ardipithecus ramidus* (the oldest known hominid fossil, 4.4 Mya), *Australopithecus anamensis* (4 Mya), and *A. afarensis, A. africanus, Paranthropus aethiopicus, P. boisei,* and *P. robustus* (various times between 3 and 1 Mya). Of these, *A. anamensis* is probably in a direct line of descent leading to *A. afarensis, Homo habilis, H. erectus,* and *H. sapiens.* About 2 Mya, *Homo erectus* emerged and soon spread to other regions including Eurasia and the Middle East where fossils as old as 1.8 My have been discovered. A transition from *H. erectus* to archaic *H. sapiens* occurred about 0.4 Mya, but the exact time is difficult to ascertain or perhaps arbitrary because of taxonomic uncertainties for fossils dating between 0.1 and 0.4 Mya. In any event, anatomically modern humans appeared at least 100,000 years ago.

Within this general framework, a major controversy in anthropology (Bräuer and Smith, 1992) has centered on three competing hypotheses for human origins (Fig. 3.4). An extreme model known as the "candelabra" hypothesis (review in Lewin, 1993) supposes that different regional hominid populations were isolated completely from one another for well over one million years. This hypothesis predicts ancient, geographically structured lineages in modern humans, and, thus, is difficult to reconcile with mitochondrial and other genetic data (Takahata, 1995). A less extreme version of the candelabra is the "multiregional" or "regional continuity" model (Wolpoff, 1989, 1992) in which anatomically modern humans arose consonantly in multiple Old World hominid populations that have been connected by persistent gene flow over the past 1.5 million years. Under the recent "out-of-Africa" or "African replacement" model (Stringer and Andrews, 1988), modern humans arose in Africa or the Middle East proba-

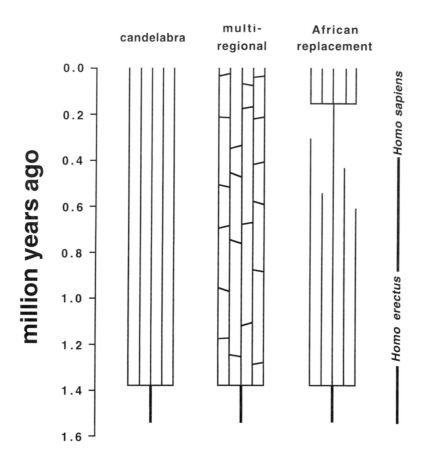

FIGURE 3.4   Three models of human evolution during the Pleistocene (after Ayala et al., 1994).

bly within the last 200,000 years, and then spread throughout the old world by replacing populations of *H. erectus* or archaic *H. sapiens*. Both the multiregional and African replacement models envision an African origin for the genus *Homo*, but differ in the supposed exodus time for the genetic lineages that led to modern *Homo sapiens*.

The mtDNA data usually have been interpreted as support for the African replacement scenario because of the shallow depth in the global mitochondrial genealogy and the suspected recent root of the gene tree in

Africa (Cann et al., 1987). However, challenges have been voiced to this molecular-based conclusion. First, it leaves unexplained a suggested morphological continuity in fossils for some regional hominid populations (see Bräuer and Smith, 1992). Second, parsimony tree-building algorithms that originally were applied to the mtDNA data may be less than definitive in support of an African placement of the gene-tree root (Maddison, 1991; Hedges et al., 1992a; Maddison et al., 1992; Templeton, 1992, 1996; see Penny et al., 1995, and Stoneking, 1997 for responses). Third, estimates of absolute time from the mtDNA genealogy have wide confidence limits due to uncertainties in molecular clock calibrations and to interpretive challenges of alternative historical demographic scenarios, including likely episodes of population extinction and restoration that are difficult to model formally (Takahata, 1995). Fourth, as emphasized repeatedly, mtDNA registers only a small fraction of hereditary pathways through a pedigree and, thus, cannot provide a full story of human genetic history.

An autosomal locus recently subjected to genealogical appraisal is β-globin (Fullerton et al., 1994), where allele-specific PCR primers have been used to amplify and sequence haplotypes from a 3-kb region in human populations worldwide (Harding et al., 1997a). Moderate sequence diversity was observed from which a unique gene tree for 326 (of the 349 assayed) sequences could be obtained without complications stemming from intragenic recombination (Fig. 3.5). Use of β-globin sequences from other primates permitted estimates of the gene-tree root and the mutation rate, from which coalescent times were calculated. Harding et al. (1997a) conclude that the MRCA for alleles at this nuclear locus lived in Africa about 800,000 years ago, a date interpreted as consistent with earlier estimates of about 200,000 years for the mtDNA Eve (because, all else being equal, mean coalescent times for autosomal genes are expected to be fourfold higher than for cytoplasmic genes). However, a rather deep separation among some Asian β-globin lineages (Fig. 3.5) was interpreted as evidence that hominids had dispersed across Asia more than 200,000 years ago.

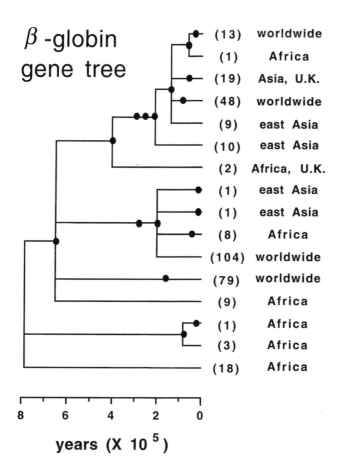

FIGURE 3.5   Time-scaled coalescent tree for 326 β-globin haplotypes observed in samples of human populations from around the world (after Harding et al., 1997a). Dots on the gene tree indicate inferred positions of recorded mutations, and numbers in parentheses are tallies of sampled individuals displaying a given haplotype.

Cladistic and phylogenetic analyses of genetic data from several other molecular systems (mini- and microsatellites, RFLPs, and others) likewise have been interpreted as consistent with a relatively recent African origin for human genetic diversity (Hill et al., 1992; Mountain and Cavalli-Sforza, 1994; Nei, 1995; Zischler et al., 1995; Armour et al., 1996; Nei and Takezaki,

1996; Hammer et al., 1997; Reich and Goldstein, 1998). For example, genetic distances based on 30 microsatellite polymorphisms from humans worldwide yielded a tree whose root, dated to ca. 156,000 years ago, separates African from non-African populations (Goldstein et al., 1995; see also Bowcock et al., 1994 and Chu et al., 1998). Prior analyses of allozyme polymorphisms suggested that the earliest divergence among ethnic groups occurred between Africa and Asia (Nei and Roychoudhury, 1982). Although available data from nuclear gene frequencies are consistent with a relatively recent out-of-Africa model, they alone may not rule out an out-of-Asia alternative.

These latter population genetic approaches are traditional in the sense that most of them provide a composite or average picture across multiple nuclear genes, rather than detailed phylogenetic histories of particular loci. Genealogical analyses have been attempted for a few nuclear loci in addition to β-globin (e.g., Rapacz et al., 1991; Xiong et al., 1991; Tishkoff et al., 1996), but the data are not yet definitive on matters of genealogical origins (Goldman and Barton, 1992; Takahata, 1995). Considering all available evidence, an intermediate scenario between the multiregional and replacement models remains viable: late-departing African (or, conceivably, Asian) groups may have interbred with rather than completely replaced archaic hominid populations elsewhere (Li and Sadler, 1992). If true, it then must be determined what fraction of nuclear gene genealogies will display relatively early lineage separations (i.e., > 200,000 years ago) that trace to particular locations in Africa, Asia, or perhaps elsewhere (Xiong et al., 1991; Harding et al., 1997a).

OTHER GENE-GENEALOGICAL STUDIES

Regional Populations

The global mtDNA genealogy of humans is exceedingly shallow in comparison to those of many species, including several higher primates (Fig. 3.6). Nonetheless, sequence analyses particularly of the rapidly evolving control region have provided many opportunities for finer-scale phylogeographic

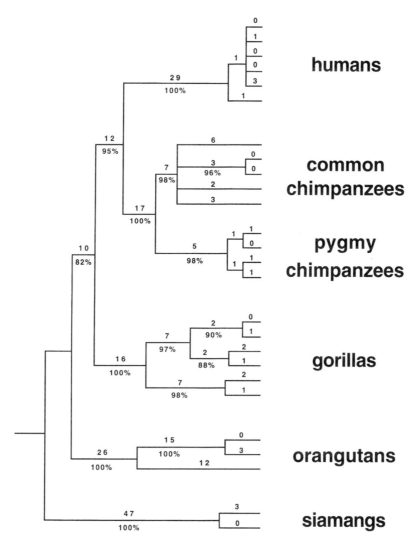

FIGURE 3.6   Phylogenetic tree for hominoids indicating approximate depths of intraspecific lineages as well as interspecies relationships (after Ruvolo et al., 1994). Shown is a maximum parsimony consensus tree from mitochondrial COII gene sequences. Branch lengths indicated are minimum numbers of inferred changes, and bootstrap support levels (> 80 percent) also are given.

analyses of regional human populations (Stenico et al., 1998). One complication is that much of total human mtDNA variation (like that in nuclear genes; Lewontin, 1972) is apportioned within rather than among populations or ethnic groups (Di Rienzo and Wilson, 1991; Ward, 1997). Thus, extant mtDNA genotypes often represent ancestral conditions retained across colonization or expansion episodes, rather than recent markers of local phylogeographic clades. Indeed, it is common to infer the geographic source and number of colonization events consistent with an observed distribution of mtDNA haplotypes in suspected donor and recipient populations. A case in point concerns the colonization of the New World (Gibbons, 1996).

PEOPLING OF THE AMERICAS

Most researchers believe that the historical affinities of Native Americans lie with Asiatic peoples who colonized the New World recently, probably about 12,000 to 14,000 years ago, via a Beringian land route between what is now Alaska and Russia (Hoffecker et al., 1993; Ward, 1997). Under debate have been the exact geographic origins of the founders, the number of colonization events, and the founding population sizes that were to lead to the great cultural and linguistic diversity displayed today among native American peoples.

A controversial suggestion that helped to motivate early genetic studies envisioned several independent colonization events that accounted for three proposed major linguistic groups in the New World: the Eskimo-Aleut speakers in the far North; the Na-Dene speakers mostly of north-central Canada, interior Alaska, and Greenland; and the Amerind speakers in most of the remainder of North and South America (Greenberg et al., 1986). Initial genetic reports identified four relatively deep mtDNA lineages or haplogroups (A-D) in North America. These displayed large frequency differences across linguistic populations, and initially were interpreted as supporting Greenberg et al.'s (1986) separate-waves-of-migration model (Schurr et al., 1990; Torroni et al., 1992). However, further mitochondrial studies of Asian and native American populations (Ward et al., 1991, 1993; Sambuughin et al., 1992; Horai et al., 1993; Shields et al., 1993; Torroni et al., 1993a, b, 1994a, b; Santos et al., 1994; Kolman et al., 1995) uncovered additional information that has prompted challenges to the separate-colonizations scenario (Merriweather et al., 1995; Kolman et al., 1996).

The distinctive New World haplogroups identified in the early mtDNA studies now appear to be distributed almost ubiquitously in the Americas, albeit often in highly different frequencies across populations. In most cases, linguistic groups formerly assigned to separate migrations merely exhibit different frequencies of the same set of haplogroups, suggesting to Merriweather et al. (1995) and Kolman et al. (1996) that genetic drift and founder events in the New World following a single migration are sufficient to account for the current mtDNA distributions. This argument is strengthened by the restricted codistribution within Asia of these same four mtDNA haplogroups, which have been observed jointly only in present-day populations in Mongolia, Tibet, and central China. Thus, Kolman et al. (1996) concluded that this area in east-central Asia (rather than Siberia, for example) is the most likely source of a single colonizing migration to the New World, with subsequent founder effects and genetic drift responsible for the variation in mtDNA haplotype frequencies among indigenous American populations.

One caveat to this interpretation is that other candidate source populations in Asia might be overlooked because they too might have lost one or more of the four haplogroups subsequent to the colonization of the Americas. Another caveat is that a rare mitochondrial lineage recently discovered in Native Americans seems to register an historical link to peoples of Europe and the Middle East rather than Asia (Morell, 1998). If so, at least a few peoples originally from these areas also may have traveled through Asia to be included among the New World's first settlers.

HISTORIES OF OTHER AREAS

Similar genealogical appraisals based on mtDNA (and, rarely, nuclear genes—Harding et al., 1997b) have been applied to the colonization histories of other regional human populations, such as those along Asian coastlines, New Zealand (Murray-McIntosh et al., 1998), and smaller islands of the South Pacific. One mtDNA marker with special phylogenetic potential in a cladistic sense is an Asian-prominent 9-bp deletion (in the intergenic region between the COII and tRNA^{Lys} genes; Wrischnik et al., 1987) that probably originated in central China (Ballinger et al., 1992). As judged by its current population distribution, this marker appears to have been carried along two major colonization routes: into North America via the Bering land bridge (Kolman et al., 1996), and along the Asian coast and then southeastward into Indonesia and onto the Pacific islands (Hertzberg et al., 1989).

However, interpretive caveats apply even to such a potentially ideal genealogical marker. First, this deletion may be polyphyletic because it appears to have arisen independently in Africa, for example (Redd et al., 1995), and because several other mtDNA size variants also are thought to have originated more than once (Cann and Wilson, 1983; Ballinger et al., 1992). Second, populations genealogically connected to ancestral stocks in Asia could have lost the deletion secondarily through founder effects and genetic drift. A possible example involves the highlanders of Papua New Guinea (PNG), who unlike their coastal counterparts lack the 9-bp deletion (Hertzberg et al., 1989; Stoneking and Wilson, 1989). Alternatively, PNG highlanders might be related more closely to native Australians, who also lack this feature (Hertzberg et al., 1989; Stoneking et al., 1990). As always, support from multiple markers is desirable before drawing firm genealogical conclusions.

LOCAL MIGRATORY AND MATING PATTERNS

Another area beginning to yield interesting results involves the study of extremely recent human history using mtDNA and Y-chromosome markers in conjunction. A few case studies will illustrate this approach.

Under the traditional marriage customs of Arab tribes in the Sinai Peninsula, women can marry into outside groups whereas men cannot. If this marriage habit has been practiced for a long time, different patterns of population variation should characterize maternally inherited mtDNA and the paternally inherited Y-chromosome. Consistent with this expectation, Salem et al. (1996) report that Sinai populations display extremely reduced levels of Y-chromosome variation (in relation to 2 nearby populations in the Nile Delta, for example) but no such reduction in mtDNA variability. This outcome was attributed to female-biased movements into the Sinai, rather than to a demographic bottleneck or founder event that probably would have reduced variation in mtDNA as well as the Y-chromosome in the Sinai population.

Similar cytonuclear studies have been conducted on Amerindians (Pena et al., 1995) and Finns (see Salem et al., 1996). In both populations, significant reductions in Y-chromosome variation were detected relative to presumed ancestral sources, but so too were some reductions in mtDNA variation. Thus, the genetic findings in these cases were attributed to past population bottlenecks that affected both males and females.

In India, the stratified Hindu caste system has dictated marriage choices of the citizenry for more than 3,000 years. Preferred marriages involve partners of the same caste, and occasional departures from this social tradition usually involve matings of lower-caste women with higher-caste men. In a large population survey, Bamshad et al. (1998) found that men in adjacent castes frequently share mtDNA haplotypes, but no such blurring of caste lines was detected in polymorphic Y-linked markers. These genetic patterns were interpreted as consistent with historical inter-caste leakage of maternally inherited genotypes via the upward social mobility of women, and caste-specific confinement of paternally inherited markers due to the extreme religious constraints on the social mobility of men.

More generally, recent genetic evidence suggests that female-mediated gene flow among human populations has exceeded by severalfold that attributable to movement by males (Seielstad et al., 1998). This conclusion was based on a global compilation of allele frequency data from single-nucleotide polymorphisms (SNPs) in mtDNA, the Y-chromosome, and autosomal loci, and from microsatellite loci in the latter two genetic systems.

Notably, the mean spatial localization of variants on the Y-chromosome and autosomes proved to be severalfold greater than for mtDNA. The authors conclude that the higher historical migration rates inferred for females than for males probably reflects "patrilocality," the tendency for a bride to leave her parent's home and move into her husband's natal household. About 70 percent of the world's cultures practice this social tradition. The genetic findings challenge one traditional view of human migrations (the "Marco Polo–Genghis Khan" scenario) in which men are perceived as the usual wayfarers and women as homebodies. In commenting upon these genetic findings, Stoneking (1998) concludes that "if we really want to understand human migrations, we must pay more attention to women's ways."

This brief discussion has merely introduced the topic of human genetic appraisals primarily involving explicit genealogical appraisals of particular loci. Another large database consists of allele frequencies at more than 100 nuclear genes in nearly 2,000 human populations. Readers interested in a detailed interpretation of this information in the context of regional and ethnic groups, linguistic and cultural associations, and historical migration pathways are referred to an extended treatment by Cavalli-Sforza et al. (1994).

The Neanderthals

Another long-standing enigma in hominid evolution has been the evolutionary status of the Neanderthals (*Homo sapiens neanderthalensis*). These morphologically recognizable hominids appeared in Europe and western Asia at least 200,000 years ago and persisted until about 30,000 to 40,000 years ago, thus overlapping temporally with modern humans (*Homo sapiens sapiens*). Whether the two taxonomic subspecies actually coexisted at particular locales, and if so whether they interbred, have been matters of considerable speculation and debate (Ward and Stringer, 1997, and references therein). A fascinating molecular discovery recently shed some light on the issue.

In a technical tour de force, Krings et al. (1997) used PCR to isolate and sequence a 379-bp mtDNA fragment from the fossilized skeleton of a Nean-

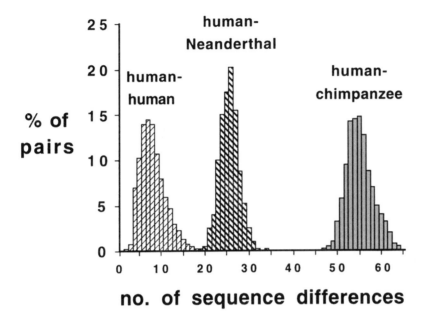

FIGURE 3.7 Mismatch frequency distributions of pairwise mutational differences in comparisons of mtDNA control-region sequences from humans, chimpanzees, and a single Neanderthal (after Krings et al., 1997).

derthal specimen who probably lived between 30,000 and 100,000 years ago. A comparison against homologous sequences from extant humans and chimpanzees indicated that the Neanderthal sequence falls essentially outside the range of variation in modern humans. Genetic distances between the Neanderthal and modern human mtDNA sequences are about one-half as large as the mean human-chimpanzee difference (Fig. 3.7). Using a molecular clock for this portion of the hypervariable control region as calibrated by the suspected time of the human-chimp split (4–5 Mya), the authors calculate that the matrilineal separation of humans and Neanderthals occurred about 600,000 years ago, a date severalfold older than conventional estimates for the matrilineal coalescent time of modern humans.

These molecular findings suggest that Neanderthals went extinct without contributing mtDNA to modern humans, and furthermore that

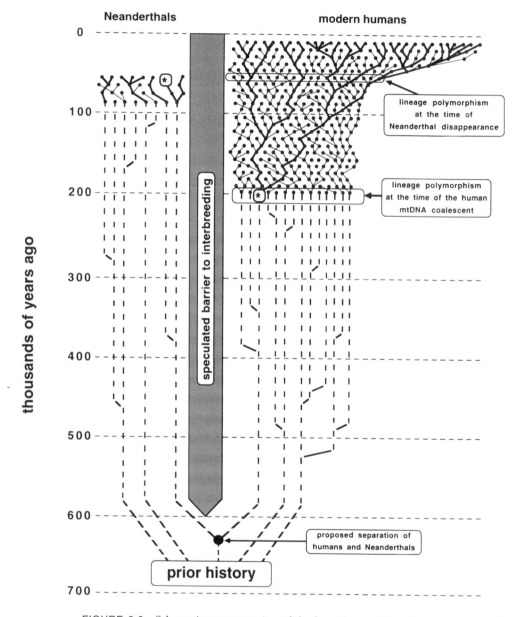

FIGURE 3.8   Schematic representation of the hypothesized long-term separation of
Neanderthals from the ancestors of modern humans, as reflected in matrilineal histories
inferred from mtDNA sequences. Stars indicate the Neanderthal specimen assayed, and
the coalescent Eve from which extant matrilines trace (heavy branches) within the last
200,000 years. Boxes indicate points of uncertainty and special interest.

for more than half a million years they may have evolved independently from the hominid populations ancestral to modern humans (Fig. 3.8). However, with current data these conclusions must remain provisional for at least two reasons. First, the scope of genetic variation in Neanderthals (as well as in early populations of *Homo sapiens sapiens*) remains unknown from direct evidence. As judged by subsequent laboratory failures to extract mtDNA from other ancient hominid specimens less well preserved (Cooper et al., 1997), such genetic information may remain difficult to obtain. Second, results do not rule out the possibility that interbreeding between Neanderthals and modern humans led to an exchange of nuclear gene lineages, some of which may have survived to the present.

This brief introduction to the literature on human phylogeography should serve to illustrate the power and conceptual novelty of molecular genealogical appraisals of intraspecific evolution. When viewed against the gross uncertainties that existed about human phylogeny as recently as a decade ago, many of the remaining points of contention in biological anthropology represent relatively refined issues, such as the precise historical demography of the human species and the genetic histories of particular regional populations or ethnic groups. This overview of human phylogeography also should warn the reader of the kinds of limitations and caveats that apply with equal or greater force to genealogical interpretations in other animal species, where available molecular data sets typically are smaller.

SUMMARY

1. With respect to molecular phylogeographic analysis, perhaps no species has received greater attention than *Homo sapiens*. By all available evidence, our species shows limited phylogeographic population structure on a global scale. For example, the shallow evolutionary depth of branches in the human mtDNA gene tree contrasts with much deeper lineage separations characteristic of many other animal species even on far smaller spatial scales.

2. Available mitochondrial and nuclear gene genealogies for humans suggest that: (a) the evolutionary effective population size of our species has been on the order of a few thousand to a few tens of thousands of individuals; (b) a pronounced population expansion occurred probably in the late Pleistocene; and (c) Africa is the likely source of what may be a surprisingly recent global expansion of our species. Whether this expansion resulted in a complete replacement of archaic humans or was accompanied by genetic introgression cannot be decided definitively by available gene-tree evidence, but preliminary indications are that an intermediate level of mixing and replacement may have been involved.

3. Although the molecular genealogies of extant humans are relatively shallow evolutionarily, fine-scale phylogeographic analyses of numerous tribal groups, cultural units, and linguistic assemblages have contributed to an understanding of the origins and spread of particular genetic lineages during the recent peopling of the planet. For example, special attention has been paid to the history of colonization of the New World. A major challenge in such analyses has been to distinguish between *de novo* mutations, homoplasy, and retention of ancestral lineages in interpreting the distributional patterns of molecular variants.

4. A recent analysis of fossil-recovered mtDNA suggests that molecular lineages in *Homo sapiens neanderthalensis* have been separated from those in *H.s. sapiens* for more than half a million years, and that the Neanderthals may have gone extinct without contributing matrilines to modern humans.

In the study of dispersal and distribution of animals, it is important to see that the physical conditions lead, and that in a more or less definite succession the flora and fauna follow; thus the fauna comes to fit the habitat as a flexible material does a mold. The time is passed when faunal lists should be the aim of faunal studies. The study must not only be comparative, but *genetic,* and much stress must be laid on the study of the habitat, not in a static, rigid sense, but as a fluctuating or periodical medium.

—Charles Adams, 1901

4

INTRASPECIFIC PATTERNS IN OTHER ANIMALS

IN COMPARISON to many other species, the phylogenetic branches in the intraspecific human pedigree (Chapter 3) are shallow and only weakly structured geographically. Human mtDNA lineages on a global scale show far lower sequence divergence (maximum $p \cong 0.006$) than do, for example, regional populations of pocket gophers in the southeastern United States (mean $p \cong 0.034$). However, as described in this chapter, a tremendous diversity of phylogeographic outcomes has been observed among animal species.

PHYLOGEOGRAPHIC HYPOTHESES

In 1987, a comparative summary of molecular phylogeographic patterns available at that time suggested that historical biogeographic factors as well as contemporary ecologies and behaviors of organisms had played important roles in shaping the genetic architectures of extant species (Avise et al., 1987a). These preliminary findings led to several intraspecific phylogeographic hypotheses (Table 4.1) deemed worthy of further evaluation as new molecular and other evidence became available. These

TABLE 4.1   Phylogeographic hypotheses for mitochondrial gene trees as formulated
             originally by Avise et al. (1987a).

I.   Most species are composed of geographic populations whose members occupy
      recognizable matrilineal branches of an extended intraspecific pedigree.

II.   Species with limited or "shallow" phylogeographic population structure have
      life histories conducive to dispersal and have occupied ranges free of firm, long-
      standing impediments to gene flow.

III. Intraspecific monophyletic groups distinguished by large genealogical gaps
      usually arise from long-term extrinsic (biogeographic) barriers to gene flow.[a]

a. This hypothesis has a series of corollaries that relate to various aspects of genealogical
concordance (Chapter 5).

hypotheses are analagous to traditional ecogeographic rules (Table 1.1) in
the sense that they suggest trends rather than state inviolable truths. More
than a decade later, these phylogeographic tendencies now can be reexamined in the light of many additional appraisals conducted within a
genealogical framework.

Several distinctive phylogeographic patterns (along a continuum of
theoretical outcomes) can be categorized for mtDNA gene trees or those of
other loci (Fig. 4.1). These five catagories will be discussed in turn.

CATEGORY I: DEEP GENE TREE, MAJOR
LINEAGES ALLOPATRIC

Category I is epitomized by the presence of spatially circumscribed haplogroups separated by relatively large mutational distances. In other words,
prominent genetic gaps distinguish deep allopatric lineages in a gene tree.
Spatial substructure also may be present among more closely related lineages within regions. Category I patterns commonly appear in phylogeographic surveys of mtDNA. Cases involving the pocket gopher and deer
mouse were described earlier, and many more examples will follow.

Long-term extrinsic barriers to genetic exchange often provide a likely
explanation for major phylogeographic discontinuities. Through time,

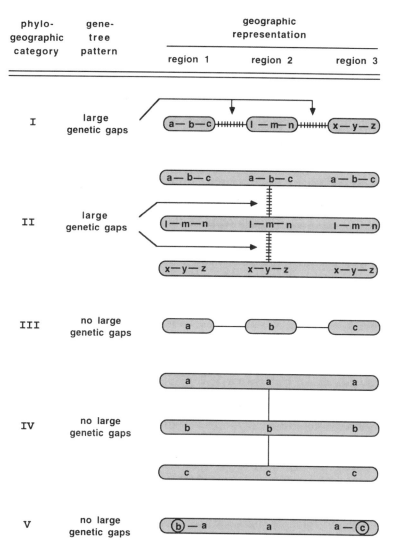

FIGURE 4.1   Five categories of phylogeographic outcome (after Avise et al., 1987a). Circles or ovals encompass particular mtDNA haplotypes (denoted by letter) or groups of closely related haplotypes whose geographic distributions are shown. Haplotypes are connected into phylogenetic networks whose branches involve one mutation step each unless otherwise indicated by slashes (number of mutation steps) along a network pathway. Genetic gaps imply much larger genetic distances between than within phylogroups.

regional allopatric populations come to occupy recognizable, deeply separated branches in an intraspecific gene tree. The genetic differences may reflect an accumulation of *de novo* mutations postdating population separation, and/or effects of lineage sorting from a highly polymorphic ancestral gene pool. Especially in broadly distributed species with low dispersal and gene flow, extinctions of intermediate haplotypes also may contribute to the appearance of pronounced phylogenetic gaps.

Ideally, confirmation of historical sundering at the population level requires concordant support from multiple genes, or agreement with other sources of biogeographic information such as historical geology or comparative systematics. Otherwise, the possibility remains that a Category I pattern in a gene tree is idiosyncratic to the locus under investigation.

CATEGORY II: DEEP GENE TREE, MAJOR LINEAGES BROADLY SYMPATRIC

This pattern is characterized by pronounced phylogenetic gaps between some branches in a gene tree, with the principal lineages codistributed over a wide area. Theoretically, this pattern could arise in a species with large evolutionary N_e and high gene flow. Then, some anciently separated lineages might by chance have been retained whereas many intermediate genotypes were lost over time by gradual lineage sorting. Balancing selection could facilitate this outcome by favoring long-term evolutionary survival of some haplotype lineages (as in the human HLA system).

However, most Category II patterns in mtDNA surveys probably involve zones of secondary admixture between allopatrically evolved populations or species (e.g., Avise et al., 1984b, 1997; Taberlet et al., 1992; Scribner and Avise, 1993; Arctander et al., 1996; Kim et al., 1998). In such situations, independent evidence (from historical geology, genetics, or morphology) often exists to suggest that the distinctive mtDNA phylogroups diverged in allopatry. These mtDNA lineages as well as nuclear loci also provide conspicuous genetic markers for detailed population-

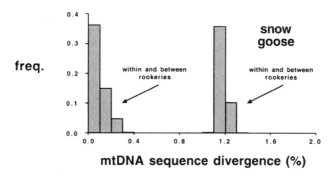

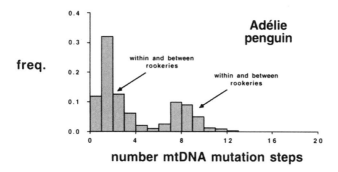

FIGURE 4.2 Bimodal mismatch distributions of mtDNA genetic distance estimates within each of two avian species that display phylogeographic Category II: *above:* snow geese as assayed by restriction-site analyses of the whole mtDNA genome (after Avise et al., 1992b); and *below:* Adélie penguins as assayed by mtDNA control region sequences (Baker and Marshall, 1997, after Monehan, 1994).

genetic dissections of possible hybridization phenomena in contact zones (Barton and Hewitt, 1985; Harrison, 1990; Avise, 1994; Arnold, 1997).

   Two examples of phylogeographic Category II with uncertain evolutionary etiology are illustrated in Fig. 4.2. In the colonially breeding snow goose (*Chen caerulescens*), two distinct mtDNA haplogroups (mean sequence divergence in the control region $p \cong 0.067$; Quinn, 1992) are present in each surveyed rookery across the trans-Canadian Arctic and Russia

(Avise et al., 1992b). Within any rookery, individuals of both mtDNA clades interbreed freely. In the Adélie penguin *(Pygoscelis adeliae)*, two distinctive mtDNA haplogroups ($p \cong 0.051$) occur within each of three geographically separated rookeries on Ross Island in Antarctica (Monehan, 1994; Baker and Marshall, 1997). The bimodal mismatch distributions of genetic distances within snow geese and Adélie penguins (Fig. 4.2) bear close resemblance to the intraspecific phylogenetic signatures presented previously for sharp-tailed sparrows and seaside sparrows (Fig. 2.12), except that the divergent mtDNA lineages within the geese and penguins are not separated geographically at present.

The current rookery sites for snow geese and Adélie penguins occur in high-latitude glaciated areas that were uninhabitable for these species as recently as 5,000–10,000 years ago. Thus, these rookeries were settled by colonizers from other breeding sites occupied during the Pleistocene. Also, these large-bodied birds are unlikely to have maintained large effective population sizes (and, hence, distinctive neutral lineages) for long periods of time. For these reasons, Category II phylogeographic patterns in these species probably reflect recent admixtures of birds from allopatrically diverged sources (Quinn, 1992; Baker and Marshall, 1997).

In populations of the toque macaque monkey *(Macaca sinica)*, Hoelzer et al. (1994) observed two highly distinct mtDNA lineages (sequence divergence > 3.0 percent) in sympatry in Sri Lanka. No explanation involving past population subdivision was obvious, and the authors wisely urged caution in using mtDNA data alone to make such inferences.

CATEGORY III: SHALLOW GENE TREE, LINEAGES ALLOPATRIC

In Category III, most or all haplotypes are related closely, yet are localized geographically. The implication is that contemporary gene flow has been low enough in relation to population size to have permitted lineage sorting and random drift (or, perhaps, diversifying selection) to promote genetic divergence among populations that nonetheless were in historical

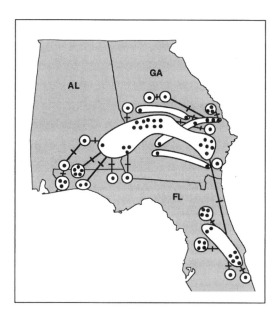

FIGURE 4.3   MtDNA parsimony network for the beach mouse superimposed over the geographic sources of collections throughout most of the species' range in Alabama, Georgia, and Florida (after Avise et al., 1983). Circles or extended ellipses encompass the documented geographic range for each of 22 haplotypes. Slashes across network branches indicate restriction site changes; solid dots are individual animals. Two haplotypes and one haplotype group are shown as unconnected to the remainder of the network because of placement ambiguities.

contact recently. Category III grades into Categories IV and V depending on historical levels of gene flow among conspecific populations that had not been sundered by firm, long-term biogeographic barriers.

Empirical examples are common. One early study involved the beach mouse *Peromyscus polionotus* (Avise et al., 1979a, 1983), a species mostly confined to coastal plains of the southeastern United States. There, it replaces the related deer mouse *P. maniculatus* that inhabits the remainder of North America. In restriction site assays, 22 different mtDNA haplotypes were detected among 68 beach mice, but all genotypes proved to be closely related and spatially localized (Fig. 4.3). The number of restriction sites assayed was small by current standards, but by these same criteria many more mutational steps distinguished regional populations of deer mice from one another (Fig. 2.21). Furthermore, the latter species appeared to be paraphyletic to *P. polionotus* in matriarchal genealogy (Avise et al., 1983), indicating that the beach mouse maintains only a small subset of the total lineage diversity in the beach mouse/deer mouse complex. The mtDNA pattern in *P. polionotus* probably reflects a relatively recent separation from ancestral stock during the species' colonization of the southeastern United States, plus pronounced restrictions on contemporary gene flow between small and scattered local populations.

Category III outcomes also characterize many within-region phylogroups in species that display a Category I phylogeographic pattern overall. For example, an assemblage of deer mice in central and western North America (Fig. 2.21) consists of closely related mtDNA haplotypes, many of which appear to be localized geographically. Similarly, within both the eastern and western mtDNA phylogroups in pocket gophers (Fig. 1.9), closely related mtDNA haplotypes display evident spatial structure.

CATEGORY IV: SHALLOW GENE TREE, LINEAGES SYMPATRIC

This pattern is expected for high-gene-flow species of modest or small effective size whose populations have not been sundered by long-term biogeographic barriers. A good example involves the American eel, *Anguilla rostrata,* where coastal samples from Maine to Louisiana failed to display statistically significant differences in frequencies of various mtDNA haplotypes, all of which were related closely (Avise et al., 1986). As mentioned in Chapter 2, this outcome probably reflects the unusual life history of this catadromous species, whose larvae disperse more or less at random from what may be a nearly panmictic spawning aggregation in the Sargasso Sea. In effect, this life cycle entails exceptionally high gene flow among freshwater populations across a vast area. In conjunction with the small evolutionary effective population size of American eels as inferred from shallow mtDNA lineage separations, the net result is a phylogeographic outcome in which multiple closely related haplotypes are broadly sympatric.

Other species already mentioned that approximate phylogeographic Category IV include humans and red-winged blackbirds. In both cases, population numbers increased explosively (as judged by historical evidence as well as by the nature of the mismatch distributions of mtDNA distances) as these species expanded their ranges in recent evolutionary time. The net result has been genealogical trees with shallow branches, many of which are dispersed widely across the species' current ranges.

For a species that has expanded in size rather recently from small or modest numbers of founders, another expected genealogical signature is a "star phylogeny." A common ancestral-like haplotype typically lies at the star's center and recent derivatives are connected to it independently by short branches. Examples from two avian species are presented in Fig. 4.4. If two long-isolated populations each underwent recent population expansions, the species should display a "dumbbell" gene tree with two starbursts connected by a longer branch (Fig. 4.5). Depending on the dispersal patterns attendant with the population expansions, such a species could display a Category I or Category II phylogeographic pattern overall. Of course, some species may consist of more than two starbursts.

CATEGORY V: SHALLOW GENE TREE, LINEAGE DISTRIBUTIONS VARIED

This pattern is intermediate between Categories III and IV, and involves common lineages that are widespread plus closely related lineages that are "private" (each confined to one or a few nearby locales). This phylogeographic outcome intimates low or modest contemporary gene flow between populations that are connected tightly in history. Common haplotypes often are the ancestral (plesiomorphic) states within such species, whereas rare haplotypes are the presumed derived (apomorphic) conditions. Thus, rare haplotypes shared by individuals or populations provide potential cladistic markers for recent clades in a gene tree.

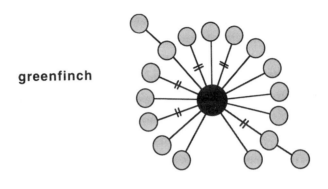

greenfinch

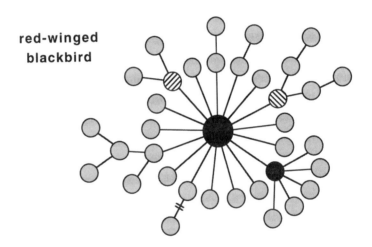

**red-winged
blackbird**

FIGURE 4.4   Star phylogeny reported in mtDNA surveys of greenfinches across Europe (after Merilä et al., 1997) and red-winged blackbirds across North America (after Ball et al., 1988). The most common and widespread haplotype(s) are in black, those with crosslines are hypotheticals not observed, and other haplotypes appear to be rare and perhaps localized in available samples. All network branches are of unit length (one mutation) except where indicated by slashes denoting two observed mutations. A starburst phylogeographic pattern is an expected signature for an abundant species that has expanded its range rather recently from small or modest numbers of founders.

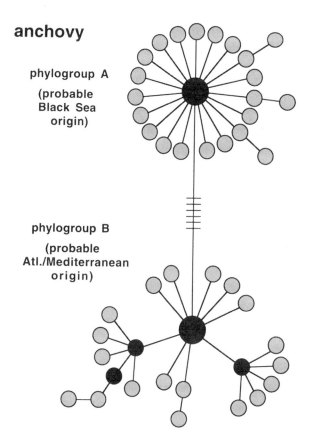

anchovy

phylogroup A

(probable Black Sea origin)

phylogroup B

(probable Atl./Mediterranean origin)

FIGURE 4.5   Modified dumbbell pattern in the empirical mtDNA network for the European anchovy, *Engraulis encrasicolus* (after Magoulas et al., 1996). Except for the long branch with the six indicated restriction site changes, all other network branches are of unit length. Common haplotypes are shown in black. The somewhat different geographic distributions of phylogroups A and B led to their proposed origins.

An example involves the bowfin fish (*Amia calva*) in Atlantic coastal and Floridian rivers of the southeastern United States (Bermingham and Avise, 1986). In this study, one mtDNA haplotype was earmarked as the probable ancestral form within the region based on four lines of evidence: (a) It was by far the most common haplotype, occurring in 30 of 59 assayed specimens; (b) It was geographically widespread, observed in nine of the ten surveyed drainages in the area; (c) It formed the core of a star phylogeny whose rays (tree branches) connected separately to seven other mtDNA genotypes within the Atlantic region; and (d) It was at least one mutation step closer (than other haplotypes in its clade) to a phylogenetically distinctive group of lineages in river drainages entering the Gulf of Mexico (the species registers a Category I phylogeographic pattern overall). In available samples, each of the eight rarer haplotypes in the eastern clade was confined to a single drainage or regional set of adjacent drainages.

With regard to the utility of rare haplotypes as synapomorphic markers for gene-tree clades, one cautionary note is illustrated by a phylogeographic survey of the mtDNA control region in the snapping turtle *Chelydra serpentina* (Walker et al., 1998b; see also Phillips et al., 1996). This species shows exceptionally low mtDNA variation and geographic differentiation across the southeastern United States. One common haplotype was widespread (60 specimens in nine states) and two rare variants differed from it by a single base transition. One variant was observed in South Carolina and at a site in Louisiana about 1,000 km away. Perhaps these variant specimens belong to a true matrilineal clade (their distribution might be the result of recent transport by humans, for example). Alternatively, perhaps their mtDNAs are alike at this nucleotide site by virtue of parallel mutations.

Thus, the potential cladistic value of rare haplotypes under Category V (or III and IV) is offset partially by a greater likelihood of homoplasy falsely uniting closely related genotypes. Such difficulties are far less likely to confound the *general* lineage assignments for anciently separated haplotypes, such as those characteristic of the deep gene-tree branches under phylogeographic Categories I and II.

Another cautionary point should be made about the usual difficulty of distinguishing cleanly between phylogeographic Categories III–V in finite molecular surveys. As longer DNA sequences are assayed and larger numbers of individuals surveyed, the power to detect statistically significant lineage structure increases, all else being equal. Thus, some species that bear a molecular footprint of high gene flow (Category IV) under low-resolution genetic assays might convert to an appearance of low or intermediate gene flow (Categories III or V) under more stringent assay conditions. Conversely, some species that under limited sampling bear signatures of low gene flow might convert with additional sampling to an appearance of moderate or high gene flow, for example if private alleles in preliminary assays later proved to be widespread. For these reasons, and because the precise boundaries between Categories III–V are somewhat arbitrary, these patterns characterized by shallow phylogenetic depth will be pooled in most of the summaries of taxonomic groups that follow.

The next section highlights empirical phylogeographic findings at the intraspecific level for numerous species of five vertebrate classes, as well as representative invertebrates surveyed for molecular genealogy.

MAMMALS

Small-Bodied Terrestrial Species

The spatially-structured phylogeographic motifs already described for the southeastern pocket gopher and deer mouse have proved to be general patterns for small- or medium-bodied terrestrial mammals. In most cases, mtDNA surveys have uncovered deep genealogical subdivisions across broad regional scales (phylogeographic Category I) in addition to evident spatial organization among closely related haplotypes in local areas. In some species with small or modest geographic range, such as *Peromyscus polionotus* in the southeastern United States (Avise et al., 1983) and *Cynictis penicillata* (the yellow mongoose) in southern Africa (Van Vuuren and Robinson, 1997), only the shallower genealogical separations characteristic of phylogeographic Categories III or V have been evident.

Category I phylogeographic patterns have been observed for small mammals in many countries and on nearly all continents. Examples from the Americas include: in Canada, the ground squirrel, *Spermophilus columbianus* (MacNeil and Strobeck, 1987); in western North America, the pocket gopher, *Thomomys bottae* (Smith, 1998), grasshopper mouse, *Ony-chomys leucogaster* (Riddle and Honeycutt, 1990; Riddle et al., 1993), cactus mouse, *Peromyscus eremicus* (Walpole et al., 1997), and various pocket mice in the genera *Perognathus* and *Chaetodipus* (McKnight, 1995; Riddle, 1995; Lee et al., 1996); in the eastern United States, regional populations of the woodrat, *Neotoma floridana* (Hayes and Harrison, 1992); in Middle and Central America, deer mice in the *Peromyscus aztecus* complex (Sullivan et al., 1997); in Costa Rica, the pocket gopher, *Orthogeomys cherriei* (Demastes et al., 1996); and in the Amazonian basin, a variety of forest-dwelling rodents and marsupials (da Silva and Patton, 1993; Patton et al., 1994, 1996; review in da Silva and Patton, 1998).

On other continents also, Category I phylogeographic patterns have been reported. Examples include a rabbit (*Oryctolagus cuniculus*) on the Iberian peninsula (Biju-Duval et al., 1991); the wood lemming (*Myopus schisticolor*) in Scandinavia (Federov et al., 1996); the field vole (*Microtus agrestis*) in northern Europe (Jaarola and Tegelström, 1995, 1996); *Sorex* shrews in central Europe (Taberlet et al., 1994); European populations of hedgehogs in the genus *Erinaceus* (Santucci et al., 1998); the woodmouse (*Apodemus sylvaticus*) in the Mediterranean region (Michaux et al., 1996); Eurasian forms of the house mouse, *Mus musculus* (Boursot et al., 1996); *Spalax* mole rats in the Middle East (Suzuki et al., 1996); the naked mole rat (*Heterocephalus glaber*) in central Africa (Faulkes et al., 1997); the springhare rodent (*Pedetes capensis*) in southern and eastern Africa (Matthee and Robinson, 1997); the rock hyrax (*Procavia capensis*) in southern Africa (Prinsloo and Robinson, 1992); the rock-wallaby, *Petrogale xanthopus* (Pope et al., 1996) and eastern barred bandicoot, *Perameles gunnii* (Robinson, 1995) in Australia; the musk shrew (*Suncus murinus*) in Asia and Indonesia (Yamagata et al., 1995); and the dormouse (*Glirulus japonicus*) in Japan (Suzuki et al., 1997). Another example of a Category I pattern observed in a small mammal is detailed in Fig. 4.6.

In many of these cases, the *major* phylogeographic gaps in mtDNA orient with independent evidence and thereby implicate long-term population isolation and divergence. For example, two major mtDNA lineages in

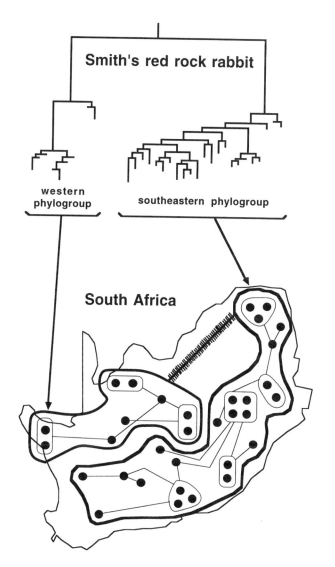

FIGURE 4.6   Category I phylogeographic pattern in Smith's red rock rabbit, *Pronolagus rupestris* (after Matthee and Robinson, 1996). *Above:* Phylogeny based on RFLP assays of mtDNA. *Below:* The matrilineal phylogeny superimposed on a map of South Africa (each black dot is a different haplotype). Note the large number of mutational steps (more than 40) distinguishing the two major phylogroups, whereas within either phylogroup only about 1–12 inferred mutational steps occurred along branches of the parsimony network.

the southeastern pocket gopher are mirrored by nearly fixed differences in frequencies of allozyme alleles. So too in some cases are the major mtDNA phylogroups in the western pocket gopher. In the desert pocket mouse as well as in the cactus mouse, two principal mtDNA lineages coincide with conventionally recognized forms in the Chihuahuan versus Sonoran deserts that differ also in nuclear-encoded proteins, karyotype, or morphology. In the Japanese dormouse, two major mtDNA phylogroups differ substantially in assayed sequence at a nuclear ribosomal gene and a Y-linked sex-determining locus. The two principal mtDNA lineages of springhares in southern versus eastern Africa characterize populations that are distinct in karyotype and morphology.

In many cases, support for historical interpretations of mtDNA phylogeographic units has come from geographic or taxonomic evidence. Two primary mtDNA lineages in the European field vole have distributions consistent with their isolation in suspected glacial refugia during the Pleistocene, as do the two major mtDNA lineages in European rabbits across Iberia. In the Costa Rican pocket gopher, two deep mtDNA lineages are separated by a physiographic feature (the Cordillera de Tilaran) that probably provided a long-term barrier to genetic exchange. Two well-differentiated mtDNA clades in the barred bandicoot distinguish populations on the island of Tasmania from those of mainland Australia. In European populations of commensal house mice, two primary mtDNA clades (Ferris et al., 1983a,b) generally conform to taxonomic units (variously recognized as subspecies or species) that are distinct in morphology and in nuclear genes but hybridize in narrow zones of contact (Hunt and Selander, 1973). On a wider geographic scale, mtDNA lineages distinguish additional regional populations of house mice, and concordantly with nuclear genes point to the northern Indian subcontinent as the likely cradle of evolutionary diversification within the assemblage (Din et al., 1996).

On the other hand, deeply separated mtDNA lineages within some species of small mammals have proved to be discordant in particular geographic areas with independent genetic characters (Boissinot and Boursot, 1997) or other evidence. For example, in parts of Denmark and Sweden a mtDNA genotype normally characteristic of one taxonomic form of house mouse is fixed in populations otherwise ascribed by nuclear genes and morphology to a second subspecies or species. Here, the discordant pattern probably evidences a postulated colonization event by *domesticus* females into *musculus* range followed by introgressive hybridization and an eventual population-level replacement by the alien *Mus* cytoplasms (Gyllensten and Wilson, 1987; Vanlerberghe et al., 1988). Similar scenarios of introgression were invoked to account for discordances between mtDNA lineages and taxonomic species boundaries in local populations of voles (*Clethrionomys rutilus* and *C. glareolus*) in northern Scandinavia (Tegelström, 1987), and between mtDNA, nuclear markers, and taxonomic boundaries in *Thomomys* pocket gophers in some areas of the western United States (Patton and Smith, 1994; Ruedi et al., 1997).

In other reported instances of phylogeographic discordance between mitochondrial and nuclear data, evolutionary factors other than secondary cytoplasmic introgression may apply. For example, geographic ranges of major mtDNA lineages in the *Sorex araneus* group track the suspected population histories of these shrews in Europe, whereas chromosomal mutations do not (Taberlet et al., 1994). These authors speculate that particular chromosomal variants arose multiple times independently (producing homoplasy in the karyotypic data), and/or that such variants were dispersed across regional mtDNA phylogroups either as retained ancestral polymorphisms or as secondary interregional transfers mediated by males (who typically are less philopatric than the highly territorial females).

Many small-bodied mammals that display deep phylogeographic subdivisions on a species-wide scale also show lineage substructure on finer spatial spectra within regions. Such substructure typically involves closely related lineages and can be attributed to limited historical dispersal and transient biogeographic sundering agents operating over meso- and microevolutionary time scales. Often, the current genetic substructure of these populations will have resulted from agents acting primarily upon ancestral lineage polymorphisms, and merely may have been supplemented by occasional *de novo* mutations that arose on site (e.g., Good et al., 1997). In one of the few direct examinations of temporal dynamics in population lineage structure, Thomas et al. (1990) used PCR assays to compare mtDNA sequences in extant populations of a kangaroo rat (*Dipodomys panamintinus*) to same-site samples preserved as museum skins from the early 1900s. Results showed between-site differences in haplotype frequencies but temporal continuity of mtDNA sequences within each location.

Lineage structure also may extend to extreme microspatial scales due to matrilineal ties among close kin. In the cotton rat (*Sigmodon hispidus*), various mtDNA haplotypes were localized within a single 3-hectare field (Kessler and Avise, 1985). So too were mtDNA genotypes in meadow voles (*Microtus pennsylvanicus*) across a 25-hectare field (Plante et al., 1989), and in grey-sided voles (*Clethrionomys rufocanus*) across a 1-hectare grid (Ishibashi et al., 1997). Such genealogical microstructure is probably kaleidoscopic in time and space as families continually arise and their spatial associations then dissipate through dispersal.

The demographic and biogeographic histories of many species are likely to produce complex mixtures of deep and shallow genealogical structures. For example, the labyrinthine arrays of mixed conifer forests across the partially fragmented montane archipelagos of the American Southwest are home to numerous mammal species whose current distributions have been explained by historical vicariant events, dispersal episodes, and various temporal combinations of these processes. An example involving an arboreal squirrel is as follows.

Three biogeographic hypotheses have been advanced for the origins of disjunct populations of *Sciurus aberti* in the American Southwest. Perhaps these populations: (a) are vicariant relicts from an ancestral form widely distributed across the more-or-less continuous coniferous forests of the Pleistocene; (b) arose via post-Pleistocene dispersal across unsuitable nonmontane habitats; or (c) reflect early vicariant events from which dispersal ensued at later times. Lamb et al. (1997) reported molecular evidence consistent with the latter view. Among 22 squirrel populations sampled rangewide, 21 different mtDNA haplotypes were observed. These arrayed into distinctive eastern and western phylogroups distributed in montane areas from, respectively, Mexico to Colorado and Utah, and from Arizona to southwestern New Mexico (Fig. 4.7). Overall, the current

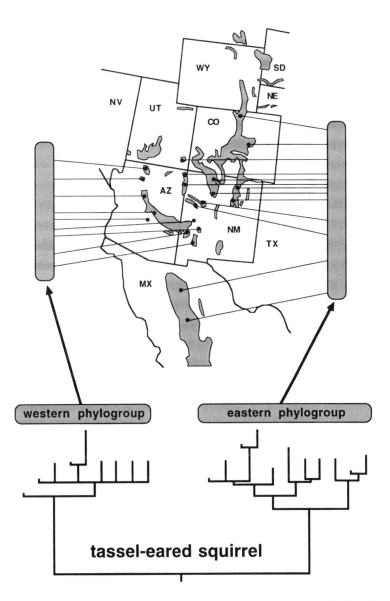

FIGURE 4.7   Phylogeographic pattern in the tassel-eared squirrel (after Lamb et al., 1997). *Below:* Matrilineal phylogeny based on mtDNA assays. *Above:* Map indicating (by shading) montane archipelagoes of ponderosa pine that provide suitable habitat for this squirrel.

spatial distributions and branch depths in the mtDNA gene tree were interpreted to indicate a population vicariant event in the early Pleistocene followed by Quaternary dispersal in conjunction with a documented northward range expansion of ponderosa pines.

Large-Bodied Terrestrial Species

Many small terrestrial mammals have limited vagility so it is not surprising that these species often display strong mtDNA phylogeographic structure. What of larger-bodied terrestrial mammals, many of which are highly mobile? A variety of phylogeographic outcomes has been observed.

Among several codistributed bovid species from the African savannahs surveyed using similar mtDNA assays, different genealogical histories have been deduced (Templeton and Georgiadis, 1996). According to the Templeton algorithms (Chapter 2), the phylogeographic pattern in the buffalo, *Syncerus caffer*, is compatible with restricted gene flow under isolation by distance (see also Simonsen et al., 1998), whereas patterns in the impala, *Aepyceros melampus*, and wildebeast, *Connochaetes taurinus*, register various mixes of restricted gene flow, population fragmentation, and occasional long-distance dispersal. In the Grant's gazelle (*Gazella granti*), pronounced mtDNA differences distinguish neighboring populations that presumably established contact recently following former vicariant separation into allopatric units (Arctander et al., 1996).

In the pampas deer (*Ozotoceros bezoarticus*) sampled from Argentina to central Brazil, the effects of isolation by distance without ancient population vicariance (phylogeographic Category III) were suggested from the mtDNA distributions (González et al., 1998). Phylogeographic patterns in some other large mammals have been intermediate between Categories I and III. For example, populations of the rhinoceros, *Dicerorhinus sumatrensis*, in Borneo showed fixed haplotype frequency differences from those in Malay and Sumatra, but the estimated sequence divergence was modest ($p \cong 0.01$) in the mtDNA control region (Morales et al., 1997; see also Amato et al., 1995).

In some large-bodied mammals, substantial phylogeographic gaps (Category I) have been reported that were unanticipated but that make retrospective sense. For example, Morin et al. (1993, 1994; see also Goldberg and Ruvolo, 1997) identified two highly divergent mtDNA lineages in chimpanzees that distinguish a geographically disjunct population in western Africa (*Pan troglodytes verus*) from populations to the east (*P.t. troglodytes* and *P.t. schweinfurthii*). Original descriptions of these subspecies were based primarily on their allopatric distributions (they are nearly identical morphologically). Morin et al. (1994) emphasize that the newly identified phylogeographic subdivision is not merely of academic interest but also carries ramifications for conservation efforts.

The previously unsuspected genetic divergence in chimpanzees also could affect interpretations of any differences that might be observed in behavioral or physiological features of captive animals traditionally assumed to be uniform genetically. Similar conclusions apply to another distinct chimpanzee lineage recently discovered in central Africa (Gonder et al., 1997). Deeply divergent phylogeographic lineages also have been identified in mtDNA surveys of other primates (Fig. 3.6) including *Gorilla gorilla* (Garner and Ryder, 1996; Saltonstall et al., 1998), the orangutan, *Pongo pygmaeus* (Ryder and Chemnick, 1993; Xu and Arnason, 1996; Zhi et al., 1996; but see Muir et al., 1998), and the pigtail macaque, *Macaca nemestrina* (Rosenblum et al., 1997).

Similar reports exist of moderate or deep mtDNA phylogeographic subdivisions in several other large-bodied mammals. The mule deer (*Odocoileus hemionus*) and white-tailed deer (*O. virginianus*) both are structured geographically in North America (Carr et al., 1986; Cronin et al., 1991a; Cronin, 1992). In the southeastern United States, for example, the latter is arrayed into three coherent phylogeographic units—in extreme south Florida, the Atlantic coast, and continental areas to the west—that were interpreted as generally consistent in position with major phylogroups in some other codistributed vertebrate species, and with the physiographic history of the region (Ellsworth et al., 1994a).

Other examples of strong mtDNA phylogeographic structure in large mammals include the following. In the leopard (*Panthera pardus*), a combination of mitochondrial and nuclear assays revealed six genetically distinct and geographically isolated assemblages—in Africa, central Asia, India, Sri Lanka, Java, and east Asia (Miththapala et al., 1996). The African wild dog (*Lycaon pictus*) is subdivided phylogenetically into at least two distinct groups in southern and eastern Africa (Girman et al., 1993; Roy et al., 1994a), as are kit and swift foxes in the *Vulpes macrotis-velox* complex on opposite sides of the Rocky Mountains in North America (Mercure et al., 1993). The Asian elephant (*Elephas maximus*) carries two highly divergent mtDNA lineages that are geographically patterned but that do not coincide with the two conventionally recognized subspecies in Sri Lanka and the Asian mainland (Hartl et al., 1996). In contrast, two highly divergent mtDNA lineages in the collared peccary (*Tayassu tajacu*) from the American Southwest agree well with earlier subspecies assignments (Theimer and Keim, 1994).

In the brown bear (*Ursus arctos*), several highly distinct mtDNA lineages have been observed both in the Old World and New World (Randi, 1993; Randi et al., 1994; Taberlet and Bouvet, 1994; Taberlet et al., 1995; Talbot and Shields, 1996; Waits et al., 1998). Viewed globally, the deepest mtDNA clades are patterned geographically and reflect population fragmentations probably related to Pleistocene climatic fluctuations. The related cave bear (*Ursus spelaeus*) went extinct about 20,000 years ago. A portion of the mtDNA control region recently was PCR-amplified from fossil remains of one specimen and the sequence compared to those of extant brown bears (Hanni et al., 1994). Results suggest that *U. spelaeus* diverged from an early offshoot of *U. arctos* at about the same time (early Quaternary) as some of the main lineage separations preserved in extant brown bears were initiated.

Sometimes, a species that appeared to conform to phylogeographic Category III or V converts upon more extensive geographic sampling to a Category I pattern overall. A case in point involves the black bear (*Ursus americanus*), which in initial surveys displayed minimal mtDNA differentiation east of the Rocky Mountains in North America (Cronin et al., 1991b; Paetkau and Strobeck, 1996). Another major phylogeographic unit later was discovered in coastal regions of western Canada (Byun et al., 1997; Wooding and Ward, 1997). The data suggest that black bear populations survived the Pleistocene in forested glacial refugia both in the Pacific Northwest and in an eastern portion of the continent (Wooding and Ward, 1997).

When deep phylogenetic splits in an intraspecific mtDNA gene tree have been observed, almost invariably they distinguish populations in separate portions of a species' range. So prevalent is this trend that rare departures from it, in which deeply separated lineages are sympatric (Category II), are of special note. Wayne et al. (1990) observed highly distinct mtDNA lineages (sequence divergence $\cong 8.0$ percent) within several contiguous populations of the black-backed jackal (*Canis mesomelas*) in central Africa. The explanation is uncertain but probably involves either secondary admixture of formerly separate populations, or the long-term maintenance of distinct gene-tree lineages in a high-gene-flow species with unusually large N_e (Wayne et al., 1990).

Several large mobile mammals reportedly display shallow or no detectable phylogeographic subdivisions in mtDNA. Extant populations of these species probably were not sundered allopatrically for long periods, and historical connections and gene flow across the assayed range were moderate to high in relatively recent time. Examples include the black rhinoceros (*Diceros bicornis*) in eastern Africa (Ashley et al., 1990), the African elephant (*Loxodonta africana*) in eastern and southern portions of the continent (Georgiadis et al., 1994), the red kangaroo (*Macropus rufus*) in Australia (Clegg et al., 1998), the bighorn sheep (*Ovis canadensis*) in the Rocky Mountains (Luikart and Allendorf, 1996), and several species of large herbivores and carnivores across large portions of North America. The latter include the moose, *Alces alces*, and the elk, *Cervus elaphus* (Cronin, 1992), muskox, *Ovibos moschatus* (Groves, 1997), coyote, *Canis latrans* (Lehman and Wayne, 1991; Lehman et al., 1991), and gray wolf, *C. lupus* (Wayne et al., 1992; Wayne, 1996).

Gray wolf populations from the Nearctic and Palearctic also have been compared in mtDNA control-region sequences to 67 breeds of the domesticated dog, *Canis familiaris* (Vila et al., 1997). Dogs displayed a level of mtDNA polymorphism comparable to that in wolves, and matrilines within the two species were phylogenetically intermixed in a gene tree that was remarkably shallow overall. Although the authors conclude that dogs arose more than 100,000 years ago, any such dating exercises are rather uninformative without secure knowledge of ancestral polymorphisms. Suppose that multiple wolf puppies with different mtDNA genotypes were integrated into a primordial dog gene pool. Then, the mtDNA coalescent in extant dogs could be much older than the domestication time. Barbujani et al. (1998) make an interesting analogy: "Suppose that some Europeans colonize Mars next year: If they successfully establish a population, the common mitochondrial ancestor of their descendants will be Paleolithic. But it would not be wise for a population geneticist of the future to infer from that a Paleolithic colonization of Mars."

In any event, among the dog breeds surveyed which ranged from toy poodles to giant mastiffs, morphological diversity is huge. The molecular results highlight a well-known point: morphological differentiation can occur rapidly under the influence of diversifying selection (artificial in this case). The molecular findings also raise a general cautionary note about the utility of morphology alone as a guide to genealogical depths in species where population separation times otherwise are unknown.

Volant Species

With their capacity for flight, bats might be expected to display minimal phylogeographic structures over wide areas. However, this has not necessarily proved to be the case. Four disjunct populations of the ghost bat (*Macroderma gigas*) in northern Australia show fixed frequency differences in mtDNA haplotypes that are quite divergent from one another (4.5 percent) in control-region sequences (Wilmer et al., 1994). Results suggest long-term female philopatry due to innate homing and/or distance barri-

ers to dispersal, and they extend direct ecological observations documenting female site fidelity in contemporary time.

Significant but less pronounced mtDNA population structure was reported between colonies of the mouse-eared bat (*Myotis myotis*) in southern Bavaria (Petri et al., 1997). In another European bat traditionally
described as *Pipistrellus pipistrellus*, two highly distinct mtDNA lineages (>
11 percent sequence divergence in the *cytb* gene) reveal what probably are
two cryptic species that live in nonmixed though sometimes sympatric
colonies (Barratt et al., 1997). Species status is supported further by the observation that the distinctive mtDNA lineages correspond to two phonic
types differing in echolocation calls. In a fruit bat (*Artibeus jamaicensis*) of
the Caribbean region, two distinct mtDNA lineages also exist, but in this
case the genotypes sometimes co-occur in colonies within which there is
no evidence for reproductive isolation. Pumo et al. (1988) suggest that the
mtDNA lineages have been dispersed by recent matings between individuals from different islands.

Marine Species

Most mammals inhabiting the marine realm (primarily pinnipeds and
cetaceans) are large and highly mobile. Not surprisingly, several of these
species sampled over vast areas display limited differentiation of mtDNA
lineages. For example, a shallow matrilineal genealogy with minimal geographic patterning was observed in Dall's porpoise (*Phocoenoides dalli*)
collected from locations up to 4,000 km apart in the North Pacific Ocean
and Bering Sea (McMillan and Bermingham, 1996). Likewise, little or no
intra-oceanic spatial genetic structure was reported among populations of
the widespread harbor porpoise (*Phocoena phocoena*) in either the Pacific
or Atlantic Oceans (Rosel, 1992; Wang, 1993; Rosel et al., 1995). Hoelzel
(1994) reviewed genetic evidence from mitochondrial and nuclear genes
suggesting a lack of deep evolutionary separations among conspecific
populations of several cetacean species with global distributions, including the fin whale (*Balaenoptera physalus*) and sei whale (*B. borealis*). A

near absence of mtDNA divergence also characterized morphologically separable populations of the spinner dolphin, *Stenella longirostris* (Dizon et al., 1991).

More often, however, conspecific populations of marine mammals display pronounced phylogeographic structure. This structure is deep in some species, and oriented geographically in patterns strongly suggestive of long-term population separations (phylogeographic Category I). In other cases, the lineage separations are shallow (Categories III or V) and indicative of lineage-shaping processes promoting differentiation over shorter ecological timescales: e.g., philopatry to natal site, fidelity to social groups, or isolation by distance. In many species such as the West Indian manatee (*Trichechus manatus*), both shallow and deep population structures have been reported in molecular phylogeographic surveys (Garcia-Rodriguez et al., 1998).

Among marine pinnipeds, examples of phylogeographic Category I include the following. Atlantic versus Pacific populations of the harbor seal (*Phoca vitulina*) show a modest net difference (> 3 percent) in mtDNA control region sequence (Stanley et al., 1996). Rookeries of the Stellar sea lion (*Eumetopias jubatus*) in Russia, the Aleutians, and the Gulf of Alaska differ markedly from those in southeastern Alaska and Oregon in a pattern suggestive of historical population isolation in separate glacial refugia (Bickham et al., 1996). In the sea lion (*Zalophus californianus*), three colonies from the Pacific coast of southern California and the Baja Peninsula are distinct in mtDNA control region sequence from a colony in the Gulf of California, suggesting regional philopatry and genetic separation over a protracted time (Maldonado et al., 1995). In the grey seal (*Halichoerus grypus*), populations in the western versus eastern North Atlantic Ocean differ in mtDNA sequence at a level indicative of a population separation initiated about 1.0 Mya (Boskovic et al., 1996).

Spatial matrilineal structure also characterizes several small cetacean species. Populations of the bottlenose dolphin (*Tursiops truncatus*) in the Atlantic versus Pacific are phylogenetically divergent in mtDNA, and further nested genetic subdivision is evident between various populations in the Atlantic (Dowling and Brown, 1993; Hoelzel et al., 1998b). In the harbor porpoise mentioned above as showing minimal intra-oceanic genetic differentiation, populations from the North Atlantic and Eastern Pacific Oceans appear in the same assays to be alternately fixed for mtDNA lineages that are highly divergent (> 2.4 percent) in sequence.

On finer spatial scales, shallower genealogical subdivisions within marine mammal species often are detected. Presumably, these patterns reflect philopatric behaviors or short-term impediments to dispersal rather than long-standing physical barriers to gene flow. For example, three spatially discrete populations of Hector's dolphin (*Cephalorhynchus hectori*) in the continuous coastal waters of New Zealand occupy different branches in a shallow mtDNA gene tree (Pichler et al., 1998). In the harbor seal mentioned above, nested within the major phylogroups in both the Atlantic and Pacific are substructured lineages distinguishing east-coast from west-coast rookeries in each ocean basin (Stanley et al., 1996). Statistically significant but shallow mtDNA population structure is displayed by the fur seal (*Arctocephalus forsteri*) in Australia and New Zealand (Lento et al., 1994). In the sea otter (*Enhydra lutris*), several populations along the North Pacific rim show pronounced differences in the frequencies of closely related mtDNA haplotypes (Cronin et al., 1996a; Scribner et al., 1997). Results suggest limited gene flow today between populations with rather tight historical connections.

Many cetacean species are matrifocal—socially organized around related females (Amos et al., 1993; Baker and Palumbi, 1996). One expected population-genetic signature is a spatial structure of matrilines. For example, humpback whales (*Megaptera novaeangliae*) within each of three ocean basins are subdivided into mtDNA stocks attributed in large part to maternally directed fidelity to particular migratory destinations (C. S. Baker et al., 1990, 1993, 1994, 1998). Site fidelity probably develops when a calf accompanies its mother on her annual migration between low- and high-latitude waters. At the population level, this results in an association between mtDNA genotype and migratory circuit (Palsbøll et al., 1995; Baker and Palumbi, 1996; Larsen et al., 1996).

Maternally directed philopatry to seasonal habitats also has been invoked to explain the matrilineal population structure observed in the beluga whale (*Delphinapterus leucas*) across the Nearctic (Brown Gladden et al., 1997; O'Corry-Crowe et al., 1997), and in the narwhal (*Monodon monoceros*) in the northwest Atlantic (Palsbøll et al., 1997). Fidelity to female social group also may account for the shallow but significant mtDNA variation detected in the sperm whale (*Physeter macrocephalus*) sampled on a global scale (Lyrholm et al., 1996; Lyrholm and Gyllensten, 1998). On the other hand, isolation by distance rather than maternally directed site fidelity to summer feeding areas was invoked by Bérubé et al. (1998) to account for the mild but statistically significant heterogeneity in frequencies of nuclear and mitochondrial alleles in fin whales of the North Atlantic.

Killer whales *(Orcinus orca)* in the Pacific Northwest exist as two genetically differentiable but sympatric assemblages with specialized feeding strategies directed either toward fishes or mammals (Hoelzel and Dover, 1991). Members of this species have predictable home ranges and tend to remain within their natal group for life. These behaviors in conjunction with socially learned feeding habits apparently facilitated the appearance of microspatial genetic group structure in this otherwise highly mobile species (Hoelzel et al., 1998a). In the common dolphin *(Delphinus delphis)*, two distinctive and often sympatric morphotypes (long-beaked and short-beaked) coexist in several of the world's oceans, but a survey of specimens from southern California showed that they consistently occupy two different albeit closely related branches in a matrilineal genealogy (Rosel et al., 1994). The authors suggest that the two morphs probably do not interbreed (at least in this area; a broader genetic survey is needed), and might warrant species designation.

In summary, mtDNA phylogeographic surveys of mammals show a wide mix of outcomes generally aligned with organismal differences in behavior, natural history, and environments historically occupied. Nearly all small and relatively immobile terrestrial mammals exhibit pronounced phylogeographic differentiation. Local lineage structure typically is evident in addition to deeper phylogenetic splits that often characterize regional assemblages of populations whose ancestors probably diverged in allopatry during the Pleistocene or earlier. Regional populations of many large-bodied mammals also show deep phylogeographic separations, but as expected for mobile animals, local differentiation typically is limited or expressed at larger spatial scales. Nonetheless, even highly mobile mammals in open environments often show at least modest genealogical population structure due to "self-imposed" limits on dispersal, such as behavioral site fidelity or group alliance.

BIRDS

Avian populations present an enigma regarding postulated levels of geographic structure and magnitudes of gene flow. On the one hand, birds (particularly males of many species) often display strong fidelity to natal sites for reproduction, implying severe constraints on contemporary gene movement. Also, many conspecific populations show obvious geographic differences in song, body size, plumage, or other phenotypic features. On the other hand, most birds have exceptional dispersal potential by virtue of their capacity for flight and proclivity to migrate. Numerous studies of allozyme allele frequencies suggest that avian populations in temperate regions typically show less genetic structure than do conspecific populations of most freshwater fishes, small mammals, reptiles, and amphibians surveyed over comparable geographic areas (reviews in Avise, 1983; Barrowclough, 1983; Ward et al., 1992). One complication in interpreting allozyme data is that currently segregating alleles often reflect ancestral polymorphisms retained from stocks predating population fragmentation (and sometimes speciations), thus compromising genealogical inferences on population history (Zink and Remsen, 1986).

In recent years, several avian species have been reexamined from a genealogical perspective, and a variety of mtDNA patterns has been uncovered (reviews in Avise and Ball, 1991; Avise, 1996c; Baker and Marshall, 1997; Zink, 1997). The phylogeographic outcomes can vary greatly even among closely related congeners. For example, the Carolina chickadee

(*Parus carolinensis*) of the southern United States displays highly divergent mtDNA clades east versus west of central Alabama and, thus, probably was separated historically into two Pleistocene refugia (Gill et al., 1993). By contrast, conspecific populations of the black-capped chickadee (*P. atricapillus*) and likewise of the boreal chickadee (*P. hudsonicus*) feature the same mtDNA haplotypes from New York to Alaska. Within the last 15,000 years, following retreat of the Wisconsin glacier, each of these species probably expanded to a continent-wide range from a single source population.

As mentioned earlier, other avian species such as the red-winged blackbird, downy woodpecker, and snow goose also show little or no contemporary mtDNA phylogeographic structure across much of North America, presumably as a result of post-Pleistocene range expansions into formerly glaciated regions. Other avian species in which common (and sometimes rare) haplotypes are distributed over moderate or large geographic areas include the northern flicker, *Colaptes auratus* (Moore et al., 1991), common grackle, *Quiscalus quiscula* (Zink et al., 1991), thick-billed murre, *Uria lomvia* (Birt-Friesen et al., 1992), song sparrow, *Melospiza melodia* (Zink and Dittmann, 1993a), swamp sparrow, *Melospiza georgiana* (Greenberg et al., 1998), chipping sparrow, *Spizella passerina* (Zink and Dittmann, 1993b), short-tailed shearwater, *Puffinus tenuirostris* (Austin et al., 1994), red knot, *Calidris canutus* (Baker et al., 1994), prairie grouse in the genus *Tympanuchus* (Ellsworth et al., 1994b), oilbird, *Steatornis caripensis* (Gutierrez, 1994), redpoll finches in the genus *Carduelis* (Seutin et al., 1995), northern pintail duck, *Anas acuta* (Cronin et al., 1996b), greenfinch, *Carduelis chloris* (Merilä et al., 1997), and bluethroat, *Luscinia svecica* (Questiau et al., 1998).

However, other avian species have displayed pronounced mtDNA phylogeographic differentiation at varying spatial scales. In addition to the Carolina chickadee and the two sparrow species described in Chapter 2, these include the Canada goose, *Branta canadensis* (Shields and Wilson, 1987; Van Wagner and Baker, 1990; Quinn et al., 1991), grey-crowned babbler, *Pomatostomus temporalis* (Edwards and Wilson, 1990; but see also Edwards 1993a,b), Australian white-eye, *Zosterops lateralis* (Degnan and Moritz, 1992), plain titmouse, *Parus inornatus* (Gill and Slikas, 1992), black-throated gray warbler, *Dendroica nigrescens* (Bermingham et al., 1992), ostrich, *Struthio camelus* (Freitag and Robinson, 1993), mountain chickadee, *P. gambeli* (Gill et al., 1993), streaked saltator, *Saltator albicollis* (Seutin et al., 1993), dunlin, *Calidris alpina* (Wenink et al., 1993, 1996), fox sparrow, *Passerella iliaca* (Zink, 1994), bananaquit, *Coereba flaveola* (Seutin et al., 1994), stonechat, *Saxicola torquata* (Wittmann et al., 1995), yellow warbler, *Dendroica petechia* (Klein and Brown, 1995), common guillemot, *Uria aalge* (Friesen et al., 1996), common chaffinch, *Fringilla coelebs* (Baker and Marshall, 1997; Marshall and Baker, 1997), Le Conte's thrasher, *Toxostoma lecontei* (Zink et al., 1997), and Adelaide's warbler, *Dendroica adelaidae* (Lovette et al., 1998). Several of these and other examples are reviewed in Helbig et al. (1995) and Avise and Walker (1998).

An instructive example of an avian species with unmistakable phylogeographic structure is the flightless brown kiwi (*Apteryx australis*) of New Zealand. Virtually every population possesses private mtDNA haplotypes, and in addition birds from the southern end of the South Island are strongly divergent in mtDNA sequence and other genetic characters (at a level probably warranting species recognition) from morphologically similar populations elsewhere on the islands (A. J. Baker et al., 1995). The conspicuous Category I phylogeographic pattern that characterizes this relatively sedentary avian species (Fig. 4.8) closely resembles the pattern typifying many small terrestrial mammals.

In avian species with deep mtDNA phylogeographic structure, the well-supported matrilineal clades usually make biological and historical sense because population-sundering events agree with morphological, geographical, or other independent evidence. For example, the two distinctive mtDNA clades in the Canada goose correspond to long-recognized large-bodied versus small-bodied subspecies whose breeding ranges in North America are mostly separate. The five major mtDNA clades in the

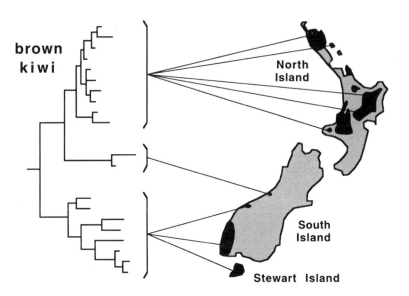

FIGURE 4.8   Phylogeographic pattern in the flightless brown kiwi (after A. J. Baker et al., 1995). *Left:* Phylogeny based on sequence assays of the mitochondrial *cytb* gene. *Right:* Map of New Zealand showing the current range of this avian species (black areas) and the sample collection sites of major mtDNA phylogroups that were supported by high (> 95 percent) bootstrap values.

dunlin distinguish regional breeding populations in Europe, eastern Siberia, central Siberia, Alaska, and Canada. (MtDNA data also show that dunlins from different breeding areas mix on migratory and wintering grounds; Wenink and Baker, 1996.) The two deeply separated mtDNA haplogroups in sharp-tailed sparrows are geographically concordant with differences in the birds' behavior and morphology, and the two major mtDNA phylogroups in the seaside sparrow are supported by independent data that implicate historical separation between populations along the coastlines of the Atlantic and Gulf of Mexico.

More such examples further document the historical nature of many of the principal intraspecific phylogroups registered in mtDNA. In the fox sparrow, four matrilineal assemblages correspond to groups defined by plumage characters. Major mtDNA lineages in the ostrich align with currently accepted subspecies' designations. In the yellow warbler, mtDNA haplogroups clearly differentiate North American migratory populations from the sedentary forms on Caribbean islands. In Le Conte's thrasher, a distinctive taxonomic subspecies displayed colorimetric as well as large mtDNA sequence differences from other populations. In the rufous-sided towhee (genus *Pipilo*), two geographic forms distinctive in mtDNA (Ball and Avise, 1992) as well as in morphology and song recently were elevated to taxonomic species status (AOU, 1995). Separate island populations of Adelaide's warbler on Puerto Rico, Barbuda, and St. Lucia are reciprocally monophyletic for mtDNA lineages with sequence differences comparable to those of some continental avian species. At least 7 of 13 Holarctic species of birds assayed for mtDNA show strong differentiation across the Beringian region separating Asia from the New World (Zink et al., 1995). About 50 percent of arid-land bird species assayed from the southwestern United States show pronounced mtDNA differentiation presumably reflective of population isolations associated with historical islands of desert habitat (Zink, 1997).

Such findings give considerable confidence that the deeper mtDNA clades within avian species often register deep historical separations at the population level (as opposed, for example, to retentions of ancient lineages within unsundered species). Thus, in well-documented reports of otherwise unsuspected mtDNA phylogeographic "breaks," provisional historical explanations should not be dismissed lightly. For example, the pronounced but unforeseen mtDNA disjunction reported by Gill et al. (1993) for the continuously distributed but morphologically conservative Carolina chickadee may indeed indicate (as the authors suggested) that populations east and west of a zone in central Alabama were separated for at least 1.0 My.

Likewise, reported failures to detect substantial matrilineal breaks should not be dismissed as uninformative about population history. For example, although valiant efforts to distinguish geographic populations of the Arctic-breeding red knot by mtDNA failed completely, due in part to a paucity of lineage variation in the species (A. J. Baker et al., 1994), the data indicate that these shorebirds likely were bottlenecked through a small population in the late Pleistocene and expanded to a current broad distribution only within the last 10,000 years (Baker and Marshall, 1997).

Some avian species such as the song sparrow, red-winged blackbird, and redpoll finch display strong geographic variation in song dialect, behavior, or morphology that is not mirrored by deep phylogeographic differences in mtDNA (see also Brawn et al., 1996). Such instances probably reflect recent historical connections among populations (e.g., via postglacial range expansion) followed by rapid onsite differentiation in other organismal attributes perhaps promoted by diversifying selection. Another possibility, not excluded in most instances, is that differences in rearing conditions or environmental circumstances may have sponsored a nongenetic (ecophenotypic) component to variation in some organismal features (James, 1983; Seutin et al., 1995).

In summary, phylogeographic analyses of mtDNA have revolutionized thought about avian population structure. Species appear to be structured matrilineally in a wide variety of patterns and presumed evolutionary timescales (Fig. 4.9). Many species show pronounced phylogeographic subdivision (Category I). This structure often is deep genealogically, concordant with differentiation in other characters, and geographically oriented in ways suggestive of long-term population isolation. In other cases, populations merely show significant differences in frequencies of closely related matrilines. A shallow genealogical structure (Categories III and V) is consistent with the limited realized vagility of most birds relative to the size of their respective species' ranges, and probably reflects isolation by distance or recent population fragmentation rather than long-term vicariant separations. In some avian species, matrilineal connections are tight across broad geographic areas. In these cases, any behavioral or morphological alterations among populations would appear to have arisen quite recently.

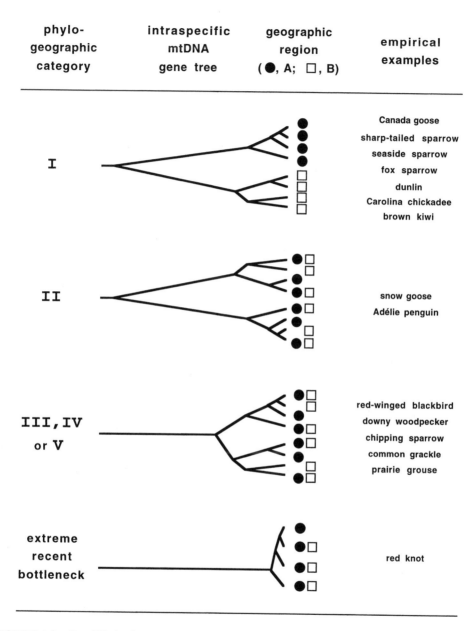

FIGURE 4.9 Simplified schematic of some common mtDNA phylogeographic patterns observed in avian species. Phylogeographic categories are those described in Fig. 4.1 and in the text.

REPTILES AND AMPHIBIANS

Many of these species are small-bodied and relatively immobile. In addition, reptiles often are specialists on patchily distributed habitats such as deserts, and nearly all amphibians are tied to discrete bodies of freshwater for reproduction. Thus, their phylogeographic patterns might be expected to resemble those of small terrestrial mammals, with pronounced structure the norm. In general, this has proved to be the case.

Terrestrial and Freshwater Species

A good example involves the desert tortoise (*Xerobates agassizi*) of the American Southwest. Molecular assays of animals throughout the species' range revealed three deep phylogeographic units differing from one another by more than 15 mtDNA restriction sites, plus additional substructure of closely related haplotypes on local geographic scales (Lamb et al., 1989). Two of the major phylogroups occurred east versus west of the Colorado River, a long-standing physical feature that probably served as a dispersal barrier for these desert-adapted animals. With regard both to local and regional population structure, the mtDNA phylogeographic pattern in the desert tortoise resembles the motif already mentioned for pocket mice from these same desert regions (McKnight, 1995).

However, any suspicion that most or all desert-adapted reptiles in the American Southwest might share a common biogeographic history were negated with subsequent findings for the desert iguana (*Dipsosaurus dorsalis*) and chuckwalla (*Sauromalus obesus*). In these species, mtDNA haplotypes showed strong geographic localization, but the intraspecific phylogenies gave no indication of having been shaped concordantly by the Colorado River or by any other singular vicariant factor of overriding multitaxon significance (Lamb et al., 1992).

The Category I phylogeographic pattern exemplified by the desert tortoise also has been reported for the related gopher tortoise (*Gopherus polyphemus*) in the southeastern United States (Osentoski and Lamb, 1995), as well as for several other freshwater turtle species in the region: *Trache-*

mys scripta (Avise et al., 1992c), *Sternotherus minor* (Walker et al., 1995), *S. odoratus* (Walker et al., 1997), and *Kinosternon subrubrum* and *K. baurii* (Walker et al., 1998a). The concordant patterns in these species will be discussed more fully in Chapter 5.

A Category I pattern also has been reported in an Australian rainforest skink (*Gnypetoscincus queenslandiae*). There, a deep mtDNA phylogenetic disjunction (17 fixed nucleotide differences) distinguished northern from southern populations that probably had been isolated in forest fragments by climatic conditions of the late Tertiary (Moritz et al., 1993b; Joseph et al., 1995; Cunningham and Moritz, 1998). Likewise, deep mtDNA phylogeographic structure has been recorded in the short-horned lizard (*Phrynosoma douglasi*) across its range in western North America (Fig. 4.10; Zamudio et al., 1997), in a boa snake (*Corallus enydris*) in northern versus southern regions of South America (Henderson and Hedges, 1995), and in populations of the bushmaster snake (*Lachesis muta*) from Central versus South America (Zamudio and Greene, 1997).

In some reptilian species, mtDNA phylogeographic structure is shallow but nonetheless conspicuous (Categories III or V) with respect to frequency differences or cladistic relationships of allied haplotypes. For example, by available evidence the marine iguana (*Amblyrhynchus cristatus*) shows little evolutionary differentiation across the Galápagos Islands in nuclear genes, and the same holds true for mtDNA sequences (Rassmann et al., 1997). Nonetheless, the mtDNA haplotypes align into island-specific lineages in an ancestral → derived pattern (determined by reference to outgroup taxa) that registers a sequential colonization of the archipelago mostly from east to west. By contrast, nuclear data from micro- and minisatellite loci could detect the effects of isolation by distance only. Brown and Pestano (1998) similarly used mtDNA sequences to infer the colonization history of skinks (genus *Chalcides*) in the Canary Islands.

In another study of microphylogeny on the Canary Islands, Thorpe et al. (1993, 1994) employed mtDNA sequences in conjunction with other genetic evidence to examine the colonization history of the lizard *Gallotia galloti*. Morphological characters then were mapped onto the molecular

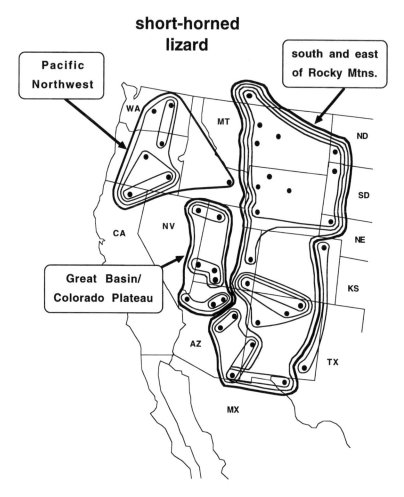

FIGURE 4.10   Matrilineal phylogeography of the short-horned lizard (after Zamudio et al., 1997). Geographic distributions of successively deeper clades in a mtDNA gene tree are encircled by concentric rings. Note the three major phylogroups.

phylogeny, and associated evolutionary features thereby deduced (Thorpe, 1996). For example, differences among island lizards in dorsal pattern were related to environmental conditions, whereas details of sexual leg markings were correlated mostly with phylogenetic history.

Similar exercises in phylogeographic character-state mapping have been conducted for body size differences and colonization histories of conspecific reptile populations in the region of Baja California (Petren and Case, 1997; Radtkey et al., 1997). As judged against a genealogical backdrop provided by mtDNA, oceanic islands in the Gulf of California were founded at least five separate times by mainland lizards in the *Cnemidophorus tigris* complex. Instances of both character relaxation and the retention of ancestral mainland conditions apparently contributed to the current distributions of body-size morphs on various islands (Radtkey et al., 1997).

Among the amphibians also, phylogeographic Category I has been recorded in mtDNA assays, as in the following examples. Geographic populations within as well as between two conventionally recognized subspecies of the California newt (*Taricha torosa*) occupy distinct branches on a matrilineal gene tree (Tan and Wake, 1995). So too do populations of the Cape clawed frog (*Xenopus gilli*) that probably were sundered by a late-Cenozoic oceanic transgression into their coastal range in South Africa (Evans et al., 1997). Interestingly, a related frog (*X. laevis*) that invaded the region recently shows no major phylogeographic subdivisions in this same area (Evans et al., 1997). In the diploid gray treefrog (*Hyla chrysoscelis*), two divergent mtDNA lineages are confined to eastern versus western portions of the species' range in eastern North America (Ptacek et al., 1994). Based on placements in a matrilineal tree, a related tetraploid form (*H. versicolor*) appears to have arisen from *H. chrysoscelis* at least three times via independent polyploidization events (Ptacek et al., 1994).

Similar breaks in mtDNA genealogy attest to the likely effects of historical biogeographic factors in sundering other amphibian populations. In the Taipei treefrog (*Rhacophorus taipeianus*), two distinctive mtDNA lineages thought to have split from a common ancestor in the early Pleistocene currently inhabit northern versus central regions of Taiwan (Yang et al., 1994). In an Australian hylid frog (*Litoria pearsoniana*), two deep clades that date to the Miocene or Pliocene occupy mountain ranges on opposite sides of the Brisbane River valley (McGuigan et al., 1998). In the canyon treefrog (*Hyla arenicolor*) from the American Southwest, two divergent mtDNA phylogroups inhabit the Colorado Plateau and the Sonoran Desert, respectively (Barber, 1996). Across Europe, four members of an assemblage of chromosomally, morphologically, and allozymically distinct crested newts (*Triturus cristatus* complex) all display divergent mtDNA genotypes, and further phylogeographic partitions in mtDNA were observed within each of two forms now recognized as taxonomic species (Wallis and Arntzen, 1989). In a salamander (*Salamandra salamandra*) inhabiting the Iberian Peninsula, two regionally distinct populations were interpreted from mtDNA and other genetic evidence to have been sundered by a Pliocene fluvial barrier (the Guadalquivir Basin) that probably played a similar role in other vertebrate speciations (García-París et al., 1998; see also Dopazo et al., 1998).

In the toad *Bufo marinus*, two highly divergent genealogical units (net sequence divergence ca. 5.5 percent) confined to opposite sides of the Andes Mountains in northern South America probably were separated by an uplift of the Eastern Andean Cordillera approximately 2.7 Mya (Slade and Moritz, 1998). This species has been translocated widely in this century and is now an important pest in Australia and Hawaii. The mtDNA findings also identified the original source of these introductions (the eastern lineage in the Americas) and confirmed, by virtue of severely diminished variation, that serial population bottlenecks were involved.

In the giant salamander (*Cryptobranchus alleganiensis*) of the eastern United States, several mtDNA lineages with sequence differences often greater than 3 percent occupy mostly allopatric or parapatric ranges (Routman, 1993; Routman et al., 1994). One lineage with extremely low intragroup sequence variation is distributed widely from Pennsylvania to the northern Ozark Mountains in Missouri in a pattern suggestive of rapid post-Pleistocene range expansion from a single refugium. Likewise, in the spotted salamander (*Ambystoma maculatum*) in the east-central United States, a closely-knit mtDNA phylogroup resides in several northeastern states as well as in the Ozark highlands (Phillips, 1994), again suggesting recent historical lineage connections across these areas. In this case, a second distinct mtDNA clade occurs from Alabama to Michigan, bisecting the range of the first phylogroup.

Overall, the diversity of phylogeographic patterns reported among salamander species is remarkable. In a continent-wide survey of control region sequences in the North American tiger salamander (*Ambystoma tigrinum*), Shaffer and McKnight (1996) observed only shallow lineage separations in this supposedly rapidly evolving portion of the molecule (however, various mtDNA haplotype clusters were localized regionally). Near the other end of the spectrum, exceptionally large genetic distances exist within a salamander (*Ensatina eschscholtzii*) of the American Pacific coast. Some geographic subspecies in California show mtDNA sequence differences exceeding 12 percent (Moritz et al., 1992), a finding that in conjunction with allozyme evidence suggests periods of population differentiation exceeding 5 My (Wake, 1997).

Unisexuals

A number of herpetofaunal (as well as piscine) taxa consist solely of females who reproduce asexually by parthenogenesis or by related quasi-sexual modes (Dawley and Bogart, 1989). Unisexual "biotypes" are not biological species in the usual sense but nonetheless are isolated genetically from related sexual species from whom they arose via hybridization. Analyses of mtDNA are of special interest for unisexual biotypes and their sexual cognates for at least two reasons: (a) They reveal the direction of the hybridization event(s) that gave evolutionary rise to the unisexuals; and (b) Within a unisexual biotype, the maternal phylogeny *is* the entire organismal pedigree (unlike the case for a sexually reproducing species).

With regard to phylogeographic assessments, several studies have taken advantage of mtDNA assays to decipher the geographic and biotic

origins of unisexual reptiles. For example, mtDNA haplotypes in nine unisexual biotypes in the *sexlineatus* group of *Cnemidophorus* lizards all stem
from one of the four nominate geographic subspecies (*C.i. arizonae*) within
the sexual ancestor *C. inornatus* (Densmore et al., 1989). In a parthenogenetic gecko (*Heteronotia binoei*), two major mtDNA lineages are present,
one in western Australia and the other from central to western regions of
the continent (Moritz, 1991). As judged by the limited mtDNA sequence
variation in the more widespread matrilineal clade (an order of magnitude
less than diversity in the codistributed sexual relative), this parthenogen
must have spread across a wide area, probably within the last few thousand years. Phylogenetic appraisals against mtDNA lineages in the sexual
cognate species identified western Australia as the geographic origin of
this widespread unisexual clade.

Similar mtDNA evidence for relatively recent colonizations also apply
to parthenogenetic house geckos (*Lepidodactylus lugubris* and *Hemidactylus
garnotii*) that now are widespread on remote Pacific Islands (Moritz et al.,
1993a). Presumably, the shallow phylogeographic pattern reflects the proclivity of these human-associated biotypes to hitch rides on boats. Many
additional molecular genealogical appraisals of the origins and ages of
unisexual vertebrates are available (review in Avise et al., 1992a).

Marine Turtles

Life histories of the 7–8 extant testudine species in the sea differ from those
of their terrestrial and freshwater counterparts by inclusion of migrational
movements over vast distances, often tens of thousands of kilometers during an individual's lifetime. Marine turtles generally are inaccessible for
close field observation, and have generation lengths measured in decades,
so numerous aspects of their behavior and rookery affinities remained unknown. Following the pioneering mtDNA studies by Bowen et al. (1989)
and Meylan et al. (1990), many gaps in knowledge have been closed by
more than 100 research publications now available (Bowen, 1996a) that
deal with population genetics, molecular evolution, and conservation

biology in marine turtles (reviews in Bowen and Avise, 1996; Bowen and Witzell, 1996; Bowen and Karl, 1997).

The first substantial mtDNA study of these organisms (Bowen et al., 1989) provides a classic resolution of a dispersal-vicariance controversy in phylogeography. A major rookery for the green turtle (*Chelonia mydas*) exists on Ascension Island, situated on the mid-Atlantic ridge halfway between Brazil and Liberia. Females that nest on Ascension otherwise occupy shallow-water feeding pastures along the South American coast. Thus, for each nesting episode (every two or three years for an individual), females embark on a 5,000-km migration to Ascension Island and back.

Two distinct scenarios exist for how Ascension turtles established such an improbable migratory circuit. Under the vicariance hypothesis of Carr and Coleman (1974), the ancestors of Ascension Island green turtles nested on islands adjacent to South America in the late Cretaceous, soon after the opening of the equatorial Atlantic Ocean. Over the past 70 My, these volcanic islands were displaced from South America by plate tectonic movements and sea-floor spreading (by about two cm per year). Carr and Coleman proposed that natally homing females gradually lengthened their migratory circuits by following a population-specific instinct to migrate to Ascension Island. This model entails a genetic separation of Ascension matrilines from those on mainland rookeries for tens of millions of years. An alternative dispersalist hypothesis suggests that Ascension Island was colonized secondarily, perhaps recently, from continental waifs.

Bowen et al. (1989) showed that the Ascension Island nesters display fixed mtDNA differences from several other conspecific rookery populations in South and Central America. However, the magnitude of the sequence divergence is small ($p < 0.002$) and thus inconsistent with the ancient vicariance hypothesis of rookery origin. The molecular data indicate that Ascension Island was colonized within the last 1.0 My (lower confidence bound: zero years bp).

Extensions of mtDNA surveys to green turtles worldwide revealed the following: (a) a fundamental phylogenetic split distinguishing assayed specimens in the Atlantic-Mediterranean from those in the Indian-Pacific Oceans; and (b) strong matrilineal substructure among rookeries or assemblages of rookeries within oceans, as evidenced by fixed or nearly fixed differences in the frequencies of closely related mtDNA haplotypes (Bowen et al., 1992; Allard et al., 1994; Encalada, 1996; Encalada et al. 1996). The first finding is consistent with a vicariant separation of tropical-restricted green turtles into separate ocean basins by the rise of the Isthmus of Panama about 3 Mya. The second finding is indicative of a propensity for natal homing by females: If each adult female normally returns to nest near her own natal rookery, colonies over time should become distinguishable in maternally inherited markers.

This latter discovery provides a clear biological illustration of the special matrilineal-demography connection emphasized in Chapter 2. Females ultimately govern the reproductive output of a rookery. Thus, natal-philopatric nesting tendencies signify a considerable *demographic* autonomy for each green turtle rookery *with regard to reproduction*. This conclusion holds in principle even if populations within an ocean basin were homogeneous for nuclear gene frequencies due to panmictic mating (empirically they are not—Karl et al., 1992). The mtDNA findings have conservation relevance for this endangered species. Because of the natal-homing propensity, recruitment of females from foreign rookeries is unlikely to compensate for mortality in heavily exploited rookeries, or to reestablish (over ecological time scales) rookeries that are extirpated by human activities or other causes.

Similar mtDNA evidence for significant matrilineal structure has been reported for rookeries of the loggerhead turtle *Caretta caretta* (Bowen et al., 1993b, 1994) and the hawksbill *Eretmochelys imbricata* (Broderick et al., 1994; Bass et al., 1996; Broderick and Moritz, 1996). Surveys of marine turtles on migratory routes and at feeding locations also have helped to reveal the rookery origins of these animals at other stages of life (Laurent et al., 1993, 1998; Avise and Bowen, 1994; Broderick et al., 1994; Norman et al.,

1994; Bowen, 1995, 1996b; Bowen et al., 1995, 1996; Sears et al., 1995; Bass et al., 1996; Norrgard and Graves, 1996; Encalada et al., 1997; FitzSimmons et al., 1997a; Bolten et al., 1998). An emerging generality is that particular assemblages of nonnesting marine turtles often derive from multiple rookery sites (Lahanas et al., 1998). Thus, with regard to mortality sources at these times, different rookeries can be jointly impacted demographically. This too can have conservation ramifications. For example, a country that permits harvest of these endangered animals at feeding arenas within its legal jurisdiction could negatively impact multiple rookeries including those under the purview of other countries.

A puzzling phylogeographic question applies to the ridley complex of marine turtles. Two species traditionally are recognized—*Lepidochelys kempi* confined to a single rookery in the western Gulf of Mexico, and *L. olivacea* distributed almost worldwide in suitable waters. However, these taxonomic species are morphologically nearly identical, and their lopsided distributions at face value make little biogeographic sense. Nonetheless, mtDNA surveys revealed that populations of *L. olivacea* from the Atlantic and Pacific Oceans are considerably less divergent from one another than either is from *L. kempi*. Furthermore, the matrilineal separation between these two taxonomic forms is slightly greater than that between *any* two conspecific populations of green turtles or loggerheads worldwide (Bowen et al., 1991, 1998).

Earlier, a biogeographic scenario that may explain the mtDNA findings (as well as the traditional taxonomy for the ridley complex) had been proposed (Pritchard, 1969). Based on a synthesis of distributional and morphological information, Pritchard suggested than an ancestral ridley population had been sundered by the rise of the Isthmus of Panama into proto-Kemp's and proto-olive forms in the Atlantic and Indo-Pacific Oceans, respectively, and that olive ridleys recently colonized the Atlantic via South Africa's Cape of Good Hope. The mtDNA phylogeographic pattern and the genetic distance estimates are consistent with this biogeographic scenario, and further suggest that the Atlantic invasion by olive ridleys occurred within the last 300,000 years.

FISHES

At any point in time, freshwater drainage basins by definition are spatially disconnected, and we might expect populations of their piscine inhabitants to be genetically structured accordingly. However, landscapes and river courses change over geological time, and adjacent drainages may be connected repeatedly by headwater stream captures or outlet mergers. Other drainages may remain isolated. One net effect of historical connections and dissociations of waterways might be a more-or-less corresponding genealogical structure for the fish populations that inhabit freshwater systems. By contrast, the marine realm at face value is less sundered, so weaker phylogeographic population structures might be anticipated for mobile species over comparable areas in the sea. These expectations often but not invariably have been met.

Freshwater Species

Numerous mtDNA surveys have been conducted on freshwater fishes (Table 4.2). The phylogenetic trees in some cases have proved to be shallow and spatially unstructured, but these instances usually involve surveys conducted within a single body of water or over a relatively small portion of a species' total range. For example, conspecific populations of several carp species show minimal phylogeographic differentiation along a 500-km stretch of the Yangtze River in China (Lu et al., 1997), as do several species of cichlid fishes within large East African lakes (Bowers et al., 1994; Meyer et al., 1996; Sturmbauer et al., 1997; but see also Verheyen et al., 1996). Sand darters of the *Etheostoma beanii-bifascia* complex also show little mtDNA differentiation (Wiley and Hagen, 1997) but their range is confined to adjacent coastal drainages in Mississippi and Alabama.

When larger geographic areas have been examined in other species, deeper phylogeographic structures often have become evident. Walleye (*Stizostedion vitreum*) populations throughout the 400-km length of Lake Erie in North America are weakly structured in mtDNA (Faber and Stepien, 1997), but a disjunct population in the southern United States differs sharply in mtDNA sequence ($p \cong 0.023$) from those in the northern states (Billington and Strange, 1995).

In widely distributed freshwater fishes surveyed broadly, deep phylogenetic subdivisions in mtDNA among regional drainage systems have been the norm. At least 27 surveyed species (56 percent) have displayed phylogeographic Category I separations, and many additional species have exhibited at least shallow (Category III or V) population structure on finer spatial scales (Table 4.2). In only a few reported cases are deeply separated mtDNA lineages broadly sympatric (Category II).

However, the allopatric versus sympatric status of principal mtDNA lineages are not always as straightforward as the summaries in Table 4.2 imply. For example, populations of the bluegill (*Lepomis macrochirus*), mosquitofish (*Gambusia affinis-holbrooki* complex), and largemouth bass (*Micropterus salmoides*) show a clear genealogical separation between eastern and western subregions of the southeastern United States, yet in each case presumed secondary overlap zones exist where the distinctive phylogroups are widely sympatric. In these areas of Georgia, the Florida panhandle, and eastern Alabama, mtDNA lineages appear to have been scrambled against alternative nuclear genetic backgrounds via introgressive hybridization (Philipp et al., 1983; Avise et al., 1984b; Scribner and Avise, 1993; Nedbal and Philipp, 1994).

Phylogeographic patterns in mtDNA often have been interpreted in conjunction with physical evidence on historical drainage patterns of the Quaternary. For example, Wilson and Hebert (1996) identified three primary mtDNA lineages in surveys of nearly 900 lake trout (*Salvelinus namaycush*) from 60 localities in North America. These lineages display mostly allopatric distributions suggestive of divergence in separate Pleistocene refugia, as well as presumed zones of secondary overlap that match dispersal corridors whose approximate ages are known (Fig. 4.11).

Throughout its native European range, the brown trout (*Salmo trutta*) displays complex patterns of variation in phenotype and life history, and this has led to disagreements about population affinities and taxonomic boundaries in the assemblage. Molecular surveys of mtDNA revealed five

TABLE 4.2   Examples of mtDNA phylogeographic patterns in freshwater and anadromous fishes.

Species	Survey Area, Major mtDNA Lineages	Reference
Phylogeographic Category I[a]		
Lepomis macrochirus	southeastern U.S., eastern vs. western drainages	Avise et al., 1984b
Amia calva	southeastern U.S., eastern vs. western drainages	Bermingham & Avise, 1986
Lepomis microlophus	southeastern U.S., eastern vs. western drainages	ibid.
Lepomis punctatus	southeastern U.S., eastern vs. western drainages	ibid.
Lepomis gulosus	southeastern U.S., eastern vs. western drainages	ibid.
Coregonus species complex	Holarctic drainages, 4 phylogeographic units	Bernatchez & Dodson, 1991, 1994
Salmo trutta	Europe, 5 phylogeographic units	Bernatchez et al., 1992
Cyprinodon nevadensis complex	N. Amer. southwest, 2 phylogeographic units	Echelle & Dowling, 1992
Poecilia reticulata	Trinidad, 2 phylogeographic units	Fajen & Breden, 1992
Salvelinus fontinalis	eastern North America, 2 phylogeographic units	Bernatchez & Danzmann, 1993; Angers & Bernatchez, 1998
Gambusia affinis/holbrooki	southeastern U.S., eastern vs. western drainages	Scribner & Avise, 1993
Micropterus salmoides	eastern U.S., Floridian vs. continental subspecies	Nedball & Philipp, 1994
Gasterosteus aculeatus	Holarctic, 2 phylogeographic units	Ortí et al., 1994
Osmerus species complex	Holarctic, 2–3 phylogeographic units	Taylor & Dodson, 1994; Bernatchez, 1997
Oncorhynchus nerka	Russia to British Columbia, 2 phylogeographic units	Bickham et al., 1995
Stizostedion vitreum	eastern U.S., 2 phylogeographic units	Billington & Strange, 1995
Cyprinella lutrensis	U.S. interior, 3 phylogeographic units	Richardson & Gold, 1995
Culaea inconstans	U.S. Great Lakes region, 2 phylogeographic units	Gach, 1996
Cyprinella venusta	southern U.S. coastal states, 4 phylogeographic units	Kristmundsdóttir & Gold, 1996
Cottus nozawae	northern Japan, 2 phylogeographic units	Okumura & Goto, 1996
Percina caprodes	central U.S., Ozark and Ouachita highlands, 3 units	Turner et al., 1996
Percina phoxocephala	central U.S., Ozark and Ouachita highlands, 2 units	ibid.
Salvelinus namaycush	eastern North America, 3 phylogeographic units	Wilson & Hebert, 1996
Salvelinus alpinus	Holarctic, 3 phylogeographic units	Wilson et al., 1996
Oreochromis niloticus	Africa, 3 phylogeographic units	Agnese et al., 1997
Agosia chrysogaster	southwestern U.S., 2 phylogeographic units	Tibbets & Dowling, 1996
Galaxias zebratus	South Africa, 4 units that might be valid species	Waters & Cambray, 1997

Phylogeographic Category II[a]

Brevoortia tyrannus	U.S. Atlantic coastline	Bowen & Avise, 1990
Hemibagrus nemurus	S.E. Asia, complex pattern incl. mixing of ancient lineages	Dodson et al., 1995
Hypophthalmichthys molitrix	China, 500 km. of Yangtze River	Lu et al., 1997
Mogurnda adspersa	Australia, 3 distinct phylogroups co-occur in a river	Hurwood & Hughes, 1998

Phylogeographic Categories III, IV, or V[a]

Salmo gairdneri	North America, Pacific northwest	G. M. Wilson et al., 1985
Alosa sapidissima	Atlantic seaboard, Florida to Quebec	Bentzen et al., 1989; Epifanio et al., 1995
Acipenser oxyrhynchus	southeastern U.S., coastal drainages	Bowen & Avise, 1990
Acipenser transmontanus	western North America, 2 drainages	Brown et al., 1993
Oncorhynchus keta	northern rim of the Pacific Ocean	Park et al., 1993
Melanochromis auratus	50 km transects within Lake Malawi, East Africa	Bowers et al., 1994
Melanochromis heterochromis	50 km transects within Lake Malawi, East Africa	ibid.
Phoxinus eos	central Ontario	Toline & Baker, 1995
Polyodon spathula	central U.S., Montana to Alabama	Epifanio et al., 1996
Simochromis babaulti	300 km transect within Lake Tanganyika, East Africa	Meyer et al., 1996
Simochromis diagramma	300 km transect within Lake Tanganyika, East Africa	ibid.
Percina nasuta	central U.S., Ozark and Ouachita highlands	Turner et al., 1996
Ctenopharyngodon piceus	China, 500 km. of Yangtze River	Lu et al. 1997
Aristichthys nobilis	China, 500 km. of Yangtze River	ibid.
Mylopharyngodon piceus	China, 500 km. of Yangtze River	ibid.
Oncorhynchus mykiss	coastal and interior California	Nielsen et al., 1994, 1997
Etheostoma beanii/bifascia	Mississippi and Alabama coastal drainages	Wiley & Hagen, 1997

a. The distinction between Categories I and III is not always unambiguous. For example, several geographic populations of *Salmo gairdneri* show fixed frequency differences among mtDNA haplotypes that are intermediate in levels of sequence divergence. Similarly, the distinction between Categories I and II is not always unambiguous. For example, the two major mtDNA clades present in *Salvelinus fontinalis* have different distributions suggestive of allopatric divergence in separate glacial refugia, yet they now occur sympatrically in some areas presumably due to joint reinvasions of northern North America following glacial retreats.

lake trout

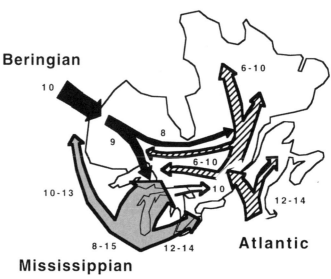

FIGURE 4.11   Schematic representation of the deduced colonization routes from Atlantic, Mississippian, and Beringian Pleistocene refugia for three major mtDNA lineages in lake trout in eastern North America (after Wilson and Hebert, 1996). Numbers
are probable dates (in thousands of years bp) for post-Pleistocene drainage connections.

major phylogroups, each spatially oriented. For example, one matrilineal
clade is fixed in all Atlantic basin samples from France to northern Scandinavia (Bernatchez et al., 1992). The sharp mtDNA differences, in conjunction with the geographic distributions of the phylogroups and generally
concordant allozyme patterns, indicate a history of ancient allopatric separations (Bernatchez, 1995). In some regions such as south-central Europe,
different phylogroups also overlap in patterns likely attributable to recent
range expansions (facilitated in some cases by anthropogenic dispersal of
this popular sport fish) from separate Pleistocene refugia (Giuffra et al.,
1994, 1996; Bernatchez and Osinov, 1995; Apostolidis et al., 1997). The

phylogenetic distinctiveness of these "evolutionarily significant units" gains added support from the reproductive isolation sometimes shown in sympatry (Bernatchez, 1995).

To exemplify further both the complexity of outcomes and the power of molecular appraisals in reconstructing historical biogeography in freshwater fishes, consider the mtDNA findings for another Holarctic complex, the *Coregonus* whitefishes. Bernatchez and Dodson (1991) surveyed 41 populations of *C. clupeaformis* (525 specimens) across North America, and identified several principal mtDNA lineages (Fig. 4.12) including: "A" and "B," confined (on this continent) to the Yukon and Alaska where they occur sympatrically; "C," a tight-knit assemblage across most of Canada; and "D," confined to southern Quebec and the northeastern United States Nucleotide diversity in the C assemblage was especially low, suggesting extremely tight genealogical connections of fishes across a vast area from the MacKenzie Delta in the west to Labrador in the east.

With the exception of central Alaska and the Yukon, which remained ice-free, the species' current range was covered repeatedly by Pleistocene glaciers. The genetic discontinuities and the strong geographic patterns of lineage distribution suggest that these four matrilineal assemblages differentiated in separate refugial areas (Bernatchez and Dodson, 1991). Clade C, for example, probably traces to the Mississippian refugium in the central and eastern United States from which many freshwater fish species are suspected to have recolonized parts of the northern continent within the last 10–20 thousand years.

Subsequent surveys of whitefish populations across the Palearctic clarified the evolutionary origins of the two sympatric mtDNA lineages in Beringia (Bernatchez and Dodson, 1994). Assemblage B is distributed throughout most northern Eurasian populations, with Alaska and Yukon its easternmost extension. Thus, this assemblage may have arisen in portions of northern Eurasia known to have remained unglaciated during the Pleistocene, such that its current geographical overlap with clade A probably reflects a recent colonization of the New World (rather than a long-term onsite retention of these lineages in Beringia). This study also identified another divergent mtDNA lineage ("E") that predominates in whitefish populations in the central alpine lakes of Europe (Fig. 4.12).

On finer spatial scales, mtDNA lineages in the whitefish have been employed in conjunction with data from nuclear genes to address the genealogical histories of distinctive sympatric morphotypes sometimes observed within a drainage basin. For example, dwarf and normal-sized whitefish coexist in the Allegash basin of eastern Canada and northern Maine where they belong to separate gene pools according to allozyme evidence (Kirkpatrick and Selander, 1979). Is this a case of sympatric speciation? If so, populations of the two morphs should be one another's closest living relatives. Alternatively, perhaps speciation was allopatric and sympatry was achieved secondarily, in which case closer genealogical ties for the dwarf and normal morphs might lie outside the area. Bernatchez and Dodson (1990) discovered that the two morphotypes in the Allegash drainage are alternately fixed for different mtDNA lineages (C and D), suggesting that their co-occurrence reflects a secondary overlap of two monophyletic groups stemming from separate Pleistocene refugia.

On the other hand, genetic analyses of sympatric dwarf and normal whitefishes in other eastern Canadian locations showed that these morphotypes do not group consistently into distinct phylogenetic clusters. This suggests more complex evolutionary scenarios and the likelihood of "sympatric divergence and multiple allopatric divergence/secondary contact events on a small geographic scale" (Pigeon et al., 1997). Similar mtDNA analyses of coexisting limnetic and benthic ecotypes in whitefish populations in the Yukon indicate polyphyletic matrilineal origins for these forms also (Bernatchez et al., 1996). Some of these co-occurrences of ecomorphs appear to reflect secondary contacts between allopatrically differentiated matrilines, whereas other instances remain ambiguous with regard to sympatric or allopatric origins.

These extensive molecular genetic appraisals of widely distributed fish species in the Northern Hemisphere exemplify the remarkable power

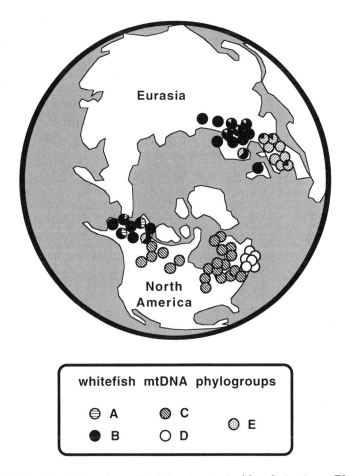

FIGURE 4.12   Distributional map (global polar view) of five distinctive mtDNA phylogroups in populations of whitefish, *Coregonus* sp. (after Bernatchez and Dodson, 1994).

of molecular tools in deciphering the phylogeographic histories of related populations in subdivided aquatic environments (Bernatchez, 1995; Bernatchez and Wilson, 1998). The interpretation of morphological and life-history differences of these fishes against an explicit genealogical backdrop has provided novel perspectives and insights into evolutionary shaping processes.

Another example of phylogeographic character-state mapping involved cave and surface-dwelling fishes in the *Amblyopsis-Typhlichthys* complex (Amblyopsidae) of eastern North America (Bergstrom, 1997). Several of the taxonomic forms show specialized features associated with cave life including severe reduction of eyes and skin pigments. By mapping these traits onto a mtDNA phylogeny, troglobytic phenotypes were shown to have evolved at least six times independently in amblyopsids of different cave systems. Similarly, in lake and stream populations of the three-spined stickleback (*Gasterosteus aculeatus*) in the Pacific Northwest, an interpretation of ecological and morphological differences against a phylogeographic backdrop from mtDNA suggests that pronounced ecotypic differences arose recently and in parallel at least twice (Thompson et al., 1997).

Marine Species

Numerous mtDNA phylogeographic surveys have been conducted on marine fishes (Table 4.3). In some cases, a Category I phylogeographic pattern has been observed wherein relatively deep matrilineal separations distinguish allopatric populations within a recognized taxonomic species. For example, disjunct populations of the black sea bass (*Centropristis striata*) in the Atlantic versus Gulf of Mexico occupy distinct branches in a mtDNA gene tree. In the bluefin tuna (*Thunnus thynnus*), a preliminary mtDNA survey suggests that Atlantic populations are divergent from those in the Pacific Ocean. In the coastal killifish (*Fundulus heteroclitus*), populations north versus south of a clinal boundary in the central Atlantic states occupy mtDNA gene-tree branches that show general phylogeographic congruence with nuclear allele frequency differences (Powers et al., 1991; Smith et al., 1998) and with respect to clade structure in a gene tree for lactate dehydrogenase (Bernardi et al., 1993). In a deep-sea fish (*Cyclothone alba*), different oceanic regions contain five highly distinct mtDNA phylogroups that may represent sibling species undistinguished by morphological differences (Miya and Nishida, 1997).

In striking contrast to the freshwater fishes, however, most (73 percent) of the marine fish species surveyed to date for mtDNA have displayed only shallow intraspecific phylogeographic structure at best (Table 4.3). Especially impressive are the spatial scales over which closely related lineages typically are distributed. Common mtDNA haplotypes in each of eight coral-reef fish species, for example, are ubiquitous in conspecific populations throughout the expanse of the Caribbean Sea (Shulman and Bermingham, 1995). These fishes are sedentary bottom-dwellers as adults but have planktonic larval stages of varying duration. The genetic data indicate considerable gene flow throughout the region (irrespective of ocean current patterns or species' life histories) and an absence of long-standing biogeographic barriers to dispersal. Similar conclusions about lower population genetic structure in most marine as compared to freshwater fishes were reached from extensive allozyme comparisons (review in Ward et al., 1994a).

The spatial scales of tight genealogical connections tend to be even greater in several widespread or cosmopolitan pelagic fishes (reviews in Graves, 1996, 1998). Populations of the striped marlin (*Tetrapturus audax*) throughout the Pacific Ocean are linked closely in a phylogenetic sense, as are populations of the white marlin (*T. albidus*) in the Atlantic. Common as well as some rare mtDNA haplotypes are shared by populations of the shortfin mako shark (*Isurus oxyrinchus*) in the North Atlantic, South Atlantic, North Pacific, and South Pacific Oceans (Fig. 4.13). In several tunas (skipjack, *Katsuwonus pelamis*; albacore, *Thunnus alalunga*; and yellowfin, *T. albacares*), mtDNA genealogical differentiation appears to be relatively low on a global scale.

To emphasize the dramatic contrast of these results with those typifying freshwater fishes, consider the mtDNA phylogeographic outcome for the yellowfin tuna globally versus that for the bowfin in adjacent freshwater drainages in the southeastern United States (Fig. 4.14). These species were chosen for this illustration because both were assayed for similar numbers of individuals (88 and 68, respectively) using similar numbers of

TABLE 4.3   Examples of mtDNA phylogeographic patterns in marine and catadromous fishes.

Species	Survey Area, Major mtDNA Lineages	Reference
Phylogeographic Category I[a]		
Centropristis striata	Atlantic vs. Gulf of Mexico	Bowen & Avise, 1990
Fundulus heteroclitus	U.S. Atlantic coast, 2 phylogeographic units	González-Villaseñor & Powers, 1990
Mallotus villosus	North Atlantic, eastern vs. western regions	Dodson et al., 1991; Birt et al., 1995
Thunnus thynnus	Atlantic vs. Pacific Oceans	Chow & Inoue, 1993
Mugil cephalus	global, several phylogeographic units	Crosetti et al., 1993
Paralabrax maculatofasciatus	Gulf of California vs. outer Pacific coast	Stepien, 1995
Pomatomus saltatrix	global, 2 phylogeographic units (one confined to Brazil)	Goodbred & Graves, 1996
Sardinops spp.	global, 3 to 5 phylogeographic units	Okazaki et al., 1996;
		Bowen & Grant, 1997
Polyprion americanus	trans-North Atlantic vs. southern hemisphere oceans	Sedberry et al., 1996
Cyclothone alba	global, 5 phylogeographic units	Miya & Nishida, 1997
Phylogeographic Category II[a]		
Makaira nigricans	2 phylogenetic units in Atlantic	Graves & McDowell, 1995;
		Finnerty & Block, 1992
Istiophorus platypterus	2 phylogenetic units in Atlantic	Graves & McDowell 1995
Engraulis encrasicolus	2 phylogenetic units in Mediterranean and Black Sea	Magoulas et al., 1996
Phylogeographic Categories III, IV, or V[a]		
Katsuwonus pelamis	Atlantic and Pacific Oceans	Graves et al., 1984
Anguilla rostrata	North American coastlines in Atlantic and Gulf of Mexico	Avise et al. 1986
Arius felis	Gulf of Mexico and U.S. Atlantic coast	Avise et al., 1987b
Bagre marinus	Gulf of Mexico and U.S. Atlantic coast	ibid.
Opsanus beta	Gulf of Mexico	ibid.
Opsanus tau	U.S. Atlantic coast	ibid.
Thunnus alalunga	Atlantic and Pacific Oceans	Graves & Dizon, 1989;
		Chow & Ushiama, 1995

Species	Location	Reference
Cynoscion regalis	U.S. Atlantic coast	Graves et al., 1992
Pseudopentaceros wheeleri	central and northern Pacific Ocean	Martin et al., 1992a
Theragra chalcogramma	Bering Sea, Gulf of Alaska	Mulligan et al., 1992;
		Shields & Gust, 1995
Lutjanus campechanus	Gulf of Mexico	Camper et al., 1993
Sciaenops ocellatus	Gulf of Mexico and U.S. Atlantic coast	Gold et al., 1993
Dicentrarchus labrax	Mediterranean Sea	Patarnello et al., 1993
Thunnus albacares	Atlantic and Pacific Oceans	Scoles & Graves, 1993;
		Ward et al., 1994b
Hoplostethus atlanticus	across southern hemisphere oceans	Smolenski et al., 1993
Epinephelus morio	Gulf of Mexico	Gold & Richardson, 1994
Pogonias cromis	Gulf of Mexico	Gold et al., 1994
Tetrapturus audax	throughout Pacific Ocean Basin	Graves & McDowell, 1994
Xiphias gladius	global	Grijalva-Chon et al., 1994;
		Bremer et al., 1995;
		Rosel & Block, 1996
Tetrapturus albidus	Atlantic Ocean	Graves & McDowell, 1995
Carcharhinus plumbeus	Gulf of Mexico and U.S. Atlantic coast	Heist et al., 1995
Stegastes leucostictus	throughout the Caribbean	Shulman & Bermingham, 1995
Ophioblennius atlanticus	throughout the Caribbean	ibid.
Abudefduf saxatilis	throughout the Caribbean	ibid.
Gnatholepis thompsoni	throughout the Caribbean	ibid.
Haemulon flavolineatum	throughout the Caribbean	ibid.
Halichoeres bivittatus	throughout the Caribbean	ibid.
Holocentrus ascensionis	throughout the Caribbean	ibid.
Thalassoma bifasciatum	throughout the Caribbean	ibid.
Sebastolobus alascanus	Pacific coast of North America	Stepien, 1995
Microstomus pacificus	Pacific coast of North America	ibid.
Gadus morhua	throughout Norway	Arnason and Pálsson, 1996
Rhizoprionodon terraenovae	Gulf of Mexico and U.S. Atlantic coast	Heist et al., 1996a
Isurus oxyrinchus	global	Heist et al., 1996b
Macquaria novemaculeata	eastern Australian coastline	Chenoweth & Hughes, 1997

a. For several of these studies, caveats like those in footnote *a* of Table 4.2 again apply.

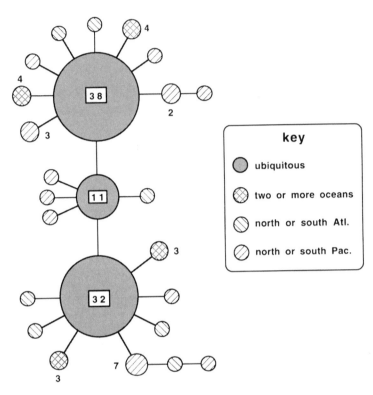

FIGURE 4.13   Phylogenetic network of mtDNA haplotypes in a global survey of the shortfin mako shark (after Heist et al., 1996b). All branch lengths are of unit length (one observed restriction site change). Shown also are numbers of individuals of each haplotype ($N = 1$ if not otherwise specified).

restriction enzymes (12 and 13). Also, identical numbers of common haplotypes (those present in multiple individuals) were observed. Not only is the total genealogical depth somewhat greater in the gene tree for the bowfin (which by standards for freshwater fish is near the *shallow* end of intraspecific matrilineal networks), but the geographic structure of bowfin variation is sharply discontinuous and displayed at a spatial scale that is vastly smaller than the distributional range of the yellowfin tuna.

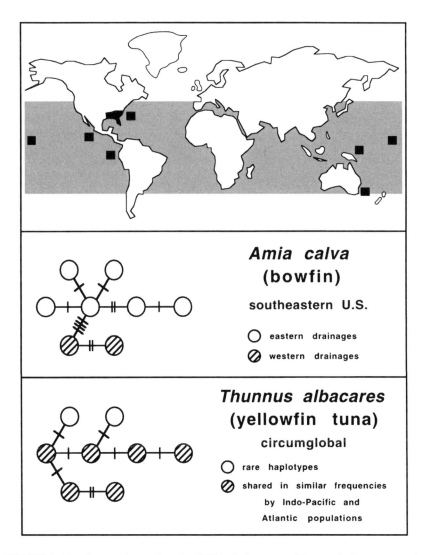

FIGURE 4.14   Contrasting scales of mtDNA phylogeographic population structure in a freshwater versus a marine fish (after Avise, 1998b). *Above:* Map showing the assayed range of the bowfin in the southeastern United States (black), and the sampling locales (black squares) and approximate total range (shaded) of the yellowfin tuna. *Below:* Parsimony networks summarizing phylogenetic relationships among mtDNA haplotypes, where slashes across branches are numbers of restriction site changes. Data for the bowfin are from Bermingham and Avise (1986), and for the yellowfin tuna are from Scoles and Graves (1993).

Some additional points should be emphasized about the shallow phylogeographic structures in many (certainly not all) marine fishes. First, tight genealogical connections do not necessarily imply genetic homogeneity across space. Genotypic frequencies often differ among locales. For example, statistically significant differences in the frequencies of closely related mtDNA haplotypes exist between and sometimes within oceans in several pelagic billfish species (e.g., Kotoulas et al., 1995), and these findings have relevance for stock identification and population management (Graves, 1996). Nonetheless, the populations appear not to be sundered deeply in a historical or genealogical sense.

Second, perceived taxonomic boundaries can influence interpretations of phylogeographic population structure greatly. For example, several of the Caribbean reef fishes surveyed by Shulman and Bermingham (1995) belong to species complexes that include related congeners in the Pacific Ocean or elsewhere, and these often display considerable mtDNA phylogeographic differentiation overall (Bermingham et al., 1997). Thus, if phylogeographically distinct members of the *Abudefduf saxatilis* species group traditionally had been named a single polytypic species, that species on a global scale would present a Category I phylogeographic pattern not unlike that of many widely distributed freshwater species such as *Coregonus clupeaformis* or *Salmo trutta.*

Another such example involves sardines, which occur in six discrete regional populations of controversial taxonomic status. These forms traditionally were recognized as *Sardinops ocellatus* in South Africa, *S. neopilchardus* in Australia, *S. sagax* in Chile, *S. caeruleus* in California, *S. melanostictus* in Japan, and *Sardina pilchardus* in Europe. On a global scale, varying levels of historical separation are evidenced in mtDNA (Okazaki et al., 1996; Grant and Bowen, 1998), with a recent (< 0.5 Mya) inferred founding of populations around the rim of the Pacific Ocean and a much deeper split between the *Sardina* and *Sardinops* lineages (see also Grant and Leslie, 1996). The taxonomic complex as a whole displays a Category I phylogeographic pattern, yet shallow lineage structures within each named species are consistent with recent historical connections probably stemming from population bottlenecks or dramatic fluctuations in population size (Grant and Bowen, 1998).

Finally, mtDNA lineage separations within regional populations of most marine fishes, including those that are currently abundant, typically appear shallow and often star-like in genealogical pattern (Shields and Gust, 1995; Grant and Bowen, 1998). In addition to some of the species listed under phylogeographic Categories III–V in Table 4.3, other examples include Atlantic herring, *Clupea harengus* (Kornfield and Bogdanowicz, 1987), the hake, *Merluccius capensis* and *M. paradoxus* (Becker et al., 1988), orange roughy, *Hoplostethus atlanticus* (Ovenden et al., 1989; C. S. Baker et al., 1995), Pacific herring, *C. pallasi* (Schweigert and Withler, 1990), Atlantic cod, *Gadus morhua* (Carr and Marshall, 1991), haddock, *Melanogrammus aeglefinus* (Zwanenburg et al., 1992), and greater amberjack, *Seriola dumerili* (Richardson and Gold, 1993). Such outcomes are further indicative of close historical connections among conspecific marine fishes, and suggest that regional populations often may fluctuate greatly in abundance such that evolutionary N_e values are relatively small.

Diadromous Species

Life cycles of many fishes include a freshwater and a saltwater phase. In catadromous species such as the American eel, spawning takes place in the ocean and larvae migrate to freshwater streams. The reverse is true in anadromous species such as many salmon, where freshwater spawning is followed by juvenile migration to the sea. Fish with diadromous life cycles are likely to have diminished contemporary population genetic structure in comparison to many primary freshwater species. In a review of the extensive allozyme literature on the subject, Gyllensten (1985) concluded that freshwater fish species typically show significantly more geographic differentiation than do marine species, and that anadromous taxa tend to display patterns more similar to those of marine forms.

In terms of mtDNA phylogeography, the near absence of population structure in eels collected from freshwater streams of North America (as discussed above) almost certainly reflects the consequences of lineage dispersal under a catadromous life cycle. However, catadromous species can show significant population structure if multiple breeding populations exist and larval dispersal is constrained. A case in point appears to be the Australian bass (*Macquaria novemaculeata*), in which significant mtDNA structure among freshwater populations was attributed to isolation by distance among marine spawning stocks (Jerry and Baverstock, 1998).

Similarly, in mobile anadromous species such as salmon, migratory behavior and a capacity to colonize new areas should tend to constrain population differentiation, in some cases perhaps overriding natal homing propensities that otherwise promote the development of population structure in the short term (Thomas et al., 1986; Shedlock et al., 1992). An empirical example involves the chum salmon, *Oncorhynchus keta*. In 42 populations sampled throughout the species' range along the northern Pacific rim from Japan to the United States, essentially no differentiation was detected in sequence assays of the complete 1-kb mtDNA control region (Park et al., 1993).

Another diadromous life cycle, amphidromy, entails a brief marine larval stage in organisms that otherwise live and breed in freshwater. All five of the indigenous "freshwater" fishes of the Hawaiian Islands (four gobies and one eleotrid) display this modified version of anadromy, and these have been the subject of recent phylogeographic appraisals (Zink et al., 1996; Chubb et al., 1998). Populations within each species proved to be nearly undifferentiated in mtDNA composition across the five main islands, presumably because the marine larvae mediate high inter-island gene flow. Such tight historical connections probably have had another evolutionary consequence for these fishes: prevention of the extensive and rapid speciations that have characterized many groups of terrestrial organisms in the Hawaiian archipelago (Wagner and Funk, 1995).

INVERTEBRATES

A great diversity of ecologies and lifestyles is represented among the invertebrates and this makes impossible any groupwide predictions on phylogeographic patterns. Mitochondrial studies have been conducted on numerous species ranging from insects (Roderick, 1996) to snails (Douris et al., 1998), and a full panoply of phylogeographic outcomes has been observed. Here, numerous examples will be reviewed to illustrate two messages: The informativeness of mtDNA genealogical approaches as documented above for conspecific vertebrates also holds for other multi-cellular animals; and historical factors as well as contemporary behaviors and natural history can be important sculptors of phylogeographic patterns in extant species.

Terrestrial and Aerial Species

Molecular genetic studies of the honeybee (*Apis mellifera*) illustrate both of these points. This species is native to Europe, Africa, and the Middle East, but in recent times has been transported worldwide by beekeepers. More than 20 taxonomic subspecies are recognized, yet a recent review of bio-geographic and morphometric evidence suggested that these should be grouped into three evolutionarily coherent assemblages native to: sub-Saharan or tropical Africa; north Africa and the western Mediterranean; and the Middle East and southeastern Europe (Ruttner, 1988). These distributions are suspected to reflect long-term isolation following separate historical colonization routes taken by the species from its hypothesized evolutionary homeland in northeast Africa or the Middle East.

Molecular surveys of honeybees throughout their native range revealed three major mtDNA lineages that correspond closely to these morphometric units (Smith and Brown, 1990; Cornuet and Garnery, 1991; Hall and Smith, 1991; Garnery et al., 1992). Estimates of mean within-group sequence divergence are about 0.3 percent whereas sequences in different groups differ by 2.0 percent or more (Smith, 1991). Concordant support for these phylogroups is provided by nuclear-gene assays (McMichael and Hall, 1996, and references therein). Cornuet and Garnery (1991) suggest that these assemblages separated between 0.3 and 1.3 Mya.

Thus, on a broad scale the honeybee phylogeographic pattern clearly falls into Category I. Against this deeper evolutionary framework, recent anthropogenic dispersal as well as the behaviors of the bees themselves have impacted current distributions of the major historical lineages. For

example, honeybees from Europe and Africa have been introduced repeatedly into the New World. Several studies have employed diagnostic mitochondrial and nuclear markers to identify the geographic sources (where
otherwise unrecorded) and outcomes of the introductions (review in
Smith, 1991).

One intriguing question is whether the New World spread of African-
like honeybees following their introduction into Brazil in the late 1950s
resulted solely from gene flow mediated by drones, or also from colony-
swarming movements led by queens. Under the former hypothesis, males
travel considerable distances and mate with the more docile honeybees of
European ancestry, thereby "Africanizing" recipient colonies without the
involvement of African matrilines. Under the alternative possibility of
colony swarming, both nuclear and mitochondrial genes would have been
dispersed geographically during the Africanization process. Molecular
studies have settled the issue to some degree by demonstrating the involvement of colony swarming: Surveyed colonies of Africanized bees in
the Neotropics typically carry African-type mtDNA (Hall and Muralidharan, 1989; Smith et al., 1989).

Nuclear and cytoplasmic molecular markers have been examined in
populations of another flying hymenopteran, the bumblebee (*Bombus terrestris*) across Europe and from isolated Tenerife Island off the coast of
western Africa (Estoup et al., 1996). At eight microsatellite loci, no significant differences were detected among continental samples, but the Tenerife population showed a large genetic distance due mostly to the presence
at each of several loci of one fixed allele drawn from among the much more
variable allelic pools on the continent. In mtDNA sequence, however,
Tenerife bees were nearly identical to those on the mainland.

At face value, the small genetic distance in mtDNA sequence might
appear to contradict the large composite genetic distance at nuclear loci,
but this would be an improper interpretation—both data sets are consistent with a population bottleneck during a recent colonization of Tenerife
Island. Thus, allele frequencies at multiple nuclear loci probably were
shifted rapidly by genetic drift, whereas a clear historical signal of close

genealogical affinity between the Tenerife and European populations is retained in the mtDNA sequence. This scenario illustrates an important distinction that sometimes arises between the phylogenetic information content of composite genetic distances based on allele frequencies at multiple loci versus the explicit genealogical assessments that can come from a single nonrecombined locus such as mtDNA.

In general, both the magnitude and pattern of mtDNA haplotype differences can be powerful sources of information on the colonization histories of isolated populations. Introduced populations of *Drosophila buzzatii* on the Iberian Peninsula in Europe are monomorphic for the most common haplotype in the species' native homeland in South America, suggesting that the colonizations were recent and perhaps channeled through relatively few individuals (Rossi et al., 1996). A contrasting outcome involves *Drosophila subobscura*. Across its broad continental range in Europe and central Russia, mtDNA haplotypes are closely similar in sequence, but a deep phylogenetic split distinguishes an isolated population in the Canary Islands (Afonso et al., 1990). The abundance of different haplotypes within the distinctive Canary Island clade, and the fact that no mtDNA genotypes were related closely to those on the mainland, suggest that this island population is old.

Many flying insects are relatively mobile and, hence, as for the bumblebees and fruitflies just described, might be expected to show minimal phylogeographic structure across wide areas (except where firm physical barriers to dispersal exist). One example that conforms to these expectations involves a migratory grasshopper (*Melanoplus sanguinipes*) that has a legendary reputation for explosive population outbreaks that devastate crops in the prairies of central North America. A mtDNA survey detected little or no differentiation among widely scattered populations in Canada and the United States (Chapco et al., 1992). Similar findings apply to several other mobile grasshopper species in the subfamily Melanoplinae (Chapco et al., 1994; Martel and Chapco, 1995).

Other surveyed insects that reportedly display little or no mtDNA phylogeographic differentiation over wide areas include the following:

swallowtail butterflies (*Papilio polyxenes* and *P. zelicaon*) across their respective broad distributions in the Western Hemisphere (Sperling and Harrison, 1994), a chrysomelid beetle (*Leptinotarsa decemlineata*) across North America (Zehnder et al., 1992), an Old World fruitfly (*Drosophila subobscura*) across its introduced range in the New World (Rozas et al., 1990), the screwworm fly (*Cochliomyia hominivorax*) in Mexico and Texas (Roehrdanz and Johnson, 1988), and the migratory monarch butterfly (*Danaus plexippus*) in North America (Brower and Boyce, 1991).

However, populations of other flying insects have displayed significant phylogeographic structure at a variety of spatial scales and temporal depths (e.g., Brown et al., 1997). In *Drosophila melanogaster*, which is commensal with humans, mtDNA patterns suggest that both long- and short-distance migration rates are too low to have resulted in a geographic homogeneity of matriline frequencies (Hale and Singh, 1987). However, the mobile nature of this species is evidenced by the fact that rare as well as common haplotypes often co-occur in widely separated localities around the world. In the fire ant, *Solenopsis invicta*, limited vagility of queens rather than long-standing environmental barriers to gene flow may account for the regional differentiation observed in frequencies of mtDNA haplotypes (and also allozyme and microsatellite markers) in South American populations (Ross et al., 1997).

Toward the other end of the scale are species that show deep phylogeographic structure probably due to long-term dispersal barriers. For example, in the tropical butterfly *Heliconius erato,* a basal split in the mtDNA gene tree (dating to about 1.5 to 2.0 Mya) distinguishes populations east versus west of the Andes Mountains in northeastern South America (Brower, 1994). More than a dozen allopatric races exist in this Müllerian mimicry complex, each characterized by a spectacular wing-coloration pattern that advertises unpalatability to predators. Yet, sequence differences *within* either of the two regional phylogroups are low, with mtDNA haplotypes often shared across races (Fig. 4.15). The molecular results evidence an ancient vicariant separation in the species apparently followed

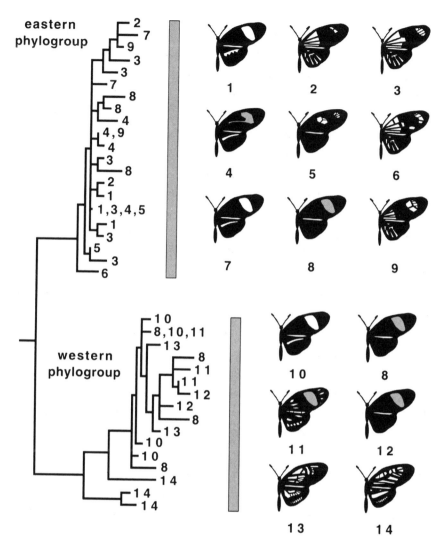

FIGURE 4.15   Matrilineal phylogeny (as estimated from mtDNA sequences) for 14 allopatric wing-color races of the butterfly *Heliconius erato* in South and Central America (after Brower, 1994). The wings display brilliant color motifs as well as the pattern differences shown here.

by rapid and often convergent evolution in wing coloration patterns due to strong selection via Müllerian mimicry (Brower, 1994).

In more sedentary insect species, pronounced phylogeographic structure often has been observed. The meadow grasshopper (*Chorthippus parallelus*) has been the subject of multiple genetic assays of geographic variation and hybrid-zone contacts across its range in Europe to Siberia (Butlin and Hewitt, 1985; Ritchie et al., 1989; Cooper and Hewitt, 1993; Hewitt, 1993; Vasquez et al., 1994). In particular, haplotype sequences from an anonymous, noncoding nuclear locus indicate the presence of five major phylogeographic units with mostly parapatric ranges: in Iberia, Italy, Greece, Turkey, and a broad area across northern Europe and western Russia (Cooper et al., 1995). These haplotype distributions were used to trace the likely post-Pleistocene expansion routes from suitable Ice Age refugia identified by independent physiographic evidence. For example, the colonizers of western Russia, northern Europe, Great Britain, and France trace to a Balkan Pleistocene refugium, and those now throughout Spain trace to a refugium in southern Iberia (Hewitt, 1996). The two forms currently meet in a narrow zone of secondary hybrid contact in the Pyrenees, as do other phylogeographic forms elsewhere on the continent.

In special cases, deep phylogeographic population structure in terrestrial invertebrates exists even across small spatial scales. Juan et al. (1996, 1997) describe an interesting example involving the darkling beetle (*Pimelia radula*) on a tiny island in the Canary Archipelago. Two ancient mtDNA phylogroups coexist on the Island of Tenerife: one confined to the northern coast and the other to the central and southern portions of the island. The authors speculate that an ancestral beetle colonized the proto-Tenerife area in the Pliocene when two or three separate volcanic islands existed, and before the archipelago was amalgamated by further volcanic activity (2.0 to 0.6 Mya). Thus, the current phylogeographic structure in mtDNA, and in an assayed nuclear ribosomal gene, appear to be genetic footprints of long-term population divergence in allopatry. Support for this scenario has come from concordant phylogeographic findings on another terrestrial inhabitant of Tenerife, the lacertid lizard *Gallotia galloti* (Thorpe et al., 1996). In that species also, populations on the northern versus southern portions of the island display a deep split (dating to 0.7 Mya) in the mtDNA gene tree.

Rarely, highly divergent mtDNA lineages within an invertebrate species are distributed as broadly sympatric genetic polymorphisms (phylogeographic Category II). Dramatic examples involve the pulmonate land snails *Cepaea nemoralis* and *Helix aspersa* (Thomaz et al., 1996), in which local conspecific populations often display mtDNA haplotypes with sequence differences exceeding 10 percent (comparable to distances described elsewhere for taxa recognized as genera or families). The authors favor an explanation that invokes long persistence times for lineages within these low-mobility species. Evolutionary effective population sizes may be exceptionally large due to the subdivision of each species into millions of partially isolated local demes connected to one another by stepping-stone gene flow.

Especially in poorly known invertebrate groups, questions routinely arise concerning taxonomic interpretations of unexpected phylogeographic outcomes. Quite often, mtDNA phylogroups are discovered that either were unanticipated or appear to be discordant with existing taxonomic assignments. Introgressive hybridization may account for some cases. For example, mtDNA phylogenetic relationships in bark weevils of the *Pissodes* species complex challenge hypotheses of species' relationships based on morphology, allozymes, and karyotype (Boyce et al., 1994). The authors suspect that introgressive hybridization contributed to the apparent genetic contradictions by moving mtDNA lineages across species' boundaries. Scenarios of historical introgression are more secure when discordances between genetic markers are confined to boundary areas between well-differentiated allopatric forms, as for example between two parapatric chromosomal forms of the grasshopper *Caledia captiva* in Australia (Marchant et al., 1988), or between two species of *Gryllus* crickets in contact zones in eastern North America (Harrison et al., 1987).

Other cases of disagreement between mtDNA lineages and taxonomy may be due to prior failures to recognize particular sibling species. In the wood roach (*Cryptocercus punctulatus*), huge genetic distances distinguish Appalachian populations from those in the Pacific Northwest. Morphologically cryptic species probably were present in the collections (Kambhampati et al., 1996). In Appalachian cave spiders of the *Nesticus tennesseensis* complex, distinct mtDNA genotypes characterize populations that had been considered conspecific, implying here that "species as morphological lineages are currently more inclusive than basal evolutionary or phylogenetic units" (Hedin, 1997). In a Neotropical pseudoscorpion currently described as a single species (*Cordylochernes scorpioides*) from Central America to Argentina, large mtDNA sequence differences (2.6 to 13 percent) exist between some geographic populations (Wilcox et al., 1997). In this case, the mtDNA patterns are matched by postzygotic reproductive incompatibilities, thus evidencing the presence of multiple biological species.

Conversely, some recognized taxonomic species display minimal mtDNA differentiation. A case in point involves the spruce budworms, *Choristoneura orae, C. occidentalis,* and *C. biennis,* which by virtue of an absence of mtDNA distinctions were judged to represent recently evolved geographic or ecological races (Sperling and Hickey, 1994). Of course, however, no arbitrary level of genetic divergence can serve by itself to define the boundaries between biological species.

Phylogeographic findings can have important consequences not only in taxonomy, but also in epidemiological and medical issues. In North America, ticks in the *Ixodes ricinus* complex are the transmitting agents of Lyme disease (*Borrelia burgdorferi*) and human babesiosis (*Babesia microti*). A survey of *cytb* gene sequences across the eastern United States revealed distinctive southern versus northern mtDNA clades that correspond to two forms of *Ixodes* previously assigned to separate species: *scapularis* and *dammini* (Rich et al., 1995). The most intense zoonotic transmission of Lyme disease is associated with *I. dammini,* whose behavioral and morphological distinctions from *I. scapularis* now can be interpreted as associated with longstanding historical separations. A range expansion of *I. dammini* in this century probably promoted recent disease outbreaks in humans.

Freshwater Species

Like freshwater fishes, invertebrate species confined to disjunct bodies of water might display pronounced phylogeographic structure, the temporal depths and patterns of which should reflect, in large measure, historical associations among drainages (sometimes confounded by recent anthropogenic dispersal). One example of a freshwater species with pronounced phylogeographic structure is the white-clawed crayfish (*Austropotamobius pallipes*). Specimens from different European rivers reportedly belong to highly divergent mtDNA lineages, and British stocks trace genetically to a recent origin in France (Grandjean et al., 1997).

To date, molecular evidence for pronounced phylogeographic population structure in aquatic invertebrates has come primarily from multi-locus allozyme assessments (e.g., Bilton, 1994; Jabbour-Zahab et al., 1997). Relatively few species have been surveyed for mtDNA, and those that have may be somewhat unusual with regard to life history pattern and phylogeographic outcome.

Particular attention has focused on *Daphnia* microcrustaceans. Results generally indicate that enormous expanses of the Holarctic realm were colonized recently by recognized forms (e.g., Taylor et al., 1996; Weider et al., 1996; Weider and Hobaek, 1997). The eggs of *Daphnia* are encased in a resistant structure known as an ephippium that is released into the water column and can be transported widely by vectors such as waterfowl, wind, and water. This highly dispersive phase of the life cycle probably accounts for near ubiquity of particular mtDNA lineages across vast areas such as northern Eurasia.

Marine Species

Most marine invertebrates spend at least part of their life cycle in open water as free-moving gametes, larvae, or adults. Thus, opportunities for moderate to high gene flow would seem to be the norm except where firm ecological or biogeographic impediments to dispersal exist. Surprisingly, however, a wide variety of phylogeographic outcomes has been observed.

Some marine invertebrates display little mtDNA differentiation over vast areas. Examples include lobsters in the genera *Jasus* (Ovenden et al., 1992) and *Panulirus* (Silberman et al., 1994), a hydrothermal vent mussel (*Bathymodiolus thermophilus*) in the abyssal depths of the Pacific Ocean (Craddock et al., 1995), and several shallow-water species of sea urchins in the genera *Heliocidaris* and *Strongylocentrotus* (Palumbi and Wilson, 1990; Palumbi and Kessing, 1991; McMillan et al., 1992) (but see Palumbi and Metz [1991] and Palumbi et al. [1997] for different phylogeographic outcomes in *Echinometra* urchins). In another sea urchin (*Echinothrix diadema*) studied for both mtDNA and allozymes, genetic markers suggest recent and massive trans-Pacific gene flow, perhaps through periodic larval transport during El Niño events across a broad expanse of otherwise uninhabitable open ocean (Lessios et al., 1998).

Such cases presumably reflect recent gene flow through natural dispersal. An ecologically troubling development mostly in this century is the anthropogenic transport of small marine organisms, often worldwide, via ballast waters in ships (Carlton and Geller, 1993; Lodge, 1993). Some of the invasions involve sibling species and, thus, may go undetected by morphological appraisals alone. Molecular phylogeographic studies can help document invasion events and pinpoint sources of the introductions. For example, Geller et al. (1997) used mtDNA assays to document cryptic recent invasions of *Carcinus* crabs into waters of California, Japan, Tasmania, and South Africa from ancestral populations of *C. maenas* and *C. aestuarii* in the Atlantic.

In any event, as is generally true for fishes, marine invertebrates often display less mtDNA phylogeographic population structure across wide areas than do related freshwater taxa on far smaller spatial scales. Among open-water copepods, for example, Bucklin et al. (1998) report higher genetic patchiness across pond populations of *Diaptomis leptopus* in Quebec than in a pelagic species, *Calanus finmarchicus*, across the North Atlantic Ocean (see also Bucklin and Kocher, 1996; Kann and Wishner, 1996).

Nonetheless, significant heterogeneity in mtDNA haplotype frequencies, usually on large oceanic scales, *has* been uncovered within several species of marine zooplankton (Bucklin et al., 1996a,b, 1997).

Significant genetic patchiness is also a common observation for marine invertebrates on local spatial scales (Burton, 1983; Hedgecock, 1986; Palumbi, 1994, 1995, 1996a). For example, despite the high dispersal potential of purple sea urchins (*Strongylocentrotus purpuratus*), populations display significant subdivision in mtDNA and allozymes along the coastlines of California and Baja California (Edmands et al., 1996). A mosaic distribution of genotypes in this and many other marine organisms might result from physical processes that restrain gametic or larval movements, behavioral mechanisms that limit effective dispersal, or spatially variable selective pressures that promote differentiation even in the face of high gene flow. Local heterogeneity in mtDNA typically is shallow (phylogeographic Category III) in the sense of involving closely related haplotypes.

On the other hand, deep mtDNA phylogeographic structure (Category I) has been uncovered in population genetic surveys of many marine invertebrates. These include: a tidepool copepod (*Tigriopus californicus*) along the California coast (Burton and Lee, 1994; Burton, 1998); an antitropical goose barnacle (*Pollicipes elegans*) from American coastlines in the North versus South Pacific (Van Syoc, 1994); a larval-brooding sea cucumber (*Cucumaria pseudocurata*) in northern versus southern portions of the northeastern Pacific (Arndt and Smith, 1998); the horseshoe crab, *Limulus polyphemus* (Saunders et al., 1986) and the American oyster, *Crassostrea virginica* (Reeb and Avise, 1990) along coastlines of the Atlantic versus Gulf of Mexico in the southeastern United States; a mussel (*Mytilus galloprovincialis*) between Atlantic waters and those of the Mediterranean (Quesada et al., 1995, 1998); asexual intertidal clams in the genus *Lasaea* along the Pacific coast of North America (O'Foighil and Smith, 1996); a planktonic gastrotrich (*Xenotrichula intermedia*) in the Mediterranean as distinct from mtDNA clades (that may register sibling species) in the Atlantic and Gulf of Mexico (Todaro et al., 1996); a marine-dispersed land crab (*Birgus latro*) from islands in the Pacific versus Indian Oceans (Lavery et al., 1996); and a starfish (*Linckia laevigata*) from these same ocean realms (Williams and Benzie, 1998).

An interesting finding is that in several of these Category I species, distributions of the preeminent intraspecific mtDNA phylogroups match well with traditional marine zoogeographic provinces as identified by range lists for multiple species, or by pronounced differences in water masses and associated ecological conditions. The unique assemblages of species characterizing zoogeographic provinces in the marine realm conventionally are thought to reflect contemporary ecological barriers to dispersal, perhaps operating in conjunction with historical factors that promoted divergence (and occasional extinctions) of regional populations. Phylogeographic breaks within species that are continuously distributed across provincial boundaries provide independent support for the importance of both historical and contemporary impacts on the genetic structures of regional biotas. Marine zoogeographic provinces appear to be important descriptors of faunal discontinuities not only with respect to species' ranges, but also to deep phylogeographic breaks within many taxonomic species whose ranges extend across such zones (discussed in Chapter 5).

Another growing realization is that pronounced genetic differences (as registered in various molecular assays) often characterize adjacent or sympatric populations of many marine invertebrates traditionally assumed by morphological evidence to be conspecific. Citing examples from the Porifera, Cnidaria, Nemertea, Mollusca, Arthropoda, Polychaeta, Echinodermata, Ascidiacea, and Bryozoa, Knowlton (1993) argues persuasively that reproductively isolated sibling species are rife in marine environments and that failure to recognize them has had a crippling effect on studies of ecological and evolutionary processes in the sea.

SUMMARY

1. Phylogeographic outcomes can be grouped for heuristic purposes into several phylogeographic categories that reflect different temporal scales and spatial aspects of population genealogical struc-

ture. Category I entails deeply separated gene-tree lineages confined to separate areas of a species' range. At the other end of the
spectrum, Category IV entails broad sympatry of lineages with
recent evolutionary connections. Other outcomes involve alternative combinations of these temporal and spatial aspects: e.g.,
lineages deeply separated but sympatric (Category II), lineages
allopatric but with shallow evolutionary separations (Category III),
and lineage distributions varied but also with shallow evolutionary
separations (Category V).

2. Molecular phylogeographic surveys, based mostly on mtDNA
   assays, have been conducted on hundreds of species in nature.
   Each phylogeographic pattern described above has been reported
   on numerous occasions, and often makes considerable sense with
   respect to a species' natural history, dispersal mode, and suspected
   biogeographic past. Such findings give confidence that mtDNA
   assays are hugely informative on genealogical matters over microevolutionary time.

3. Mammals display a wide variety of phylogeographic outcomes.
   Nearly all small and relatively immobile terrestrial species show
   deep phylogeographic differentiation at the intraspecific level,
   whereas vagile large-bodied mammals typically display less pronounced phylogeographic structure. Many species of volant and
   marine mammals, though highly mobile, nonetheless exhibit at
   least modest genealogical population structure due to self-imposed
   dispersal restraints such as behavioral site fidelity or allegiance to
   social groups.

4. Phylogeographic analyses of mtDNA have revolutionized thought
   about avian population structure. Conspecific populations can be
   structured at a variety of evolutionary timescales, ranging from
   deep lineage separations that often are concordant with differences
   in other characters such as behavior, morphology, or biogeographic
   area, to shallow separations attributable to recent population fragmentation and limited gene flow in contemporary time. In some of
   these latter cases, behavioral or morphological alterations appear

to have arisen against a framework of relatively shallow genealogical differentiation.

5. Many reptiles are small and relatively sedentary, and are habitat specialists. Nearly all amphibians are associated tightly with particular bodies of water during at least part of the life cycle. Not surprisingly, mtDNA analyses of herpetofauna have uncovered evidence for strong phylogeographic differentiation among conspecific populations, with general patterns often reminiscent of those typifying small terrestrial mammals. The origins and ages of lineage separations have been determined for numerous sexual species as well as asexual biotypes, and other character state distributions sometimes have been mapped onto these intraspecific phylogenies. Marine turtles, by contrast, are highly mobile, but even here both shallow and deep phylogeographic structures are evident. These appear to be attributable, respectively, to behavioral fidelity to natal site, and to historical vicariant and dispersal events on a global scale.

6. A notable difference exists between freshwater and marine fishes with respect to levels and patterns of phylogeographic divergence. Most freshwater fishes display pronounced phylogeographic structure, in patterns often relatable to historical isolations and mergers of drainage basins. By contrast, many (but certainly not all) marine fishes show minimal phylogeographic population structure across sometimes vast areas. Diadromous species often show genetic patterns more closely resembling those of marine species. All of these trends generally relate to the different physical structures (both contemporary and historical) of the marine and freshwater environments inhabited by fishes.

7. Analyses of intraspecific mtDNA patterns in invertebrates have yielded similar phylogeographic insights: Historical biogeographic factors as well as contemporary behaviors and ecologies have sculptured the genetic architectures of species in generally interpretable ways. The great diversity of lifestyles and vagilities in

invertebrate taxa has provided many opportunities for assessing phylogeographic outcomes in a comparative context.

8. For both vertebrate and invertebrate animals, phylogeographic hypotheses originally formulated more than a decade ago have been supported abundantly by subsequent molecular genetic research.

GENEALOGICAL CONCORDANCE: TOWARD SPECIATION AND BEYOND

Much of the effort of empirical phylogeography summarized in Part II involved molecular examinations of the genealogical histories of single genes (usually mtDNA) in particular species, one at a time. The third section of this book extends such approaches to multiple genes, multiple codistributed species, and comparisons with more traditional kinds of biogeographic data. Such extensions into comparative phylogeography introduce several distinct aspects of genealogical concordance germane to evolutionary inference at the scale of landscape history. Interpretations of phylogeographic population structure at the intraspecific level often grade into broader issues of species concepts and taxonomic assignments, so this concluding section also addresses these issues in greater detail. In particular, phylogeographic studies can be highly informative about population demography, geography, and temporal durations of the speciation process.

The species phylogeny is more like a statistical distribution, being composed of
various trees (the gene trees), each of
which may indicate different relationships.

—Wayne Maddison, 1995

5

GENEALOGICAL CONCORDANCE

GIVEN THE INHERENT stochastic elements of genetic transmission
under Mendelian inheritance—random segregation of alleles and independent assortment of unlinked loci through extended organismal pedigrees—each piece of DNA within a species might be expected to trace an
idiosyncratic genealogical history. Furthermore, given the great diversity
of ecological and evolutionary factors likely to have impinged upon the
historical demographies and genetic architectures of extant faunas, an
idiosyncratic phylogeographic outcome also might be expected for each
species. If the noise of genealogical idiosyncrasy is taken to be a general
null expectation, then any pronounced departures from the null should assume special significance with regard to phylogeographic signal. Such departures are described generically as "genealogical concordance" (Avise
and Ball, 1990).

Four distinct aspects of genealogical concordance of relevance to
phylogeography can be distinguished (Table 5.1; Fig. 5.1). These will be
described briefly as a prelude to empirical case studies from regional faunal assemblages.

TABLE 5.1   Four aspects of genealogical concordance in phylogeographic inference
(after Avise, 1996b). See text for further explanation.

I   *Concordance across sequence characters within a gene.*
     Relevance: yields statistical significance for putative gene-tree clades.

II   *Concordance in significant genealogical partitions across multiple genes within a species.*
     Relevance: establishes that gene-tree partitions register phylogenetic partitions at
     the population or species level.

III *Concordance in the geography of gene-tree partitions across multiple codistributed species.*
     Relevance: implicates shared historical biogeographic factors in shaping
     intraspecific phylogenies.

IV   *Concordance of gene-tree partitions with spatial boundaries between traditionally
     recognized biogeographic provinces.*
     Relevance: implicates shared historical biogeographic factors in shaping
     intraspecific phylogenies and organismal distributions.

ASPECT I: AGREEMENT ACROSS
CHARACTERS WITHIN A GENE

Nearly by definition, a "deep" phylogenetic split deduced in the intraspecific mtDNA tree of any species will have been registered concordantly by
multiple sequence characters. If this were not the case, such matrilineal
separations would not be evident in the data analyses, nor would they
have received significant phylogenetic support by criteria such as boot-
strapping (Felsenstein, 1985a). Typically, at least three or four diagnostic
characters (uncompromised by homoplasy in the broader data) are required for robust statistical recognition of a putative gene-tree clade in
most phylogenetic appraisals.

   All species described in Chapter 4 as belonging to phylogeographic
Categories I or II display a mtDNA gene tree whose major branches were
earmarked by multiple diagnostic characters (nucleotides or restriction sites)
within the mitochondrial genome. These genetic markers are independent
of one another in the sense of having had separate mutational origins,
although once jointly present within a matriline they are cotransmitted as

Genealogical Concordance

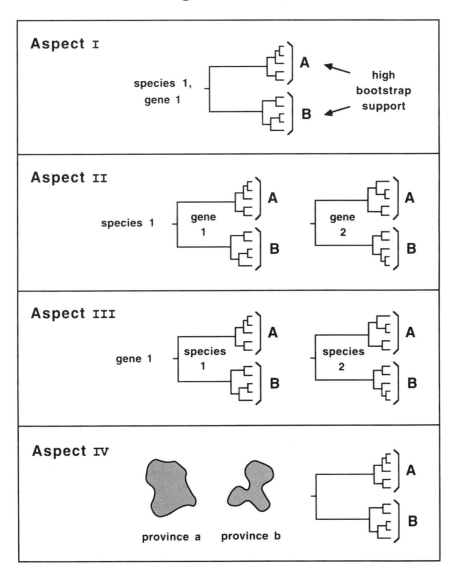

FIGURE 5.1 Schematic presentation of four distinct aspects of genealogical concordance (see text and Table 5.1 for further explanation). "A" and "B" are distinctive phylogroups in a gene tree.

a linked supergene. Similar statements would apply to tightly linked sequence characters within any nuclear gene.

Thus, this first aspect of genealogical concordance merely confirms the identity and distinctiveness of the principal branches or phylogroups within a gene tree. Furthermore, empirical experience shows that the major mtDNA phylogroups within a species nearly always show a strong geographic arrangement. Thus, such gene-tree branches are prime but provisional candidates for constituting major historical population units especially worthy of further consideration in a biogeographic context.

ASPECT II: AGREEMENT ACROSS GENES

Suppose that haplotypes were isolated, assayed, and gene trees estimated at each of several unlinked nuclear loci where intra-genic (inter-allelic) recombination had been absent within the species of interest. Suppose further that deep branch separations were apparent in each gene tree (concordance Aspect I), and that these major genealogical branches characterized the same sets of geographic populations. Aspect II of genealogical concordance ideally refers to such phylogenetic agreement across independent loci. The occurrence of concordance Aspect II demonstrates almost conclusively that particular partitions in the gene trees accurately register fundamental phylogenetic subdivisions at the population or species level. Several statistical approaches have been suggested to assess phylogenetic agreement among partitioned data sets from, for example, empirical gene trees from two or more loci (Day, 1983; Page, 1990, 1994; Bull et al., 1993; Farris et al., 1994; Lyons-Weiler and Milinkovitch, 1997).

An important theoretical qualification arises when Aspect II of genealogical concordance is addressed with respect to nuclear versus mitochondrial gene trees. Recall that under neutrality, the expected time for lineage sorting to reciprocal monophyly in two isolated populations is fourfold longer for nuclear lineages than for mitochondrial lineages, all else being equal (see Birky, 1991 and Hoelzer, 1997 for exceptions to this generality). In other words, due to a fourfold larger effective population size, the coalescent process in theory normally proceeds more slowly for alleles at nuclear genes than for those at mitochondrial loci.

This expectation has given rise to the so-called three-times ($3x$) rule, which predicts the level of phylogenetic concordance across nuclear gene loci as a function of relative branch lengths in a mitochondrial gene tree (Palumbi and Cipriano, 1998). The logic is that if a matrilineal tree for two isolated populations has just barely achieved a status of reciprocal monophyly at time x, then on average about $3x$ more time is required for a typical nuclear gene tree to achieve the same status through lineage sorting. The x in these comparisons can be estimated from the mean intrapopulation mtDNA diversity (e.g., as measured by pairwise sequence divergence, p), because p scales with N_e and provides a lineage-specific estimate of the time required for lineage coalescence. If the actual time of population separation is between x and $4x$, a mitochondrial gene tree likely will have achieved reciprocal monophyly whereas a nuclear gene will not. Thus, in theory, only when times of population separation fall outside the $3x$ window is it probable that major clades in many nuclear gene-tree lineages will display concordance with those in a mitochondrial gene tree (Fig. 5.2).

As discussed in Chapter 2, technical and biological complications have conspired to inhibit explicit genealogical appraisals of nuclear genes within and among populations of most species. These include the procedural difficulties of isolating nuclear haplotypes one at a time from diploid tissues, the challenge of identifying nuclear loci whose sequences have evolved rapidly enough to provide informative phylogenetic markers over recent evolution, and a requirement that the sequences have been mostly free of interallelic recombination (and/or gene conversion) over the historical timescale of interest. At the time of this writing, these hurdles have been overcome in relatively few instances. Recall, for example, the genealogical studies of human Y-linked loci and an autosomal globin gene (Chapter 3). These loci generally supported mtDNA data in suggesting a small evolutionary N_e and a relatively recent global population expansion for *Homo sapiens*.

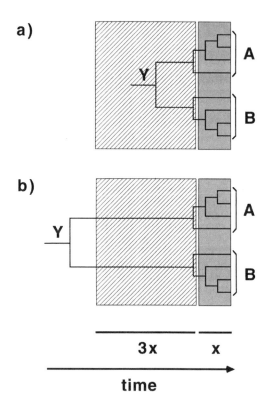

FIGURE 5.2 Diagrammatic representation of the 3x rule (after Hare, 1998). Shown are mtDNA gene trees whose major branches, stemming from node Y, characterize populations A and B. The mean lineage divergence time within populations (as might be estimated by intra-population mtDNA sequence diversity) is denoted by *x*. *(a)* A case in which node Y is within the 3x window. *(b)* The same pair of populations at a later point in time in which node Y now falls outside (is earlier than) the 3x time window. Only in the latter situation is it likely in theory that monophyletic clades at most nuclear loci will show concordance with those in a mtDNA gene tree.

Two studies that explicitly searched for (and found) phylogeographic concordance between mitochondrial and nuclear gene trees at the intra-specific level are summarized in Fig. 5.3. In the North American killifish, *Fundulus heteroclitus,* a pronounced phylogenetic distinction between northern and southern populations along the Atlantic Coast is registered

Fundulus heteroclitus

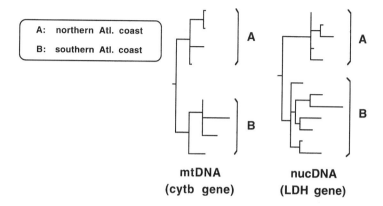

A: northern Atl. coast

B: southern Atl. coast

mtDNA
(cytb gene)

nucDNA
(LDH gene)

Tigriopus californicus

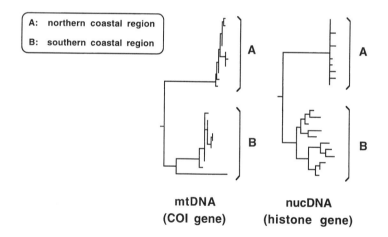

A: northern coastal region

B: southern coastal region

mtDNA
(COI gene)

nucDNA
(histone gene)

FIGURE 5.3 Empirical examples of Aspect II of intraspecific genealogical concordance. These cases involve general agreement between deep phylogeographic topologies in mitochondrial and nuclear gene trees. *Above:* Northern versus southern populations of a coastal killifish, *Fundulus heteroclitus* (after Bernardi et al., 1993). *Below:* Northern versus southern populations of a coastal copepod in California, *Tigriopus californicus* (after Burton and Lee, 1994; for a description of additional phylogeographic structure in this species, see Burton, 1998).

in a sequenced nuclear gene (encoding lactate dehydrogenase) as well as in mtDNA. Similarly, in a copepod, *Tigriopus californicus*, along the Pacific Coast, a phylogenetic split that distinguished northern from southern populations was detected both in mitochondrial and nuclear gene trees. Both of these cases exemplify what normally is implied by Aspect II of genealogical concordance: a general agreement in the principal phylogeographic units registered in gene trees from multiple unlinked loci.

Few such examples currently are available. In the absence of explicit genealogical evidence from nuclear genes to compare directly against phylogeographic data from mtDNA, surrogate information can be employed. This may come from phylogenetic appraisals of traditional population genetic data: e.g., allelic frequencies and genetic distances, preferably from multiple nuclear loci. Alternatively, the evidence may rest on geographic appraisals of organismal phenotypes—morphological or behavioral characters that might be presumed to register notable genetic differences among the populations monitored. As discussed in Chapter 4, the deepest phylogeographic partitions in intraspecific mtDNA gene trees often agree with traditional taxonomic partitions as reflected in subspecies' designations.

ASPECT III: AGREEMENT ACROSS CODISTRIBUTED SPECIES

Imagine that each of several codistributed species with comparable natural histories or habitat requirements proved to be phylogeographically structured in similar fashion. In particular, divergent branches in the intraspecific gene trees might map consistently to the same geographic regions. Aspect III of genealogical concordance presumably would reflect shared historical elements in the evolutionary or ecological factors that had shaped the intraspecific phylogeographic architectures within this regional fauna (Rosen, 1975; Platnick and Nelson, 1978; Cracraft, 1988). Concordance Aspect III has been documented for several regional biotas, as discussed later.

ASPECT IV: AGREEMENT WITH OTHER
BIOGEOGRAPHIC DATA

The most important discoveries of molecular phylogeography often emerge when molecular genealogical data are integrated with independent biogeographic and systematic information. For example, historical vicariant events might have facilitated morphological or behavioral divergence between taxonomic subspecies with disjunct distributions. Or, the traditional data may pertain to collective information about historical biogeographic factors suspected to have shaped distributional boundaries in multiple species of a regional biota.

Major phylogeographic branches in a gene tree are most likely to register long-standing separations at the population level when they are concordant with morphological differences, or are consistent in position with obvious historical geographic or geologic barriers to dispersal. For more subtle cases, formal phylogenetic procedures have been suggested for evaluating correspondence between a molecular phylogeny's branching pattern and spatial relationships of physiographic regions (Wiley, 1988; Brooks, 1990; Brooks and McLennen, 1991).

With respect to multispecies patterns, biogeographers long have sought to identify unique or distinctive biotic assemblages at various spatial scales. Zoogeographic provinces, subprovinces, and areas of endemism traditionally are described from analyses of faunal lists, from which it is apparent that the distribution of the Earth's biodiversity is nonuniform and shaped in large measure by historical biogeographic factors. To pick an obvious example, the distinctive mammalian fauna of Australia (as well as its avifauna; Sibley and Ahlquist, 1986; Sibley, 1991) registers a long-standing isolation of the continent beginning in the early or middle Tertiary. An emerging generality from molecular phylogeographic studies is that deeply separated phylogroups at the intraspecific level often are confined to biogeographic provinces or subprovinces as identified from traditional faunal lists. This kind of agreement has proved to be another common element of Aspect IV of genealogical concordance.

REGIONAL PERSPECTIVES: THE SOUTHEASTERN UNITED STATES

Regional phylogeographic analyses consider the joint spatial patterns of intraspecific genealogies across multiple codistributed taxa. All four aspects of genealogical concordance can be illustrated by reference to the regional phylogeographic surveys currently available.

The first large collection of molecular phylogeographic appraisals involved faunas in the southeastern United States. Historical biogeographic factors played a major role in shaping the contemporary genetic architectures of numerous species in this region. A brief introduction to the physical environment of this area over the past few million years follows, indicating the types of physiographic factors that likely impinged on species' demographies and phylogeographic population structures.

Environmental Background

Freshwater Realm. About a dozen major rivers and numerous smaller streams currently traverse the southeastern United States, with eastern drainages entering the Atlantic Ocean and western drainages entering the Gulf of Mexico (Fig. 5.4). During the high sea-stands of the Pliocene and the moderate sea-stands of Pleistocene interglacials, smaller coastal streams likely were flooded, and freshwater faunas probably were isolated in the upper reaches of the larger rivers and perhaps in lakes and rivers of the Floridian peninsula itself (Wright, 1965). At these times, any inter-drainage transfers of strictly aquatic species must have occurred via lateral stream captures. During the low sea-stands associated with the glacial episodes that dominated much of the Pleistocene, adjacent drainages also may have coalesced periodically as they meandered across the broader coastal plains. Such histories of drainage isolation and connection (the details of which remain poorly known) undoubtedly influenced the phylogenetic histories of aquatic populations. For example, if freshwater fishes periodically were isolated in separate refugial areas, the geographic strongholds involved and the subsequent range expansions should have been impacted by historical patterns of drainage connections.

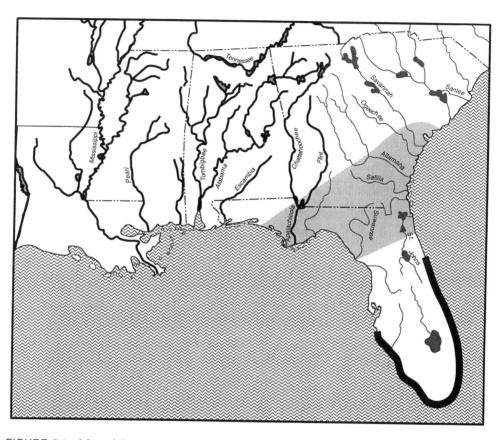

FIGURE 5.4   Map of the southeastern United States showing major environmental features relevant to phylogeographic interpretations (from Avise, 1996b). For freshwater fish, the darker and lighter lines indicate, respectively, rivers in the distinctive western and eastern faunal zones as evidenced in a compilation of species' ranges (Swift et al., 1985). For terrestrial organisms, the shaded area represents a region of presumed secondary contact and hybridization between many endemic Floridian forms and their continental counterparts as described by Remington (1968). In the maritime realm, the heavy line along the Florida coast indicates the distribution of a tropical faunal province as contrasted with the cooler zoogeographic zone further north (Briggs, 1974).

About 250 species of freshwater fishes inhabit coastal drainages of the southeastern United States, with the number of species per drainage ranging from 20 to 157. Swift et al. (1985) detailed these species' ranges and uncovered a great diversity of distributional patterns (as might be expected across a fauna this diverse). Nonetheless, the authors were able to define

about nine faunal provinces that implicated significant historical and contemporary barriers to dispersal. One analysis grouped river drainages according to species' composition. Based on a presence-absence matrix
across all fish species, the most basal split in a cluster phenogram distinguished all western drainages (western Georgia to Louisiana) from those
to the east (Atlantic coast and throughout the Floridian peninsula) (Fig.
5.4). Thus, the geographic distributions of an unusually high proportion of
fish species are confined to distinctive western or eastern freshwater
provinces.

Terrestrial Realm.   The Floridian peninsula (particularly at its southern extreme) is climatically, physiographically, and ecologically distinct from the
more temperate continental realms to the north and west. Thus, many taxa
that are adapted to tropical or subtropical conditions in the eastern United
States are confined to southern and central Florida. During the Pleistocene
glacial episodes, when sea levels were lower, the Floridian peninsula was
much larger than it is today and may have served as an important refugium
for temperate species whose ranges were compressed southward by climatic changes associated with glacial advances. Earlier in the Tertiary (as
recently as the Oligocene), an uplift area in west-central Florida existed as
one or more large islands separated from the mainland by a shallow marine incursion in the current area of northern Florida and southern Georgia
(Webb, 1990; Randazzo and Jones, 1997).

Remington (1968) first emphasized that a diverse array of endemic
plants and animals in the Floridian peninsula show morphological or
other distinctions from their respective continental near-relatives. Remington observed geographically concordant contact regions of known or
suspected hybridization between numerous Floridian forms (50 in his list)
and their continental counterparts along a "suture zone" between the
peninsula and mainland (Fig. 5.4). This secondary contact region was afforded a status equal to that of only five other major suture zones in North
America. This suture zone demarcates a distinctive biotic province in
peninsular Florida.

Maritime Realm.   The Floridian peninsula now protrudes southward into subtropical waters, separating some but not all temperate faunas into allopatric units on the Atlantic coast and Gulf of Mexico. During the many glacial advances and retreats of the Pleistocene, climatic changes and sea-level fluctuations probably had great impact on coastal species in the area (Bert, 1986; Felder and Staton, 1994). Climatic cooling associated with glacial expansions pushed temperate populations southward and perhaps increased the opportunity for contact of Atlantic and Gulf populations around south Florida (Cronin, 1988). However, sea levels during the Ice Ages also were lower (by as much as 150 meters), exposing great expanses of the Floridian peninsula. At such times, the peninsula was more arid than now, and perhaps bordered by fewer intermediate-salinity estuaries and salt-marsh habitats favored by many coastal species. Thus, an enlarged Floridian peninsula during glacial advances may have promoted a physical separation between Atlantic and Gulf coastal populations of some maritime species.

Opposing influences on species' distributions also may have been at work during periods of climatic warming, when sea levels were higher and the Floridian peninsula perhaps was bordered by more extensive estuaries and salt marshes. During these interglacial periods (as now), some strictly temperate species probably were separated into disjunct Atlantic and Gulf populations by tropical conditions of southern Florida. Conversely, some eurythermal and estuarine-adapted species may have expanded out of the Gulf region to regain increased contact with Atlantic populations around the southern tip of a smaller Floridian peninsula. Which of these effects was dominant for the maritime fauna as a whole remains unclear. At present, marine currents moving out of the Gulf of Mexico contribute to the Gulf Stream that hugs the southeastern coast of Florida and may facilitate transport of Gulf-derived pelagic gametes or larvae into the southern Atlantic.

Faunal specialists have summarized distributional records for various marine taxa in the area (review in Briggs, 1974). The most striking pattern is a disjunction between the warm-temperate and tropical maritime faunas of southern Florida (and sometimes the Gulf of Mexico) versus the cooler temperate biotas farther north, particularly along the Atlantic coast (Fig. 5.4). For example, a temperate molluskan species assemblage extending down the northern third of Florida is replaced gradually by more tropical-adapted species to the south. A faunal transition for octocorals similarly occurs in eastcentral Florida. Many fish species also have northern or southern range limits in this area (Briggs, 1958). Thus, the eastcentral coastline of Florida describes a transitional boundary between distinctive Atlantic and Gulf faunal provinces in the maritime realm.

Genetic Findings

Freshwater Fishes. Bermingham and Avise (1986) used mtDNA restriction-site assays to examine four fish species collected from major river drainages across the southeastern United States: the bowfin (*Amia calva*) and three sunfish species (*Lepomis punctatus, L. microlophus,* and *L. gulosus*). Extensive variation and differentiation were detected, and the intraspecific matrilines clearly were nonrandom in geographic position. Notably, a major mtDNA phylogenetic break was evident within each species that distinguished nearly all individuals in eastern drainages (mostly along the Atlantic coast and peninsular Florida) from those to the west (entering the Gulf coast). The estimated magnitudes of inter-regional genetic separation in the intraspecific mtDNA gene trees differed considerably across the four species (Fig. 5.5), but invariably far surpassed genetic distances among haplotypes within either region. Geographic distributions of the two major mtDNA phylogroups within each species are summarized in Fig. 5.6.

   Similar mtDNA phylogeographic analyses have been conducted on the bluegill sunfish, *Lepomis macrochirus* (Avise et al., 1984b), largemouth bass, *Micropterus salmoides* (Nedbal and Philipp, 1994), and a *Gambusia* mosquitofish complex (Scribner and Avise, 1993). A fundamental split in each mtDNA gene tree distinguished populations in the Floridian peninsula and along the Atlantic coast from those to the north and west (Fig. 5.5). Allozyme data also have been gathered for these three species (Avise and Smith, 1974; Philipp et al., 1983; Wooten and Lydeard, 1990; Scribner

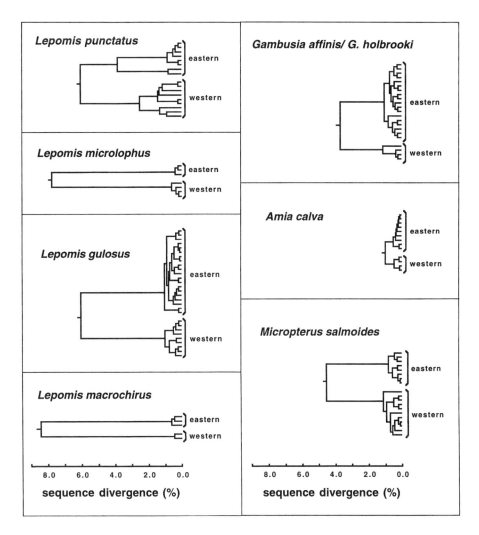

FIGURE 5.5   Phylogenetic relationships among mtDNA haplotypes within each of seven species (or species complexes) of freshwater fishes in the southeastern United States. For comparative purposes, all diagrams are plotted on the same scale of estimated mtDNA sequence divergence. "Eastern" and "western" refer to general drainage regions occupied by the two major phylogroups observed within each species (see Fig. 5.6).

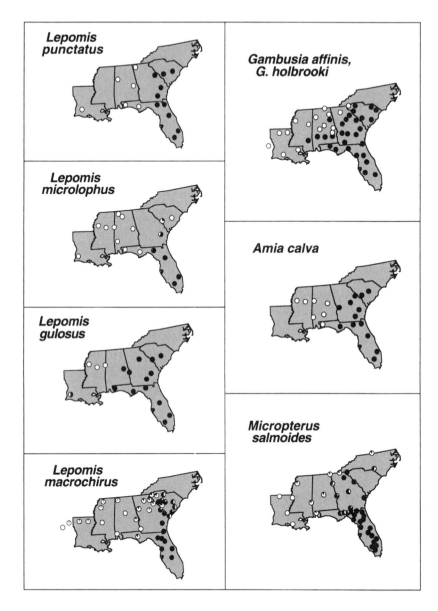

FIGURE 5.6   Pie diagrams summarizing the geographic distributions of the two primary phylogroups (Fig. 5.5) observed within each of seven freshwater fish species of the southeastern United States. Diagrams for five of the species refer to frequencies of diagnostic mtDNA haplotypes, whereas those for *Lepomis macrochirus* and *Micropterus salmoides* refer here to frequencies of diagnostic allozyme alleles.

and Avise, 1993), and in each case nearly fixed differences at several marker loci distinguished eastern from western populations. The molecular data further indicate that the phylogroups introgressively hybridize across parts of the Carolinas, Georgia, Alabama, and the Floridian panhandle. Some of these secondary contacts probably were facilitated by human-mediated transplantations (Nedbal and Philipp, 1994).

Overall, at least seven freshwater fish species or species-complexes have been assayed extensively for genetic composition across Atlantic and Gulf coastal drainages of the southeastern United States. Without exception, fundamental genealogical partitions distinguished eastern from western populations (Fig. 5.6), thus strongly suggesting shared historical biogeographic influences. The Floridian peninsula invariably is the geographic stronghold of eastern forms, whereas Gulf coastal drainages from Alabama westward house the western forms. The Mississippi or the Alabama-Tombigbee drainages may have been historical cradles for the western phylogroups (see Mayden, 1988).

In summary, all four aspects of genealogical concordance are documented abundantly in these freshwater fishes. Aspect I is evidenced by high sequence divergences between major branches in the intraspecific mtDNA gene trees. In three species surveyed also for allozyme variation, Aspect II is suggested by the general agreement between the geographic positions of the deep mtDNA phylogroups and those registered by dramatic shifts in frequencies of nuclear alleles. Aspect III of genealogical concordance appears in the similar geographic distributions of the two major phylogroups within each species. Concordance Aspect IV is evidenced by a close agreement between the geographic arrangement of intraspecific phylogroups vis-à-vis the major biogeographic provinces as identified from leading trends in the regional distributions of fish species (Fig. 5.7).

The magnitudes of the sequence divergence estimates between the major intraspecific phylogroups showed much poorer agreement across these fish species, ranging from $p \cong 0.006$ in *Amia calva* to $p > 0.050$ within each of the surveyed species of *Lepomis* (Fig. 5.5). Several hypotheses for this heterogeneity can be advanced. Perhaps the populations were sundered by

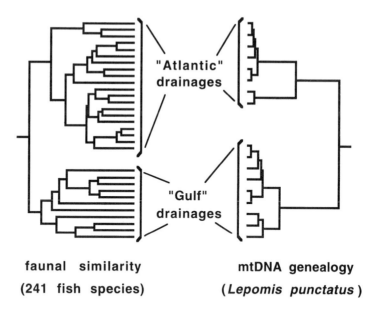

faunal similarity   mtDNA genealogy

(241 fish species)   (*Lepomis punctatus*)

FIGURE 5.7   A clear example of Aspect IV of genealogical concordance (after Walker and Avise, 1998). Geographic distributions of the two major branches in the mtDNA gene tree for the spotted sunfish, *Lepomis punctatus* (Bermingham and Avise, 1986) agree perfectly with the two principal southeastern faunal regions for freshwater fishes (Fig. 5.4).

the same historical vicariant event but rates of mtDNA evolution varied among taxa. Or perhaps different levels of ancestral polymorphism were available for conversion to differences between phylogroups. All else being equal, larger ancestral populations or those structured more strongly in space will tend to contain gene-tree lineages with deeper historical separations, and some such lineages may be those that were destined to characterize sister populations arising from a vicariant event (Fig. 5.8).

Another possibility is that the times of population separation truly differed. Climatic changes of the Pliocene and Pleistocene were repeated many times, so successive cycles in some taxa might have overridden or erased (via population extinctions and recolonizations) the phylogeographic

ancestral populations

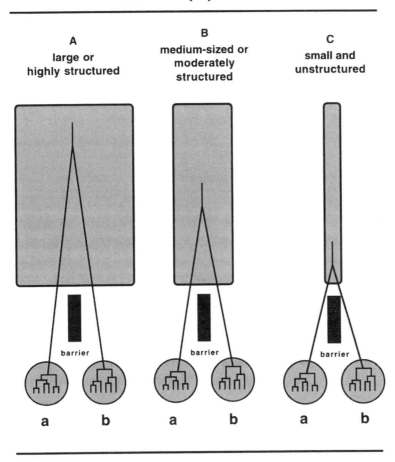

daughter populations

FIGURE 5.8   Illustration of how the intraspecific gene trees of codistributed taxa might show similar geographic patterns yet varying temporal depths as a result of differences in the sizes or degrees of spatial structuring of ancestral populations. All else being equal, lineage separations in a gene tree tend to be deeper when ancestral populations are larger or spatially structured.

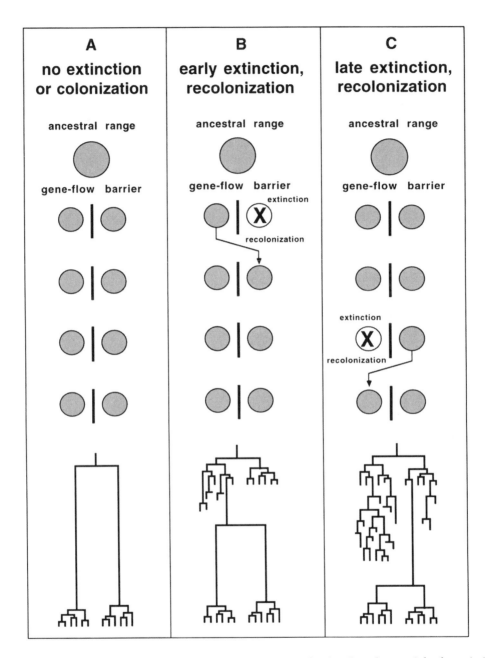

FIGURE 5.9   Illustration of how the intraspecific gene trees of codistributed taxa might show similar geographic patterns yet varying temporal depths as a result of cyclical vicariant events coupled with regional population extinctions and recolonizations (after Cunningham and Collins, 1998).

signals of earlier cycles (Fig. 5.9). Such recurring episodes of vicariance and interregional dispersal would produce a phylogeographic record that shows spatial concordance but considerable temporal heterogeneity across species.

The vicariant scenarios depicted in Figs. 5.8 and 5.9 also illustrate why the distributions of intraspecific phylogroups might tend to mirror biotic provinces as identified by species' distributional limits (e.g., Fig. 5.7). Suppose that in each of several species, population extinctions on one side or the other of the zoogeographic boundary were *not* followed by interregional recolonizations. Then, the range of each of these species today would be confined to one geographic area (either eastern or western in the case of freshwater fishes). An accumulation of such historical outcomes would produce a concentration of species' ranges that terminate at the boundaries between traditional zoogeographic provinces or subprovinces.

Maritime Species.   All four aspects of genealogical concordance also have been uncovered in molecular phylogeographic surveys of maritime taxa in the southeastern United States. The major phylogeographic patterns described next are summarized in Fig. 5.10.

MtDNA genotypes were assayed from seaside sparrows (*Ammodramus maritimus*) representing seven named subspecies (Avise and Nelson, 1989). Most noteworthy was a substantial phylogeographic break (estimated 1.0 percent net sequence divergence) that distinguished Atlantic coast birds from all specimens collected along the Gulf of Mexico. This phylogeographic pattern had been anticipated by Funderburg and Quay (1983), who speculated from distributional and other zoogeographic information that these sparrow populations were split historically into Atlantic and Gulf units. Currently, there is a pronounced hiatus in the geographic range of the species with no extant populations in southeastern Florida.

Another coastal species with disjunct Atlantic and Gulf populations is the black sea bass (*Centropristis striata*). Here too, Atlantic and Gulf forms conventionally recognized as different subspecies proved to be quite distinct in mtDNA composition, with an estimated net nucleotide

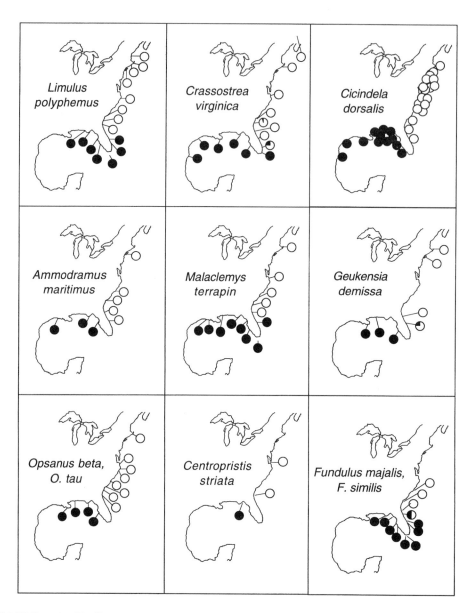

FIGURE 5.10   Pie diagrams summarizing geographic distributions of the two most fundamental phylogroups observed within each of several maritime species or species complexes along coastlines of the Atlantic and Gulf of Mexico in the southeastern United States. Diagrams for seven species refer to frequencies of diagnostic mtDNA clades (see text), whereas those for *Geukensia demissa* and the *Fundulus* species refer to frequencies of diagnostic allozyme alleles. The lower right panel exemplifies a common geographic pattern wherein one species is distributed primarily along the Gulf of Mexico and southeast Florida and a sister species occurs further north along the Atlantic coast.

quite distinct in mtDNA composition, with an estimated net nucleotide sequence divergence of 0.7 percent (Bowen and Avise, 1990).

Two other coastal vertebrates that show an Atlantic-Gulf disjunction in mtDNA phylogeny involve the toadfish complex *Opsanus* (Avise et al., 1987b) and the diamondback terrapin *Malaclemys terrapin* (Lamb and Avise, 1992). In the former assemblage, two related species (*O. tau* and *O. beta*) mostly are confined to the Atlantic and Gulf coasts, respectively, and differ by an estimated 9.6 percent mtDNA sequence divergence. In the latter species, only one fixed mtDNA restriction-site difference was detected, but it also distinguished Gulf samples from those along the Atlantic coast north of Florida.

Similar phylogeographic patterns in mtDNA have been documented for three invertebrates with quasi-continuous distributions along the coasts of the Atlantic and Gulf of Mexico. In the tiger beetle, *Cicindela dorsalis* (Vogler and DeSalle, 1993, 1994a), horseshoe crab, *Limulus polyphemus* (Saunders et al., 1986), and American oyster, *Crassostrea virginica* (Reeb and Avise, 1990), Atlantic versus Gulf populations proved highly distinct genetically. In the latter two species, the ranges of the Gulf forms extended into southeastern Florida (Fig. 5.10). In all three species, the level of genetic divergence between Atlantic and Gulf populations far surpassed observed differences within either region. In the American oyster, subsequent restriction-site assays of several anonymous nuclear loci added support for an Atlantic-Gulf population distinction (Karl and Avise, 1992; Hare and Avise, 1996). This distinction was not, however, apparent in molecular assays of other nuclear loci in this species (Buroker, 1983; McDonald et al., 1996; Hare and Avise, 1998). Hare (1998) concluded that the population separations in oysters may have occurred within the $3x$ temporal window where a split in a mtDNA gene tree is expected to be mirrored by only some modest fraction of nuclear genes.

Population studies based on allozymes or other genetic assays have uncovered evidence for similar Atlantic-Gulf distinctions in additional coastal invertebrates and vertebrates. In the ribbed mussel, *Geukensia demissa* (Sarver et al., 1992), and in a complex of killfishes (*Fundulus majalis* and *F. similis*; Duggins et al., 1995), a pronounced genetic discontinuity demarcated Atlantic from Gulf populations that the authors in both cases recommended be named separate species. The respective Gulf forms again extended along the southeastern coast of Florida in patterns reminiscent of those for the American oyster, horseshoe crab, and diamondback terrapin (Fig. 5.10). Similar genetic discontinuities also have been reported for some coastal invertebrates with disjunct Atlantic and Gulf ranges: *Hydractinia* hydroids, and crabs in the genera *Pagurus*, *Sesarma*, and *Uca* (Cunningham et al., 1991, 1992; Felder and Staton, 1994).

On the other hand, several genetically assayed coastal species have failed to show clear evidence for Atlantic-Gulf phylogenetic separations (Gold and Richardson, 1998). Examples among fishes include the red drum, *Sciaenops ocellatus* (Bohlmeyer and Gold, 1991), hardhead catfish, *Arius felis* (Avise et al., 1987b), a menhaden complex in the genus *Brevoortia* (Bowen and Avise, 1990), and the American eel, *Anguilla rostrata* (Avise et al., 1986). As described earlier, an absence of regional mtDNA divergence in the American eel probably is due to the species' catadromous life cycle. Lack of pronounced regional differentiation in the other groups probably is due to current gene flow around southern Florida or to retention of ancestral lineage polymorphisms by Atlantic and Gulf populations that separated recently.

Nonetheless, across a surprising diversity of species ranging from coastal invertebrates to maritime fishes and salt-marsh tetrapods, molecular genetic data reveal fundamental phylogenetic discontinuities and considerable concordance in the geographic distributions of the major lineages (Fig. 5.10). By these same yardsticks, little evidence exists for dramatic historical-genetic partitions within either the Atlantic or Gulf regions. Collectively, all four aspects of genealogical concordance have been evidenced: concordance across multiple mtDNA sequence characters in defining the principal mtDNA phylogroups; concordant support from nuclear genes in some of the species surveyed; general agreements across taxa in geographic boundaries between the intraspecific phylogroups; and a close agreement in the ranges of molecular phylogroups with traditional faunal provinces. Although the Atlantic-Gulf distinction is far from universal, the level of phylogeographic concordance across multiple taxa strongly implicates shared historical influences on the genetic architectures and gene-flow regimes of a substantial fraction of this maritime fauna.

Such historical influences probably have operated in collaboration with contemporary ecological conditions to influence the present-day distributions of genotypes. For example, the genetic discontinuities in several species localize to the eastcentral Florida coastline where a pronounced transition occurs between subtropical and cooler temperate waters. This ecological transition zone is mediated in part by warm waters of the Gulf Stream which flow out of the Gulf of Mexico into the coastal regions of southeastern Florida.

Such ecological conditions probably have two kinds of distributional impact on the genetic architectures of these maritime species. First, the ecological transition itself may generate distinctive selective pressures north versus south of eastcentral Florida that may further inhibit gene flow between historically differentiated Gulf and Atlantic phylogroups. Second, for some species such as the American oyster and horseshoe crab with mobile larvae, the Gulf Stream itself might have contributed to a "leakage" of Gulf genotypes into the Atlantic coast of southern Florida. These speculations emphasize a broader point. Present-day ecological and behavioral circumstances, as well as historical factors, impact the genetic architectures of all extant species. Thus, a full appreciation of why lineages occur where they do requires consideration of both contemporary and historical causation.

Freshwater and Terrestrial Turtles. The phylogeographic hypotheses in Table 4.1 and the four aspects of genealogical concordance in Table 5.1 were motivated largely by the findings just described on freshwater fishes and maritime faunas of the southeastern United States. Thus, these phylogeographic concepts were considered provisional pending further comparative evaluations. One faunal group subsequently assayed in explicit tests of phylogeographic concepts and concordance principles involved turtles (order Testudines) native to freshwater and terrestrial habitats of this continental region.

Thirty-five living species of Testudines inhabit the southeastern United States, of which about 20 have been subjects of molecular phylogeographic appraisal (Table 5.2). Nine assayed species have broad distributions across the area. Eight of these showed pronounced phylogeographic population structure in mtDNA, as predicted for relatively sedentary organisms under phylogeographic Hypothesis I (Table 4.1). Furthermore, highly divergent branches in the mtDNA gene trees were observed within these species, and the mtDNA clades invariably distinguished populations on a regional spatial scale (Fig. 5.11).

TABLE 5.2   Genetic summary statistics for several species of freshwater and terrestrial turtles examined for mtDNA phylogeographic patterns across the southeastern United States.

Species	Major Phylogeographic Units	Net Sequence Divergence between Units[a]	Reference
Sternotherus minor	A, B	0.032	Walker et al., 1995
Sternotherus odoratus	A, B, C	0.014–0.028	Walker et al., 1997
Kinosternon subrubrum	A, B, C, D	0.027–0.070	Walker et al., 1998a
Kinosternon baurii	A, B	0.010	ibid.
Trachemys scripta	A, B	0.006	Avise et al., 1992c; Walker & Avise, 1998
Graptemys (10 species)	A, B, C	0.010–0.028	Lamb et al., 1994
Gopherus polyphemus	A, B	0.021[b]	Osentowski & Lamb, 1995
Chelydra serpentina	A	—	Walker et al., 1998b
Macroclemys temminckii	A, B, C	0.017–0.028[b]	Roman et al., 1999
Deirochelys reticularia	A, (B + C)	0.043[b]	Walker & Avise, 1998

a. All sequence divergence estimates within phylogeographic units were considerably smaller.

b. From mtDNA control-region sequences. All other values in this column are based on RFLP assays.

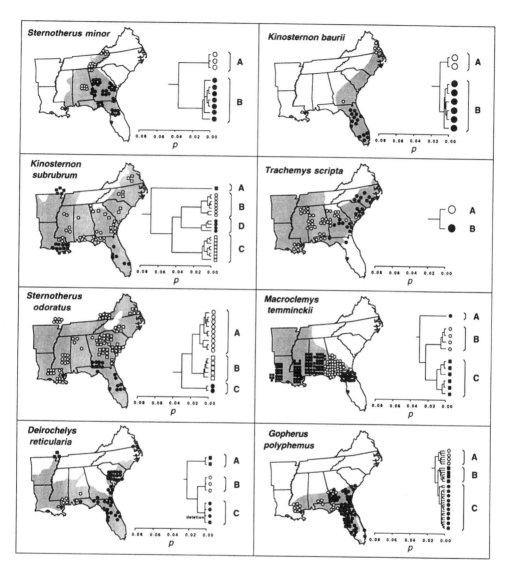

FIGURE 5.11 Intraspecific phylogeographic patterns in mtDNA for freshwater and terrestrial turtles in the southeastern United States (after Walker and Avise, 1998). Note that all diagrams are plotted on the same scale of genetic distance between haplotypes. On the maps, each symbol denotes an individual and the mtDNA phylogroup to which it belongs. Species' ranges are indicated by shading.

The evidence bearing on phylogeographic Hypothesis III for these turtles (that the major mtDNA phylogroups reflect historical population separations) consists of three aspects of genealogical concordance. First, the
gene-tree separations themselves typically involved many mtDNA sequence or restriction-site characters (concordance Aspect I). Second, the
mtDNA phylogroups often coincided reasonably well across multiple
species (Aspect III). Thus, for seven species depicted in Fig. 5.11, populations in peninsular Florida and/or along the Atlantic coast belonged to
mtDNA gene-tree branches that were highly distinct from those characterizing conspecific populations in coastal plains and Piedmont areas to the
west. These phylogeographic patterns suggest genealogical separations
between regions that are far older than those within regions.

Third, these intraspecific phylogeographic patterns agree quite well
with traditional evidence on the major Testudine faunal provinces in the
region (Aspect IV). Thus, a basal distinction in species' composition between Atlantic and Gulf zones (Fig. 5.12) parallels the intraspecific phylogeographic trends in mtDNA. Furthermore, these faunal provinces for
turtles bear a remarkable resemblance to those described previously for
freshwater fishes. All of these lines of evidence suggest that historical
biogeographic factors had cardinal influences on the population genetic
architectures and species' distributions of aquatic and semi-aquatic organisms in the southeastern United States.

OTHER MULTISPECIES REGIONAL APPRAISALS

Small Mammals of Lowland Amazonia

The lowland forests of the Amazonian Basin contain the world's richest
biota. Several hypotheses have been advanced to explain recent speciations in this region. The Refuge Model (Haffer, 1969; Cracraft and Prum,
1988) states that populations were sundered by habitat vicariance associated with cyclical expansions and contractions of forests and savannahs
during alternating wet and cool-dry Pleistocene episodes. Ecological models (Endler, 1982; Tuomisto et al., 1995) suggest that divergence has been

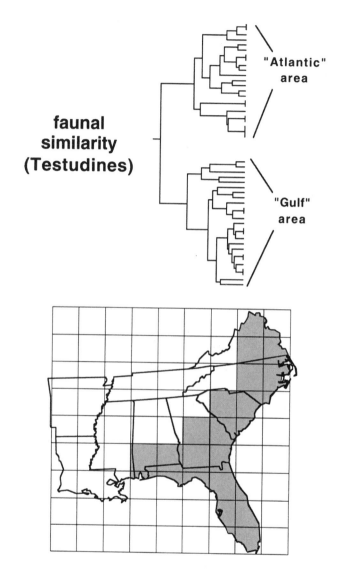

FIGURE 5.12   Faunal provinces for turtles and tortoises in the southeastern United States based on a composite assessment of species' distributions in the Testudines (after Walker and Avise, 1998). The map shows the two basal regions (shaded, "Atlantic"; unshaded, "Gulf") identified in a cluster analysis (above) of faunal similarity coefficients among areas demarcated by a grid superimposed on the map.

driven by selection pressures associated with high environmental and ecological heterogeneity in the region. The Riverine Barrier Model (Wallace,
1849; Ayres and Clutton-Brock, 1992) posits that large rivers promoted
genetic divergence in terrestrial organisms by blocking inter-regional
gene flow. These hypotheses are not mutually exclusive and multiple
factors may apply (M. B. Bush, 1994). Nonetheless, these and related possibilities provide useful frameworks for biogeographic evaluations in
the Neotropics (Simpson and Haffer, 1978; Prance, 1982; Whitmore and
Prance, 1987).

For example, tests of divergence processes in Amazonia have been analyzed from mtDNA patterns within several nonvolant mammal species
(Patton and Smith, 1992; da Silva and Patton, 1993; Patton and da Silva,
1997; Patton et al., 1994, 1996, 1997; Peres et al., 1996). An immediate goal
was to test the Riverine hypothesis of population vicariance by examining
conspecific populations along the 1,500-km length of the Rio Juruá, a major tributary of the Amazon River in western Brazil. More than a dozen
species of marsupials and rodents have been assayed (da Silva and Patton,
1998). The Riverine Model predicts historical genetic separations of populations on opposite sides of the river, but this expectation was not met by
any of the species. Instead, highly divergent mtDNA phylogroups were
observed for upstream versus downstream regions in many of the assayed
taxa (examples in Fig. 5.13). Sequence divergence estimates between these
intraspecific phylogroups ranged from 4 percent to 14 percent across
species, compared to genetic distances within clades that were typically
less than 1 percent.

Although small sample sizes prohibited strong conclusions, the large
genetic distances between phylogroups and the concordant geographic
placements of the phylogenetic breaks supported the notion of common
vicariant elements in the origin and history of this fauna. One intriguing
possibility raised by da Silva and Patton (1998) is that Amazonia is composed of several historical drainage sub-basins lying in different tectonic
settings, each separated by geological arches due to Andean uplifts of the
mid- to late Tertiary. One of these major arches cuts perpendicularly across

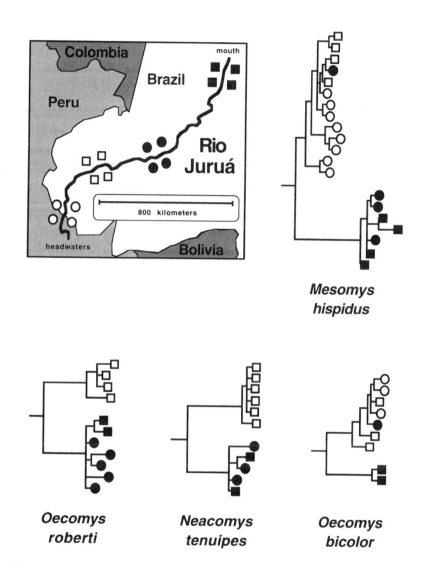

FIGURE 5.13   Phylogeographic concordance of major mtDNA phylogroups for four species of Amazonian rodents along the Rio Juruá in western Brazil (after Patton et al., 1994; da Silva and Patton, 1998; and Patton, pers. comm.). Members of *Oecomys* are semi-arboreal rice rats, *Mesomys* are spiny tree rats, and *Neacomys* are terrestrial spiny mice.

the central part of the Rio Juruá in a position coincident to and perhaps causally associated with the contemporary phylogeographic breaks observed in several of the species assayed (Fig. 5.13). If these paleobasins do constitute major historical centers for the diversification of recent Amazonian biotas, similar genetic patterns may emerge as additional species in the area are analyzed phylogeographically.

South American Cats

Another molecular genealogical study in the Neotropics revealed considerable phylogeographic concordance across species (Eizirik et al., 1998). In assays of the mtDNA control region, the margay (*Leopardus wiedii*) and ocelot (*L. pardalis*) each proved to be subdivided into 3–4 major phylogenetic units that agree closely in geographic placement (Fig. 5.14). For example, Central American populations in each species are highly distinct in mtDNA sequence from conspecific assemblages of haplotypes in northern and southern regions of South America. These genetic patterns are relevant to historical biogeographic reconstructions and also to conservation efforts for these cats.

Plants of the American Pacific Northwest

Given the relative paucity of molecular phylogeographic analyses at the intraspecific level in the botanical literature (Schaal et al., 1998), it is surprising that one of the earliest regional treatments of multiple codistributed taxa involved plants. Studies of restriction-site variation in chloroplast (cp) DNA have been conducted on several species native to the Pacific Coast of North America (Soltis et al., 1989, 1991, 1992b; Strenge, 1994). Aspects I, III, and IV of genealogical concordance emerged.

The evidence consists of more-or-less coincident distinctions in cpDNA clades between northern versus southern populations (Soltis et al., 1997). In six of the seven plant species assayed—three herbaceous perennials (*Tolmiea menziesii, Tellima grandiflora, Tiarella trifoliata*), a shrub (*Ribes bracteosum*), a tree (*Alnus rubra*), and a fern (*Polystichum munitum*)—deep separations in an intraspecific cpDNA gene tree distinguished molecular

(a) ocelots   (b) margays

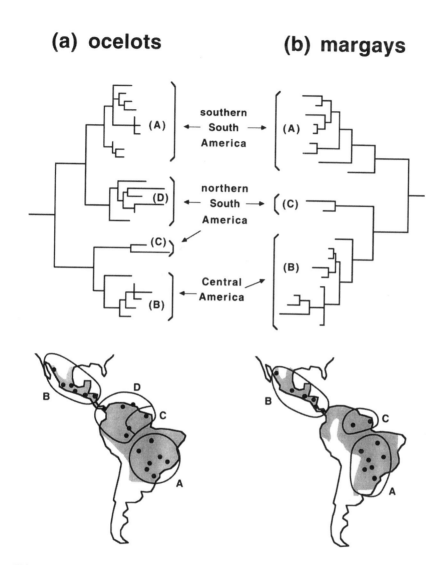

FIGURE 5.14   Concordant phylogeographic patterns within two species of Neotropical cats (after Eizirik et al., 1998). Shaded areas in the maps are species' ranges, and black dots are collecting sites for mtDNA haplotypes whose phylogenetic relationships are depicted in the maximum parsimony trees (*above*).

phylogroups centered in areas generally north versus south of the Oregon-Washington boundary (Fig. 5.15). In some of these species, "southern" genotypes also were observed in disjunct northern sites. The composite pattern suggests historical population separations tracing to previously known Pleistocene glacial refugia in the Pacific Northwest (Pielou, 1991; Soltis et al., 1997). Similar phylogeographic patterns for some animal species in the region (e.g., stickleback fishes [O'Reilly et al., 1993; Ortí et al., 1994] and brown bears [Talbot and Shields, 1996]) may register population differentiation in these same refugia (review in Byun et al., 1997).

However, assays of nuclear genes (ribosomal DNA and/or allozymes) failed to identify clear phylogenetic separations in several of these plant species (Soltis et al., 1997). This might be attributable to either greater discriminatory power inherent in assays of cytoplasmic genomes (due to smaller effective population sizes for uniparentally-inherited genes, all else being equal), or higher levels of gene flow for nuclear genes (via pollen movement) than for cytoplasmic cpDNA (via seeds). In any event, cytoplasmic genomes appear to have recorded signal biogeographic events in the histories of many codistributed plant (and animal) species of the Pacific Northwest.

Marine Organisms and the Trans-Arctic Interchange

For most of the Cenozoic Era, the region that is now the Bering Strait was a land bridge between North America and Asia that effectively prevented any exchange of cold-water marine faunas between the North Pacific and the North Atlantic. This barrier to marine dispersal vanished abruptly about 3.5 Mya when the Earth's temperatures warmed and a sea channel opened (Fig. 5.16). Fossil evidence and current distributions of marine taxa point to an ensuing faunal interchange characterized mostly by uni-directional invasions of high latitude species from the North Pacific into the North Atlantic (Vermeij, 1991a,b). The initial opening of the Bering Strait was followed by Northern Hemisphere glaciations (beginning about 2–3 Mya) that probably influenced both regional extinction patterns and likelihoods of further biotic exchanges across the sub-Arctic region.

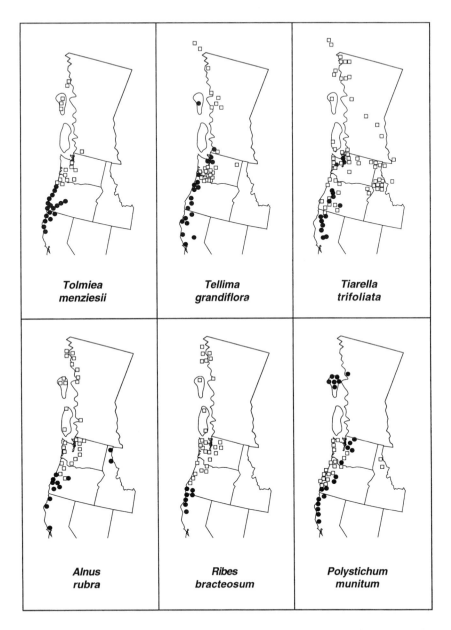

FIGURE 5.15   Phylogeographic concordance of major cpDNA phylogroups within each of six plant species along the Pacific coast of North America (after Soltis et al., 1997). In each case, open squares and closed circles denote the two principal cpDNA clades within a species.

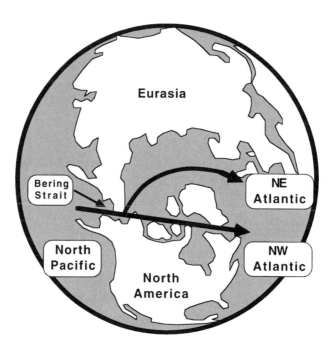

FIGURE 5.16   Circumpolar map showing avenues and suspected directions of gene flow (heavy arrows) from the North Pacific to the North Atlantic following the opening of the Bering Strait about 3.5 Mya (after Cunningham and Collins, 1998).

Molecular phylogeographic analyses have been conducted on several marine taxa in these high latitudes, and results often support the invasion scenarios described above. For example, a mtDNA-based phylogeny for rocky-shore snails in the genus *Nucella* displayed a paraphyletic pattern for North Pacific forms with respect to those in the Atlantic (Collins et al., 1996). Thus, Atlantic populations carry only a nested subset of the lineage diversity in the Pacific. The restricted position of the sole North Atlantic species within the broader phylogeny of *Nucella* snails in Pacific waters is consistent with a single successful trans-Arctic invasion of the Atlantic. Similarly, a molecular phylogeny for *Littorina* snails supports fossil-based hypotheses of two independent trans-Arctic invasions of the North Atlantic from the Pacific (Zaslavskaya et al., 1992; Reid et al., 1996).

A compilation of such case studies for high-latitude marine taxa has re-vealed a variety of phylogeographic outcomes that can be grouped into four major categories (Fig. 5.17). Classes *A* and *B* involve a secondary cessation of gene flow between the Pacific and Atlantic soon after the original opening of the Bering Strait, whereas Classes *C* and *D* involve recent genetic connec-tions between these oceans. Classes *A* and *C* entail little or no recent genetic contact between populations in the Northwest Atlantic and Northeast At-lantic, whereas Classes *B* and *D* involve recent genetic connections between these two regions. Empirical examples of each phylogeographic outcome have been documented in genetic appraisals of Holarctic marine taxa rang-ing from fishes to mollusks, crustaceans, and algae (Table 5.3).

For example, a Class *B* pattern is suggested by a close phylogenetic re-lationship between populations of the alga *Acrosiphonia arcta* across the North Atlantic and a deep separation from populations in the Pacific. On the other hand, a Class *D* history is evidenced by the close phylogenetic connections between mussel populations in the *Mytilus edulis* complex across all three oceanic regions. Detailed inspection of phylogeographic patterns in multiple species (Table 5.3) led Cunningham and Collins (1998) to conclude that regional population extinctions, recolonization events, and vicariant processes have sculptured extant diversity in these trans-Arctic taxa. Furthermore, historical factors proved to be better predictors of phylogeographic outcomes than did intrinsic organismal vagility, in-cluding mode of larval dispersal.

Vertebrates in Fragmented Australian Rainforests

Wet forests were widespread across the Australian continent during the Miocene before they became much reduced during the mid- to late Ter-tiary and Quaternary. Today, rainforests in eastern Australia are confined to a fragmented arc of uplands and coastal lowlands mostly in Queens-land (Fig. 5.18). These forest fragments house numerous endemic species as well as those with more catholic habitat tastes. Considerable interest has centered on the extent to which vicariant events dating to the Pleistocene or earlier contributed to contemporary biotic distributions via effects on

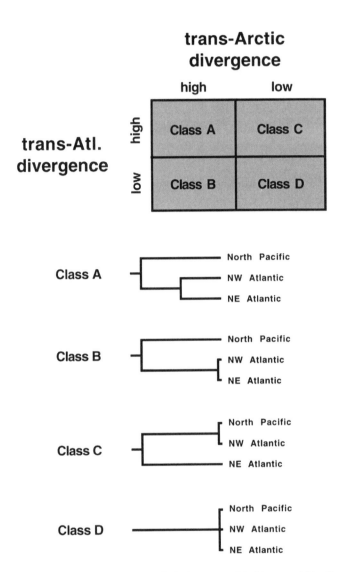

FIGURE 5.17   Four observed classes of phylogeographic history of Northern Hemisphere marine taxa as defined by degree of genetic divergence between regional populations (after Cunningham and Collins, 1998). See Table 5.3 and text.

TABLE 5.3   Marine taxa showing various classes of phylogeographic history as
deduced from genetic divergence patterns across the Arctic and North
Atlantic (see Fig. 5.17 and text).

Class A   Cessation of gene flow between Pacific and Atlantic soon after opening of the
Bering Strait; continuous residence on both sides of Atlantic with little or no
recent genetic connections across Atlantic.
Pagurus acadianus-bernhardus complex of hermit crabs (Cunningham et al.,
1992).

Class B   Cessation of gene flow between Pacific and Atlantic soon after opening of the
Bering Strait; contemporary gene flow and/or recent colonization events
across Atlantic.
Nucella lapillus complex of snails (Collins et al., 1996); *Semibalanus balanoides*
barnacles (cited in Cunningham and Collins, 1998); *Acrosiphonia arcta* algae
(van Oppen et al., 1994).

Class C   Successive trans-Arctic invasions with recent genetic connections between
Pacific and Northwest Atlantic and little or no contact across Atlantic.
Strongylocentrotus droebachiensis sea urchins (Palumbi and Wilson, 1990);
Phycodrys rubens red algae (van Oppen et al., 1995); *Macoma balthica* clams
(Meehan, 1985; Meehan et al., 1989); *Osmerus* smelt fishes (Taylor and
Dodson, 1994).

Class D   Recent genetic connections between the Pacific, Northwest and Northeast
Atlantic.
Strongylocentrotus pallidus sea urchins (Palumbi and Kessing, 1991);
Gasterosteus aculeatus stickleback fishes (Haglund et al., 1992; Ortí et al., 1994);
mussels in the *Mytilus edulis* complex (Varvio et al., 1988; McDonald et al.,
1991; Rawson and Hilbish, 1995).

---

rainforest fragmentation (Joseph and Moritz, 1994). In particular, two areas unsuitable for wet-forest species (the Black Mountain Barrier and the
Burdekin Gap) have been implicated as important historical-biogeographic
features in patterning the current distributions of forest-restricted species.

Mitochondrial DNA patterns were summarized for seven avian (Joseph
et al., 1995) and six herpetofaunal (Schneider et al., 1998) species in eastern
Australia. The most dominant and consistent phylogeographic pattern
was a deep split between northern and southern populations. For example, two rainforest-endemic birds (*Poecilodryas albispecularis* and *Orthonyx
spaldingii*) showed deep and concordant phylogeographic breaks precisely
at the Black Mountain Barrier (BMB), as did four herpetofaunal taxa

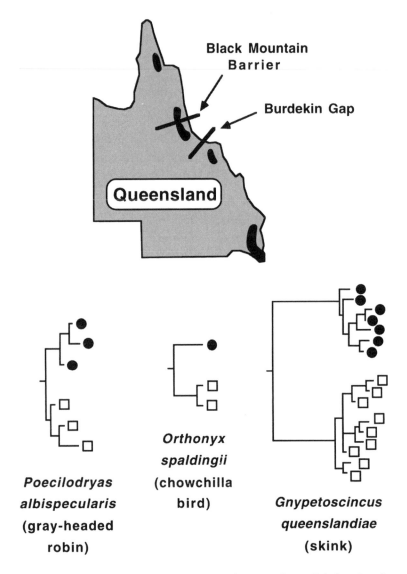

FIGURE 5.18   *Above:* Map of Queensland (northeastern Australia) showing the current distribution of rainforest fragments (black areas) and two biogeographic barriers suspected to have played important roles in structuring current species' distributions. *Below:* Intraspecific mtDNA gene trees for two avian and one reptilian species that show concordant phylogeographic breaks north (black dots) versus south (open squares) of the Black Mountain Barrier. (After Joseph and Moritz, 1994 and Joseph et al., 1995.)

(examples in Fig. 5.18). North-south phylogenetic breaks in the other forest-endemic species were associated less obviously with the BMB. Two bird species (*Sericornis citreogularis* and *S. magnirostris*) not restricted to rain-forests showed phylogeographic diversity apportioned mainly on either side of the Burdekin Gap. On the other hand, significant phylogeographic population structure was not detected in the forest-restricted *S. keri* or in the widespread *S. frontalis*.

Overall, the empirical results for these rainforest taxa show partial but incomplete congruence in phylogeographic patterns across species (con-cordance Aspect III), and between molecular phylogroups and biogeo-graphic provinces identified by traditional nonmolecular evidence (Aspect IV). The authors conclude that much of the observed phylogeographic structure is consistent with Pleistocene-refuge models for rainforest frag-ments, but also that evolutionary histories in these taxa are complicated by idiosyncrasies of local population extinctions and recolonizations.

Geminate Marine Taxa across the Isthmus of Panama

Sometimes, vicariant barriers to dispersal are so firm and well dated that they provide critical geographic frameworks for evaluating molecular pat-terns (rather than the usual converse in which molecular data inform or-ganismal biogeography). A case in point involves the rise of the Isthmus of Panama about 3 Mya, an event that sundered populations of many tropical marine organisms into geminates (Jordan, 1908) now found in the eastern Pacific and western Atlantic Oceans (Rubinoff and Leigh, 1990). Molecular data gathered for many of these pairs have been put to service in evaluat-ing evolutionary clocks for allozymes and mtDNA in organisms ranging from shrimp (Knowlton et al., 1993) and sea urchins (Lessios, 1979, 1981; Bermingham and Lessios, 1993) to tropical fishes (Vawter et al., 1980; Grant, 1987; Bermingham et al., 1997).

Two general trends have emerged from the information published to date. First, as expected, appropriate genetic assays cleanly distinguish most (but not all) of the geminate taxa in the tropical Atlantic versus Pa-cific. Second, the genetic distances show great heterogeneity across the

presumptive geminate taxa (Fig. 5.19). Thus, at face value, the phylogeographic footprints of the vicariant event on extant faunas appear similar in
geographic pattern but perhaps different in temporal depth. This outcome
is reminiscent of trends discussed previously for phylogroup separations
in faunas of the southeastern United States.

Yet, even in the geologically ideal Panamanian setting, important
qualifications apply to any conclusions about molecular evolutionary
rates. First, some of the presumed geminates might not be sister taxa, and
failure to appreciate this would bias genetic distances and inferred molecular rates upward, all else being equal (Bermingham et al., 1997). Second,
gene lineages that were destined for alternative fixation in extant pairs of
some of the true sister taxa might have separated long before the vicariant
event itself (Fig. 5.8), again resulting in inappropriate upward biases on estimates of evolutionary rate (Knowlton et al., 1993). Third, post-vicariant
genetic contact (Fig. 5.9) between some geminate taxa may have occurred
via recent circumtropical gene flow perhaps through the Indian Ocean. If
unrecognized as such, this secondary genetic contact would yield a false
impression of decelerated molecular evolution.

Bermingham et al. (1997) suggested criteria for identifying pre- and
post-vicariant contacts from genetic data on extant geminate taxa, and further suggested that such instances be eliminated (or corrected) before calibrating molecular clocks. Adoption of these protocols probably increases
the realism of resulting calibrations, but it also entails a risk of circular reasoning. For example, if a lack of appreciable genetic divergence between a
pair of geminate taxa is taken axiomatically to imply recent circumtropical
gene flow, then no accumulation of such examples could reveal what alternatively could be decelerated molecular evolution following a true 3 Mya
vicariant separation. Bermingham et al. (1997) conclude in reference to the
Panamanian studies that "Insufficient data exist to test alternative hypotheses of rate variation."

The Hawaiian Volcanic Conveyor Belt

Another favorable geological setting for evaluating phylogeographic patterns is the Hawaiian archipelago, where evolutionary diversification

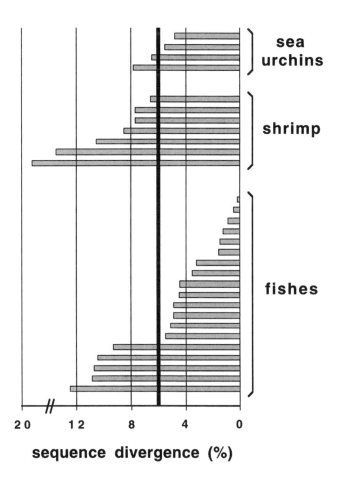

FIGURE 5.19   Differences in mtDNA sequence between presumptive geminate marine taxa separated by the Isthmus of Panama. Shown are mean genetic distances (Kimura, 1980) for four pairs of sea urchins (Bermingham and Lessios, 1993), seven pairs of shrimp (Knowlton et al., 1993), and nineteen pairs of marine fish (Bermingham et al., 1997). The heavy vertical bar indicates the expected genetic distance under a uniform molecular clock ticking at 2 percent sequence divergence per My, assuming that all geminates are true sister taxa that were sundered 3 Mya. See text for interpretations and qualifications regarding rate estimates.

within numerous taxonomic groups has occurred rapidly and recently (Wagner and Funk, 1995). The Hawaiian Islands form as the Pacific Plate moves over a volcanic hot spot that periodically extrudes island-building magma. The islands thus arose on a tectonic conveyor belt, and their times of origin are known from geological dating. These range from 0.4 My for the most recent island (Hawaii) in the southeast, to 5.1 My for the oldest major present-day island (Kauai) in the northwest of the chain (Fig. 5.20).

Molecular phylogenetic appraisals of several species-groups of birds and arthropods on the archipelago have yielded "serial area cladograms" in which the sequential order of clades in a gene tree generally parallels the linear physical (and, hence, temporal) arrangement of the islands (Rowan and Hunt, 1991; Tarr and Fleischer, 1993; Kambysellis et al., 1995; Wagner and Funk, 1995; Roderick and Gillespie, 1998). Such patterns have been interpreted as an expected result of successive lineage colonizations of younger islands. If these colonizations usually took place shortly after island formation, then ages of the islands provide a needed denominator of absolute time for assessments of molecular evolutionary rates.

Fleischer et al. (1998) used this approach to calibrate evolutionary clocks for several mitochondrial and nuclear genes in Hawaiian insects and honeycreeper birds (subfamily Drepanidinae). Results suggest that sequence divergence rates within some genes have been roughly constant over the past 4 My: e.g., 1.6 percent per My for the mitochondrial *cytb* gene in the honeycreepers, and 1.9 percent per My for the *Yp1* nuclear gene in *Drosophila* (Fig. 5.20). Nonetheless, as stressed by the authors, these calibrations remain provisional because several assumptions are entailed beyond the requirement that each island was colonized soon after its origin. Among these additional assumptions are that the geological dates of the islands are accurate, that the topology of a gene tree truly reflects serial island colonizations, and that levels of DNA sequence variation in extant populations (used as a correction factor for calculating between-island genetic distances) are similar to those in the ancestral populations.

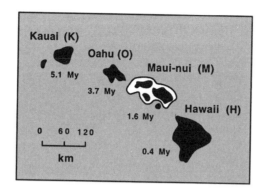

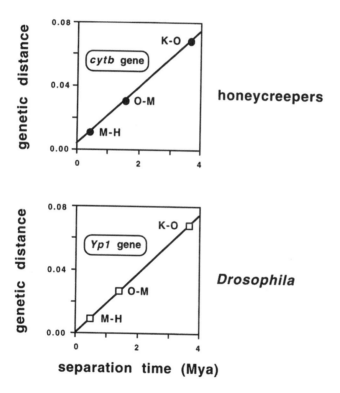

FIGURE 5.20   Molecular clock calibrations in Hawaiian birds and arthropods (after
Fleischer et al., 1998). Shown in the map are major islands or island groups whose origination times are known from potassium-argon dating. Molecular data for *Drosophila*
were from Kambysellis et al. (1995). See text for background and assumptions regarding
rate estimates in the two graphs.

Additional Examples

At the time of this writing, few other molecular studies are available in *comparative* intraspecific phylogeography at regional levels. McMillan and Palumbi (1995) observed striking genealogical concordance in two monophyletic species groups of Indo–West Pacific butterflyfishes (*Chaetodon*): in each case, a genetic break (approximately 2.0 percent in mtDNA sequence) clearly separated individuals from the Indian versus Pacific Oceans. On the other hand, Turner et al. (1996) examined mtDNA and allozyme patterns in populations of five codistributed species of darter fishes (Etheostomatini) from the Ozark and Ouachita highlands of the southcentral United States and failed to discern phylogeographic concordance across species. Nonetheless, species-specific histories were recovered in the assays of these darters.

Similar conclusions regarding species-idiosyncratic population structures but absence of between-species phylogeographic concordance were drawn for the following: assayed herpetofaunas in deserts of the American Southwest (Lamb et al., 1992); some avian species in Central America (Brawn et al., 1996) and in the Caribbean (Bermingham et al., 1996); stream fishes in the central and eastern highlands of North America (Strange and Burr, 1997; but see Bergstrom 1997 for a different interpretation); a variety of marine invertebrates along the west coast of the United States (review in Burton, 1998); and freshwater fishes in Central America (Bermingham and Martin, 1998). In the latter study, close inspection of the molecular patterns led the authors to suggest that lower Central America was colonized by fishes from northwestern Columbia in perhaps three distinct waves of invasion dating respectively to the late Miocene, mid-Pliocene (coincident with the rise of the Panamanian Isthmus), and the Pleistocene.

Zink (1996, 1997) summarized mtDNA phylogeographic patterns in five avian species with broad ranges across North America (Fig. 5.21). Little or no phylogeographic structure was detected in the red-winged blackbird (*Agelaius phoeniceus*), chipping sparrow (*Spizella passerina*), or song sparrow (*Melospiza melodia*). On the other hand, the Canada goose (*Branta canadensis*) and fox sparrow (*Passerella iliaca*) displayed two and four deep

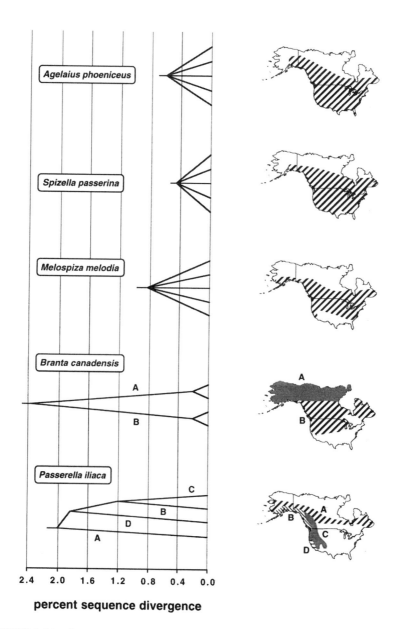

FIGURE 5.21   Approximate extent of breeding ranges and diagrammatic summaries of mtDNA gene trees for each of five avian species with wide distributions across North America (after Zink, 1996).

mtDNA clades, respectively, with strong but nonconcordant geographic orientations. In another continent-wide comparative summary, Taberlet et al. (1998) found little concordance in molecular phylogeographic patterns across 10 European taxa ranging from mammals and amphibians to arthropods and plants. Thus, in contrast to results for several regional biotas reported earlier in this chapter, these studies on broader continental scales have given little indication that populations of codistributed species have had similar phylogeographic histories.

GENEALOGICAL DISCORDANCE

Apparent and often real discrepancies among multiple gene trees, or between gene trees and traditional systematic characters, also are reported routinely in a phylogeographic context. Several aspects of genealogical discordance can be of interest in historical reconstructions.

Aspect I: Disagreement across Characters within a Gene

In principle, tightly linked sequence characters have experienced the same history of transmission through an extended organismal pedigree. Thus, any genealogical discrepancies among nonrecombined characters must in some sense represent phylogenetic "noise."

When (as is often the case) small numbers of different sequence characters earmark different but nonoverlapping clades in a gene tree, no explanations need be invoked other than the idiosyncrasies of mutational origin. The observed mutations considered individually merely may denote particular branches in a less-than-fully-resolved gene tree, without overt conflict in clade delineation. Thus, particular clades that happen to be revealed in an empirical molecular survey may vary according to vagaries of mutational origin and effects of finite genetic sampling. The net result is an estimated gene tree with unresolved branches but no topological discrepancies from the true gene tree.

On the other hand, different sequence characters within a gene may suggest conflicting or overlapping clades. If the locus was historically free of interallelic recombination (as normally is true for mtDNA), then the agents responsible must entail evolutionary homoplasy (convergence, parallelism, or reversal in particular character states). In phylogenetic reconstructions based on parsimony, the extent of homoplasy is estimated routinely as the number or fraction of extra steps in the gene tree beyond those that distinguish haplotypes in the matrix of raw data.

Sometimes, multiple sequence characters in different regions of a gene display overt disagreement with respect to inferred placement of haplotypes in a gene tree. For a nuclear locus, historical intragenic recombination (or, perhaps, gene conversion) could be responsible. Alleles that arose via intragenic recombination consist of amalgamated stretches of sequence that truly had distinct phylogenetic histories within a species. Cases are easiest to document empirically when recombination was between highly divergent alleles, and when recombination events were infrequent such that the odd recombinant haplotype can be seen to consist of two portions with distinct placements in a broader gene tree.

Aspect II: Disagreement across Genes

Some degree of phylogenetic discordance across gene trees is an inevitable consequence of Mendelian inheritance and the vagaries of lineage sorting at unlinked loci through a sexual pedigree. An example involving cytonuclear comparisons was discussed earlier under the $3x$ rule. This rule relates to the differing theoretical probabilities, at intermediate times of population separation, that nuclear versus mitochondrial lineages will have achieved reciprocal monophyly via random lineage sorting from a polymorphic ancestor. Indeed, genealogical discordances reported in the literature often involve a greater population structure inferred from cytoplasmic than from nuclear genes (although this also could result from a higher mutation rate in cytoplasmic genes).

However, other causes of cytonuclear discordance also are known (Palumbi and Baker, 1996; Rawson and Hilbish, 1998). In many species, males or their gametes tend to be more dispersive than females. Consistent gender biases in historical gene flow can result in distinctly different signatures of population structure in nuclear versus cytoplasmic loci. For example, macaque monkeys (Melnick and Hoelzer, 1992), green sea turtles (Karl et al., 1992; FitzSimmons et al., 1997b), and humpback whales (Palumbi and Baker, 1994; Baker et al., 1998) all display less geographic structure in nuclear genes than in mtDNA haplotypes, probably due at least in part to greater interpopulation movements or matings by males. In many plants, pollen are far more dispersive than seeds, one net consequence being a greater opportunity for the spread of nuclear alleles than of maternally transmitted cytoplasmic alleles (McCauley et al., 1996; Latta and Mitton, 1997).

In some secondary hybrid zones also, gender-based asymmetries have been documented with respect to contemporary (Lamb and Avise, 1986) or historical (Dowling et al., 1997) mating patterns and gene flow regimes (Arnold, 1993; Mukai et al., 1997). Often in collaboration with selective influences (Boissinot and Boursot, 1997), sex-based asymmetries can have pronounced effects on cytonuclear associations and the spatial extents of introgression across hybrid zones (Arnold, 1993, 1997; Harrison, 1993). Ancient episodes of introgressive hybridization also can produce "ghosts of hybrids past" (Wilson and Bernatchez, 1998) sometimes recognizable as pronounced incongruences among gene trees (DeSalle and Giddings, 1986; Dowling and Demerais, 1993; Rieseberg et al., 1996; DeSalle et al., 1997; Dowling and Secor, 1997; Rieseberg, 1997; Bagley and Gall, 1998).

In nonhybrid settings also, various forms of natural selection may produce distinct phylogeographic patterns among unlinked loci. At some loci, balancing selection (review in Mitton, 1997) might inhibit the extinction of allelic lineages, thereby generating a misleading impression of moderate or high gene flow among historically isolated populations. One likely example involves an aspartate aminotransferase gene in the deer mouse (*Peromyscus maniculatus*). Notwithstanding strong evidence from mtDNA, karyotype, and morphology for pronounced historical restraints on gene flow in this species, populations across North America all maintain similar frequencies of the same two protein electromorphs at this locus (Avise et al., 1979c; Aquadro and Avise, 1982). Similar observations of geographic uniformity in allozyme allele frequencies, despite evidence from mtDNA and nuclear RFLPs for restricted historical gene flow, led to speculation that balancing selection may have operated on some protein-coding loci in the American oyster (Karl and Avise, 1992). In general, any strong and consistent heterogeneities in population allele-frequency variances across loci are suggestive of selective influence of some sort because alleles at all neutral loci should paint similar pictures of population structure (Lewontin and Krakauer, 1973). Perhaps balancing selection contributes to geographic uniformity of allele frequencies at some genes, or diversifying selection contributes to geographic heterogeneity at other loci.

At the nucleotide sequence level, single-copy nuclear DNA in animals often evolves slower than mtDNA, and this provides another reason why pronounced population structure detected in mtDNA assays may remain unregistered in some nuclear assessments. Conversely, nuclear genes underlying some phenotypic features traditionally used in subspecies' taxonomies, such as pelage color in mammals or plumage features in birds (Barraclough et al., 1998; Magurran, 1998; Price, 1998), may evolve rapidly under diversifying or sexual selection. This could produce the appearance of strong population structure notwithstanding shallow lineage separations that might be registered in neutral or selectively constrained molecular characters (Hillis, 1987; Avise, 1994).

In general, disparate classes of molecular and organismal features experience varied selective regimes that can generate heterogeneities in phylogeographic outcomes. Consider the spectacular radiation of cichlid fishes in some of the African Rift Valley lakes (notably Lake Victoria). There, as gauged by the extremely shallow mtDNA genealogies and other evidence (Meyer et al., 1990), many species with diverse morphological and natural-history adaptations appear to have arisen within the last few thousand years, probably under intense sexual selection by females for differently colored males (Seehausen et al., 1997; Galis and Metz, 1998). Near the other end of the spectrum of evolutionary patterns, horseshoe crabs that have been morphologically conservative for tens of millions of years display (in molecular assays) considerable phylogenetic depth both within and among species (Saunders et al., 1986; Avise et al., 1994).

At higher taxonomic levels also, gene trees can disagree with one another and with a species tree. Discussion of this expression of discordance will be deferred to Chapter 6.

Aspects III and IV

Evolutionary processes that can generate varied phylogeographic patterns across codistributed species (Aspect III of discordance) are as numerous and varied as the historical factors that may have impinged on the population demographies of different species over an extended time. Gene trees may conflict with other biogeographic information (Aspect IV of genealogical discordance) for several reasons, including chance historical events leading to phylogeographic patterns that at face value appear inconsistent with historical geography. For example, the cattle egret (*Bubulcus ibis*) is a common wading bird in the Americas, yet it recently colonized the New World from its ancestral homeland in Africa probably by way of a storm in the late 1800s. Such colonization events are on a long list of "stochastic" historical factors that can produce species-eccentric phylogeographic outcomes.

Nonetheless, even the most species-idiosyncratic of phylogeographic architectures can inform taxonomic decisions, historical reconstructions, and conservation efforts for the particular organisms investigated. This statement holds whether or not concordant phylogeographic patterns exist across taxa, and even when recognizable genetic lineages fail to map to traditional biogeographic provinces.

CONCORDANCE AND PHYLOGEOGRAPHIC DEPTH

The primary rationale for promoting genealogical concordance concepts is that various aspects of congruence are the best and perhaps only means to distinguish between temporally deep versus shallow population genetic architectures. In nearly all species, typical dispersal distances of individuals are far below the total geographic range occupied by that species. Thus, the existence of spatial genealogical structure is virtually axiomatic (whether or not it happens to have been detected in a particular genetic assay). Through genealogical concordance principles, attention is shifted from the mere genetic diagnosis of populations to an enlightened concern about the relative magnitudes of population genetic differentiation, and of the historical processes that have shaped the more salient of the biotic partitions.

Conservation Relevance of Phylogeographic Depth

Both shallow and deep genealogical separations within species can be informative in a conservation context. Furthermore, both idiosyncratic and concordant phylogeographic patterns across genes and across taxa can have relevance for conservation efforts, depending on the particular issue and biological setting. These sentiments follow from the theoretical ties between population demography and intraspecific genealogy discussed in Chapter 2, and now they can be revisited with the force of empirical examples.

SHALLOW SEPARATIONS: MANAGEMENT UNITS

Conspecific populations not clearly delimited by large phylogenetic gaps must be connected genealogically through ongoing or recent gene flow. Nonetheless, contemporary dispersal in many cases may be far too low to promote appreciable demographic connections between geographic demes. In genetic analysis, the logic underlying the concept of a "management unit" (MU) is as follows: Any population that exchanges so few migrants with others as to be genetically distinct from them normally will be demographically independent at the present time. In the literature of commercial fisheries, MUs traditionally are referred to as "stocks" toward which harvesting quotas and other management plans are directed (Avise, 1987; Ryman and Utter, 1987; Ovenden, 1990). For any species, populations that are demographically autonomous should qualify as distinct MUs.

Provisional MUs can be identified by significant divergence in allele frequencies at neutral loci, regardless of depth in the gene tree (Moritz, 1994b). Mitochondrial haplotypes are especially powerful for identifying MUs because of their typical fourfold smaller effective population size (compared to haplotypes at autosomal loci), and because of their special relevance to demographic and reproductive connections among populations. Even shallow matrilineal subdivisions can be relevant to conservation efforts. Demographically autonomous populations, if overexploited or extirpated by humans or other causes, are unlikely to recover via natural recruitment of foreign females over ecological timescales relevant to immediate management interests (Avise, 1995).

DEEP SEPARATIONS: EVOLUTIONARILY SIGNIFICANT UNITS

In concept, an "evolutionarily significant unit" (ESU) is one or a set of conspecific populations with a distinct, long-term evolutionary history mostly separate from other such units (Ryder, 1986). As such, ESUs are the primary sources of historical genetic diversity within a species (Moritz, 1995). The ultimate goal of conservation biology is to preserve biodiversity, an important currency of which is genetic diversity (Ehrlich and Wilson, 1991). A widely held sentiment in conservation biology is that added value or worth should be attached to organismal assemblages that are more rather than less distinctive phylogenetically (Vane-Wright et al., 1991; Barrowclough, 1992; Faith, 1992, 1994; see Erwin, 1991 for an opposing view). Various measures have been suggested to summarize this phylogenetic component of biodiversity (reviews in Crozier, 1992, 1997; Krajewski, 1994; Vogler and DeSalle, 1994b; Humphries et al., 1995). Conservation plans for conspecific populations would do well to recognize ESUs and interpret them as worthy of special consideration.

Issues of phylogenetic distinctiveness often arise in the context of triage deliberations about how best to apportion finite conservation resources on species-level or higher taxa. However, similar concerns also arise at the intraspecific level. Under the U.S. Endangered Species Act, for example, legal protection is afforded not only to listed species but also to "subspecies" and to "distinct population segments." The concept of ESUs provides a phylogenetic framework for deciding which population units are most distinctive.

Operational criteria ranging from broad to detailed have been suggested for recognizing intraspecific ESUs. A general suggestion is that ESUs must contribute substantially to the overall genetic diversity of a species (Waples, 1991). A more explicit recommendation is that ESUs be identified as groups of populations "reciprocally monophyletic for mtDNA alleles and also differ[ing] significantly for the frequency of alleles at nuclear loci" (Moritz, 1994a). Any such empirical suggestion is arbitrary to some extent because there can be no clean line of demarcation along the continuum of possible magnitudes of population genetic differentiation or inferred temporal depths of population separation.

Consider, for example, the criteria noted above that were intended to answer "how much difference is enough?" for ESU qualification. Under any such guidelines, uncertainties remain: How many or what fraction of nuclear loci must show significant allele frequency differences for populations to warrant ESU designation?; How many or what fraction of individuals who carry heterotypic mtDNA lineages would disqualify a population from ESU status?; What magnitude of nucleotide sequence divergence between mtDNA clades is necessary for ESU recognition? Any universal definition that demarcates ESU status from non-ESU status is arbitrary to some extent and fails to concede that some situations truly are intermediate.

Similar practical difficulties arise in the formal application of concordance principles in ESU recognition. For example, under Aspects I and II of genealogical congruence, only arbitrary minimum guidelines can be invoked in response to questions such as: How many concordant sequence characters are required to establish a meaningful separation in a gene tree? (at least a few, but the more the better); How many gene trees must show concordant phylogenetic partitions to register important historical separations at the population level? (ditto); How deep must be the branch separations in a gene tree for ESU recognition? (the deeper the better); and, How

damning to ESU status are mild departures from perfect genealogical concordance across multiple characters? (less departure is better).

Such practical difficulties notwithstanding, the ESU concept (as a
corollary of the recognition that population separations can be of varying
evolutionary depth) remains one of the most important and revisionary of
perspectives to have stemmed from intraspecific phylogeography. Its relevance to conservation efforts applies in the context of single-species issues,
and to those of regional biotas (Avise, 1989a; Moritz, 1994b; Bernatchez,
1995; Avise and Hamrick, 1996; Smith and Wayne, 1996).

Conservation Issues in Individual Species.   Consider an example involving
the bananaquit (*Coereba flaveola*), a common land bird in Central and South
America and on several Caribbean islands (Seutin et al., 1994). A gene tree
was estimated for 58 different mtDNA haplotypes detected in this species
(Fig. 5.22). Six major mtDNA phylogroups (putative ESUs) were observed,
from which stemmed several suggestions relevant to conservation efforts
(Bermingham et al., 1996). From the perspective of maintaining phylogenetic diversity within the species, it was deemed unwise to expend extraordinary resources in protecting small bananaquit populations from each
of the northern Lesser Antilles. Birds on these small islands are close genealogically, as they are to populations on some of the larger islands such as
Guadaloupe. If a bananaquit population became extinct, the phylogeny
also could guide any proposed reintroduction program intended to restore
the original genetic condition. For example, islands in the Lesser Antilles
should not be repopulated with bananaquits from Jamaica because of a suspected long history of phylogenetic separation of birds from these regions.

This avian example involved genealogical inferences from a single locus, mtDNA, in conjunction with the geographic orientations of the inhabited islands. Ideally, proper identification of ESUs within any species
should rest on at least two facets of genealogical concordance. Concordance Aspect I merely identifies likely candidates for ESU status. Confirmation then requires support from independent genetic characters
(Aspect II), spatial agreement in phylogenetic partitions across taxa (As-

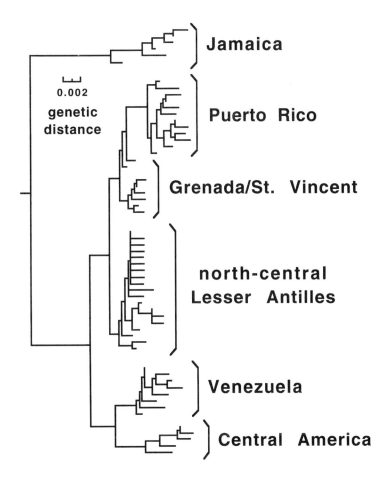

FIGURE 5.22   Mitochondrial gene tree for 58 haplotypes observed in a survey of 170 bananaquit birds, *Coereba flaveola,* in Central and South America and the Caribbean (after Seutin et al., 1994).

pect III), or congruence of spatial pattern with that from other independent biogeographic evidence (Aspect IV).

As detailed in Chapter 4, many species display deep separations in an intraspecific mtDNA gene tree. In such cases, the major genealogical branches by definition are registered concordantly by multiple mtDNA characters. These branches represent potential ESUs worthy of additional evaluation. As further detailed in Chapter 4, gene-tree phylogroups often display spatial distributions consistent with independent evidence on historical shaping processes. When such concordant support is available from multiple lines of biogeographic evidence, the putative phylogroups then warrant recognition as securely documented ESUs (or suggested terminological alternatives, such as phylogenetic subspecies [O'Brien and Mayr, 1991] or species [Frost and Hillis, 1990]).

Table 5.4 summarizes several empirical molecular studies that identified (or in some cases sought and failed to identify) ESUs within endangered or other species of special conservation concern. The phylogeographic outcomes in some instances challenged and in other cases bolstered the conventional taxonomies upon which conservation programs had been built. As noted by Daugherty et al. (1990), good taxonomies "are not irrelevant abstractions, but the essential foundations of conservation practice." Whether or not immediate conservation issues are involved, well-documented ESUs are relevant to lower-level taxonomic decisions.

Conservation Issues in Regional Biotas. Notwithstanding the phylogeographic idiosyncrasies expected across taxa, several of the comparative studies summarized above documented impressive levels of genealogical concordance in various regional biotas. An exciting application of molecular phylogeography in conservation biology is defining particular geographic regions within which multiple species display phylogenetically distinctive populations or ESUs. Such areas are special candidates for high conservation priority.

It remains to be seen how many regional biotas display the kinds of multi-taxa phylogeographic concordance currently best documented, for example, in the faunas of the southeastern United States, small mammals in Amazonia, vertebrates in rainforest fragments of northeastern Australia, or floras of the American Pacific Northwest. However, the prospect that disproportionate fractions of the Earth's biodiversity may be concentrated in recognizable historical biogeographic areas has obvious implications for conservation efforts at regional or ecosystem scales.

Analytical procedures have been introduced to formally assess genealogical data as a basis for prioritizing particular geographic areas for conservation focus. For example, Faith (1992, 1994; Faith and Walker, 1996) introduced a quantitative phylogenetic diversity measure that incorporates historical data from multiple taxa and can be adapted to summarize underlying "feature diversity" of geographic communities (Moritz and Faith, 1998). Such approaches can complement appraisals that seek to identify distinctive biogeographic provinces and subprovinces on the basis of species' richness or the distributions of taxa (Mittermeier et al., 1998; Olson and Dinerstein, 1998). They also can dovetail with a recent approach in conservation biology known as "gap analysis" wherein ranges of threatened species are compared in assessing priority locations for biological reserves (Scott and Csuti, 1997).

Regional hotspots of biodiversity or exceptional endemism often are promoted as special targets for focused conservation efforts (Scott et al., 1987; Margules et al., 1988; Myers, 1988, 1990; Dinerstein and Wikramanyake, 1993; Pressey et al., 1993; Bibby, 1994; Kerr, 1997). Governmentally sponsored "biogeographic reserves" or "biodiversity parks" could be designed to protect significant fractions of regional biotic diversity, important components of which clearly are historical. Private organizations such as the Nature Conservancy and a few countries such as Costa Rica already have championed similar ideas. A well-implemented system of biogeographic reserves also would promote public awareness of conservation issues, much as the current National Park System in the United States serves both to educate the public and to preserve exceptionally beautiful or unique features of the continent's geological history (Avise, 1996b).

Maiden analyses in comparative molecular phylogeography have yielded an encouraging observation of conservation relevance: a tendency toward spatial concordance (Aspect IV) between intraspecific phylogroups and biogeographic provinces or subprovinces as identified by species' distributions and historical physiography. Comprehensive molecular examinations of regional biotas are expensive and labor intensive, and thus can be contemplated only in model circumstances. Nonetheless, if trends from available phylogeographic studies can be generalized, an encouraging

TABLE 5.4 Examples of molecular phylogeographic appraisals of threatened or endangered species and their taxonomic relatives.

Taxon	mtDNA Gene-Tree Outcome	Concordant Support from . . .	Reference
Geomys colonus (pocket gopher)	no major phylogenetic distinction from problematic common congener	allozymes, karyotype, multivariate morphology	Laerm et al., 1982
Ammodramus maritimus (seaside sparrow)	two major phylogeographic units not coincident with subspecies' designations	phylogeographic partitions in codistributed coastal taxa, and historical biogeographic considerations	Avise & Nelson, 1989
Felis concolor (puma)	two phylogenetic units co-occur in an endangered population in South Florida	nuclear DNA markers, and macro-geographic ranges of ESUs	O'Brien et al., 1990
Lepidochelys kempi (Ridley sea turtle)	clear phylogenetic distinction from problematic congener	a biogeographic scenario from morphology and geology	Bowen et al., 1991, 1997
Canis rufus (red wolf)	close phylogenetic connections to gray wolf and coyote	microsatellite nuclear loci	Wayne & Jenks, 1991; Roy et al., 1994b
Chelonia mydas (green, black sea turtles)	two major phylogeographic units	geographic ranges (confined to Atlantic versus Indo-Pacific)	Bowen et al., 1992
Lycaon pictus (African wild dog)	three major phylogeographic units	geographic ranges (confined to distinct African regions)	Girman et al., 1993

Gopherus polyphemus (gopher tortoise)	at least two major phylogeographic units	phylogeographic partitions in codistributed taxa, and historical biogeographic considerations	Osentowski & Lamb, 1995
Macrotis lagotis (desert bandicoot)	no strong phylogeographic structure in Australian interior	microsatellite nuclear loci	Moritz et al., 1997
Sternotherus depressus (freshwater turtle)	clear phylogenetic distinction from problematic congeners	morphology	Walker et al., 1998c
Melanotaenia eachamensis (Australian rainbowfish)	clear phylogenetic distinction from a problematic congener	morphology, microsatellite nuclear loci	Zhu et al., 1998
Polioptila californica (Calif. gnatcatcher bird)	clear phylogenetic distinction from problematic congeners	morphology	Zink & Blackwell, 1998
Oreailurus jacobita (Andean mountain cat)	clear phylogenetic distinction from the ocelot and margay	morphology	Johnson et al., 1998

conservation message emerges: Much of the information needed to design historical biogeographic reserves already may be available in the form of traditionally recognized biogeographic provinces, subprovinces, or eco-regions (e.g., Scott et al., 1990; Abell et al., 1998). Thus, conservationists need not await complete genetic reexaminations of the biological world before embracing regional perspectives (in addition to traditional species-focused efforts) in preservation initiatives.

Absolute Times of Phylogroup Separation

Principles of genealogical concordance provide a conceptual framework for identifying the deeper phylogeographic units within species. What are the chronological times of the phylogroup separations? Temporal esti-mates are fraught with uncertainties, but provisional conclusions can be reached by appealing to molecular clock calibrations.

Avise and Walker (1998) summarized the mitochondrial literature on inferred separation times for intraspecific phylogroups in birds. Among 63 avian species surveyed for mitochondrial population structure across ma-jor portions of their respective ranges, 37 species (59 percent) displayed a Category I phylogeographic pattern: i.e., they were sundered into two or more significant (bootstrap supported) mtDNA phylogroups with a strong geographic orientation. In most cases, the assignment of a species to phylogeographic Category I was straightforward. Two examples involv-ing *Ammodramus* sparrows are illustrated in Fig. 5.23.

For each such species that was phylogeographically subdivided for mtDNA, net sequence divergence between major phylogroups (corrected for within-phylogroup sequence heterogeneity) was converted to an esti-mate of population separation time using a conventional avian mtDNA clock: 2 percent sequence divergence between a pair of lineages per My (Klicka and Zink, 1997). The resulting histogram of estimated phylogroup separation times (Fig. 5.24) shows that 76 percent of the 37 inferred phylo-group separations date to the Pleistocene, and most of the remainder to the late Pliocene.

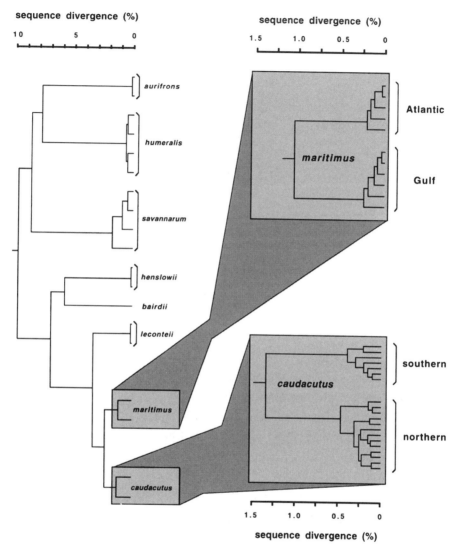

FIGURE 5.23   MtDNA phylogenies in *Ammodramus* sparrows (from Avise and Walker, 1998). *Left:* Matrilineal phylogeny for eight congeneric species as estimated by Zink and Avise (1990). *Right:* Magnified view of matrilineal relationships within each of two species surveyed across their ranges (*A. maritimus* from Avise and Nelson, 1989; *A. caudacutus* from Rising and Avise, 1993).

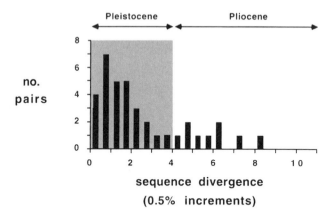

intraspecific phylogroups (birds)

FIGURE 5.24   Histogram of estimated mtDNA sequence divergences (and inferred separation times) between pairs of major intraspecific phylogroups in avian species (after Avise and Walker, 1998).

Thus, the mitochondrial evidence is consistent with conventional wisdom that late Pliocene and Pleistocene events had great impact on the phylogeographic architectures of extant birds. Indeed, authors of the original studies often invoked explicit "Pleistoscenarios" to account for particular phylogeographic outcomes. For example, from an integration of mtDNA data with evidence from morphology and behavior, Rising and Avise (1993) hypothesized glacial refugia and subsequent range expansions of two major phylogroups within *Ammodramus caudacutus*. These forms subsequently were recognized as distinct taxonomic species (AOU, 1995).

Evolutionary separation times for primary intraspecific phylogroups similarly have been summarized in species representing other vertebrate classes (Avise et al., 1998). Among 189 nonavian vertebrates surveyed for mtDNA population structure across major portions of their respective ranges, 103 species (54 percent) displayed a Category I phylogeographic pattern (examples in Fig. 5.25).

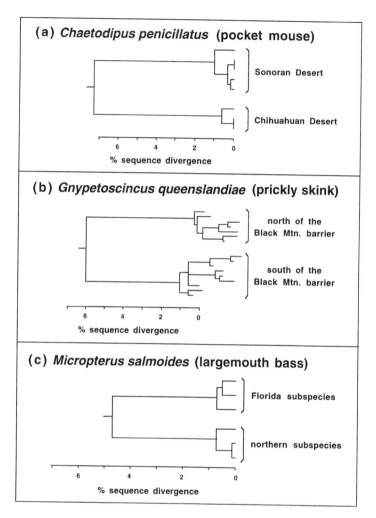

FIGURE 5.25 Typical examples of phylogeographic Category I as reported in mtDNA surveys of a mammal, reptile, and fish (after Avise et al., 1998). *(a)* cluster phenogram for the pocket mouse (Lee et al., 1996); *(b)* neighbor-joining tree for the prickly skink (Joseph et al., 1995); *(c)* cluster phenogram for the largemouth bass (Nedbal and Phillipp, 1994).

intraspecific phylogroups (mammals)

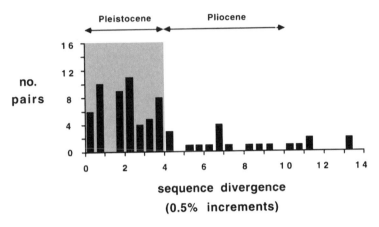

FIGURE 5.26   Histogram of estimated mtDNA sequence divergences (and inferred separation times) between pairs of major intraspecific phylogroups in mammalian species (after Avise et al., 1998).

Within the mammals (Fig. 5.26), 52 of the 72 inferred phylogroup separations (72 percent) date to the Pleistocene, and most of the remainder date to the Pliocene. These percentages are similar to those for avian taxa. For the other vertebrates (Fig. 5.27), interpretations are complicated by suspected slower mtDNA clock calibrations for some of these taxa (Avise et al., 1992c; Adachi et al., 1993; Martin and Palumbi, 1993; Rand, 1993, 1994; Martin, 1995). Under the standard mtDNA clock employed for birds and mammals, 27 of the 47 inferred phylogroup separations in amphibian and reptilian species (57 percent) date to the Pleistocene. In the fishes, 19 of 26 inferred phylogroup separations (73 percent) date to the Pleistocene. However, these fractions for herpetofauna and fishes drop to 15 percent and 31 percent, respectively, under a fourfold slower mtDNA clock calibration.

Three cautionary points should be made about these comparative analyses. First, estimates of absolute separation times are highly sensitive to the clock calibrations employed. By hard criteria, mtDNA evolutionary

intraspecific phylogroups

(a) amphibians and reptiles

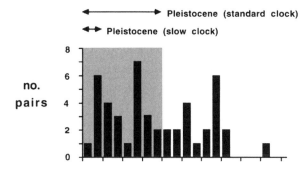

(b) fishes

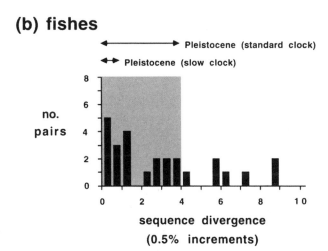

sequence divergence
(0.5% increments)

FIGURE 5.27 Histograms of estimated mtDNA sequence divergences (and inferred separation times based on two alternative mtDNA clock calibrations) between pairs of major intraspecific phylogroups in the indicated vertebrate assemblages (after Avise et al., 1998).

rates are poorly characterized for most animal groups, yet are suspected to differ by severalfold across some lineages. Second, a large variance clearly exists in the genetic distances (and inferred times of separation) across phylogroup pairs, and much of this probably reflects true heterogeneity in phylogroup separation times.

Third, a severe bias operates against the detection of any phylogroup separations that might date to less than about 200,000 years ago. About 500 bp of mtDNA sequence per individual was assayed in a typical study. Under a standard mtDNA clock, only about one nucleotide substitution in a sequence of this length is expected to distinguish two matrilines that separated 100,000 years ago. However, several concordant substitutions are required for statistical support of a putative clade in phylogenetic reconstructions. Thus, available data do not rule out the likelihood that late-Pleistocene and Recent events also initiated separations among many shallower phylogroups that remain poorly characterized with conventional laboratory efforts.

SUMMARY

1. Four distinct aspects of genealogical concordance are relevant to phylogeographic interpretations. These involve agreement in phylogenetic patterns: (I) across multiple sequence or other character states within a gene; (II) among multiple gene trees at unlinked loci; (III) across two or more codistributed species with similar ecologies or natural histories; and (IV) between molecular genetic data and more traditional classes of biogeographic evidence. Aspect I establishes that a gene tree displays significant phylogenetic subdivision. Aspect II confirms that gene-tree partitions register phylogenetic separations at the population or species level. Aspects III and IV implicate shared historical biogeographic impacts on regional biotas.

2. Regional phylogeographic surveys consider spatial patterns in the historical lineages of multiple codistributed taxa. All four aspects

of genealogical concordance have been registered in one or another of the regional phylogeographic studies conducted to date. Notable examples include: freshwater and maritime faunas in the southeastern United States; mammals in the Neotropics; marine invertebrates of the trans-Arctic; vertebrates in fragmented Australian rainforests; geminate marine taxa that were sundered by the rise of the Isthmus of Panama into Atlantic and Pacific units; and diverse plants of the American Pacific Northwest.

3. Genealogical discordances also are relevant to phylogeographic interpretations. These involve disagreement in phylogenetic depths or patterns: (I) across multiple sequence or other characters within a gene; (II) among multiple unlinked gene trees; (III) across two or more codistributed species with similar ecologies or natural histories; and (IV) between molecular genetic data and traditional classes of biogeographic evidence. Aspect I of discordance often reflects homoplasy in gene-tree data. Aspect II can result from stochastic lineage sorting through sexual pedigrees, or from other biological factors including variable rates of sequence evolution, different effective population sizes for alleles at autosomal versus cytoplasmic (and sex-linked) loci, gender-based asymmetries in historical dispersal and gene flow, and several forms of natural selection. Aspects III and IV suggest that many species-idiosyncratic biogeographic factors have impinged upon regional biotas.

4. Concepts of genealogical concordance and phylogeographic depth are related intimately and carry special relevance in conservation biology. They have led to recognition of a distinction between management units (extant populations that are genealogically close but demographically independent now) and evolutionarily significant units (populations with longer-standing evolutionary separations). The nature of evidence for deep versus shallow phylogeographic separations, and the importance of such distinctions, are increasingly appreciated in ecology, systematics, and conservation biology.

5. Sidereal separation times for intraspecific phylogroups can be estimated from sequence differences and suspected evolutionary rates
   in mtDNA. Of more than 100 surveyed species of mammals and
   birds showing significant phylogeographic population structure,
   about 75 percent of the major phylogroups within species provisionally date to Pleistocene separations under a standard mtDNA
   clock for homeotherms. Similar conclusions apply to intraspecific
   phylogroups in reptiles, amphibians, and fishes under the same
   clock calibration. However, severalfold older separation times
   for poikilothermic vertebrates are implicated under the slower
   mtDNA clocks sometimes suspected for various of these taxa.

Species are merely those strongly marked races or local forms which, when in contact, do not intermix, and when inhabiting distinct areas are. . . . incapable of producing a fertile hybrid offspring. But as the test of hybridity cannot be applied in one case in ten thousand, and even if it could be applied, would prove nothing, since it is founded on an assumption of the very question to be decided. . . . it will be evident that we have no means whatever of distinguishing so-called "true species" from the several modes of [subspecific] variation . . . into which they so often pass by an insensible series.

—Alfred Russell Wallace, 1865

SPECIATION PROCESSES AND EXTENDED GENEALOGY

IN PREVIOUS chapters, gene trees and their relationship to the demography and geography of conspecific populations were addressed. A central conclusion was that phylogenetic principles and approaches traditionally applied at supraspecific levels also have relevance, with appropriate modification, to genealogical issues within species. These top-down extensions of phylogenetic concepts to microevolution, motivated primarily by studies of mtDNA and by the rise of coalescent theory, have forged novel connections between population genetics and systematics, two major subdisciplines of evolutionary biology that otherwise had separate traditions and orientations. In this chapter, a bottom-up approach is taken to examine how concepts of microevolutionary gene trees and the coalescent might inform phylogeographic analyses of the speciation process and of higher taxa.

Species phylogenies are a conglomerate "central tendency" for the historical-descent pathways of multiple loci in the genome. Phylogenetic topologies of individual gene trees can differ from the composite species phylogeny for several reasons (Fig. 6.1), some with specifiable likelihoods (see beyond). Some of the variance across gene trees within an organismal

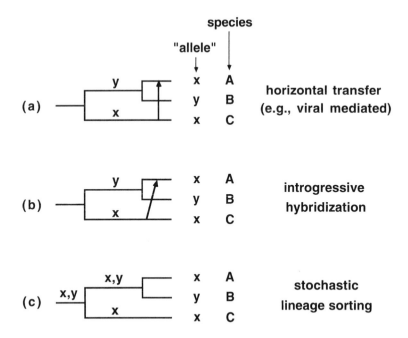

FIGURE 6.1   Three biological factors that can produce true genealogical discordance between the topology of a gene tree and a species tree. In each case, A and B are sister species but a gene lineage ("allele") in species A is genealogically closer to its homologue in C than to its homologue in B. Additional factors illustrated in Avise (1994, p. 355) can produce the false appearance of such discord.

pedigree arises inevitably from lineage sorting processes under Mendelian inheritance. Any resulting topological disagreements across genes do not mean that particular gene trees are incorrect (although they might be estimated incorrectly in an empirical appraisal). Rather, some discordances reflect the fact that unlinked genes in sexually reproducing organisms have traversed different transmission pathways through extended pedigrees.

Thus, conventional tree-like depictions of the speciation process or of species phylogenies are mere summaries that fail to convey the true fuzziness inherent in genealogical history. As stated by Maddison (1997), "A simple phylogenetic tree diagram with sticklike branches represents only the mean or mode of a distribution. Phylogeny has a variance as well, represented by the diversity of trees of different genes . . . . Given the centrality of genetics in our explanation of evolutionary diversity, we need to confront the composite, cloudlike nature of genetic history." Indeed, multigene approaches are receiving increased attention in models of speciation (e.g., Gavrilets, 1997) as well as in phylogenetic reconstructions at higher taxonomic levels (e.g., Kumar and Hedges, 1998; Pennisi, 1998).

O'Hara (1993) has likened the challenge of phylogenetic summary in biology to that of cartographic representation in geography. Phylogenies and maps alike are simplifications of reality, generalized representations with events selectively deleted according to the level of detail required. An interstate road map of the United States may be helpful in driving cross-country but is of no use in navigating the Freedom Trail in Boston, for which a local map provides appropriate resolution. Phylogenetic summaries also show varying resolution of the riverways and streams of heredity that make phylogeny, and a given depiction should be matched to the problem at hand.

PHYLOGEOGRAPHY AND THE ORIGIN OF SPECIES

The Phylogenetics of Speciation

Geography and population demography have played key roles in most speciation scenarios (Mayr, 1942, 1963). For example, under allopatric models, geographic populations begin to differentiate when environmental barriers or distance curtail genetic exchange. Extrinsic impediments to gene flow typically are envisioned as prerequisite for genetic divergence and for the eventual evolution of intrinsic (genetic) reproductive isolating barriers (RIBs) that are the hallmark of the popular biological species concept (BSC). Thus, biological speciation traditionally is viewed as the process by which a reproductive community of organisms (a field for gene recombination) is sundered into two or more reproductively isolated units.

POPULATION OR SPECIES-LEVEL PHYLOGENIES

The initial genetic sundering may involve large geographic populations (Fig. 6.2a), small founding populations outside the species' former range (Fig. 6.2b), or local syntopic populations separated by microhabitat (Fig. 6.2c). In each case, environmental obstructions to interbreeding enable population genetic divergence, a process promoted by varied selection pressures in different habitats and by founder effects and genetic drift when populations are small (Giddings et al., 1989). RIBs may arise as a nonadaptive byproduct of genetic differentiation, and in some cases they may be reinforced by selection favoring homotypic matings if the diverged populations regain sympatry (Butlin, 1989). Biological speciation also can take place suddenly in small populations via reproductive sundering agents such as polyploidization, chromosomal rearrangements, or changes in mating system (Avise, 1994, p. 258).

Most geographic and demographic scenarios for speciation initially result in paraphyletic taxa when reproductive isolation forms the basis for species definition (Fig. 6.2). This conclusion pertains with special force to subdivided gene pools shortly following their original separation. As time progresses, population extinctions often convert a paraphyletic pattern into an outcome wherein, viewed retrospectively, extant forms appear to be sister species that stemmed from a dichotomous split (Fig. 6.2, bottom).

Thus, each geographic-demographic model of biological speciation yields logical predictions about the coarse-focus phylogeny for extant taxa. Consider another case involving five hypothetical species (A-E) in which several phylogenetic outcomes are illustrated (Fig. 6.3). Here, D and E are sister species comprising a clade that is a sister group to A-C. However, the widely distributed species C that recently spawned a reproductively isolated peripheral isolate A, or a syntopic species B, is paraphyletic to these latter taxa at the outset. Furthermore, from the view of a future observer, any then-extant pair of these five taxa would appear as sister species if the remainder had gone extinct.

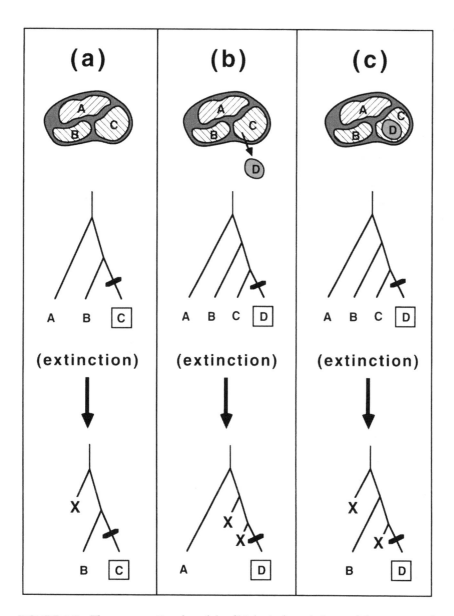

FIGURE 6.2 Three conventional models of biological speciation and the corresponding paraphyletic trees for populations A-D. A slash crossing a tree branch indicates the evolutionary origin of an intrinsic RIB. *(a)* Allopatric speciation via vicariance; *(b)* peripatric speciation via founder-effect (Mayr, 1982); *(c)* sympatric speciation (Bush, 1975, 1994). Note in each case that the initial condition of paraphyly for the ancestral species (Patton and Smith, 1989) can convert through time, via population-level extinctions (each denoted by an X), to a pattern wherein only two apparent sister species remain.

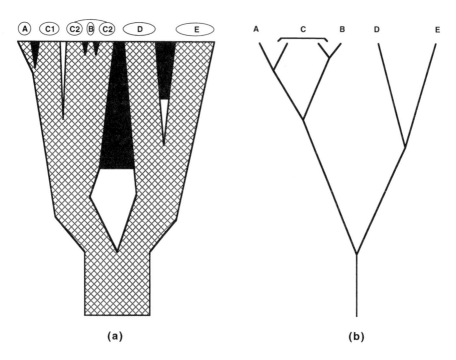

(a)   **(b)**

FIGURE 6.3   *(a)* Phylogeny for five extant biological species (A-E) and two geographically separated populations (C1 and C2) of C (after Avise and Wollenberg, 1997). The widths of branches
(cross-hatched channels) are proportional to population sizes through time, and also indicate a geographic orientation. Thus, A is a peripheral isolate from C1, and B arose within the range of C2. The
population sundering agents are extrinsic barriers to gene flow (white areas), intrinsic RIBs (black areas), or both in temporal order of appearance (white then black). *(b)* Simplified "stick" representation
of the phylogeny in *(a)*.

GENE GENEALOGIES

Within any such coarse-focus species phylogeny are multitudinous gene
trees, each with its own branching structure. These gene genealogies also
are impacted in generally predictable fashion by the geographic and
population-demographic nature of speciation events (as well as by various
operations of natural selection; Wang et al., 1997). Hence, phylogeographic
patterns in gene trees can be informative with respect to speciation mode.

For example, when speciation involves two large geographic populations (such as D and E in Fig. 6.3), most gene trees initially appear polyphyletic in the newly-arisen sister species: Some but not all lineages in D
comprise gene-tree clades with some but not all lineages in E. Through
time, lineage sorting converts most gene trees to a status of paraphyly and
then reciprocal monophyly ("exclusivity" for the two species; de Queiroz
and Donoghue, 1988; Graybeal, 1995), at rates (for neutral loci) influenced
by effective population size. The lineage-sorting processes are identical to
those described in Chapter 2 for geographically isolated populations, except that for biological species the RIBs by definition are genetic rather
than geographic alone. Divergent selection pressures may accelerate the
pace of lineage sorting to reciprocal monophyly for any two species in a
gene tree, whereas balancing selection can impede the process relative to
neutrality expectations.

If a speciation event involves a small population (A or B in Fig. 6.3) isolated from larger ancestral stock (C), the polyphyletic phase of the gene-
tree progression normally is bypassed and the temporal transition of
phylogenetic outcomes proceeds from paraphyly to eventual reciprocal
monophyly. Other geographic-demographic modes of speciation and their
anticipated phylogenetic consequences are illustrated in Fig. 6.4.

Recently separated species often display paraphyletic gene-tree patterns (DeSalle et al., 1987; Satta and Takahata, 1990; Brown et al., 1996).
Empirical cases depicted in Fig. 6.5 illustrate various explanations for paraphyletic outcomes. The annual sunflowers *Helianthus petiolaris* and *H.
neglectus* are "good" biological species because a chromosomal sterility
barrier confers reproductive isolation between them. *H. petiolaris* is distributed across much of North America, whereas *H. neglectus* is confined to
southeastern New Mexico and adjacent Texas. In this case (Fig. 6.5a), the
paraphyletic status observed in molecular data for *H. petiolaris* is hardly
surprising given the localized geographic origin of the sterility barrier that
recently gave rise to *H. neglectus*.

In a second example, the brown bear *Ursus arctos* is paraphyletic to the
polar bear *U. maritimus* in mtDNA genealogy (Fig. 6.5b). This pattern

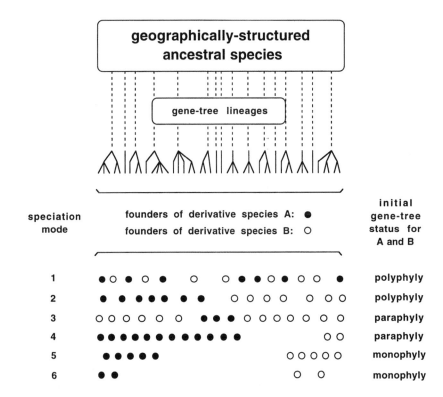

FIGURE 6.4 Alternative geographic-demographic modes of speciation and their likely phylogenetic consequences for gene trees within neophyte species A and B. Founders for the two sister species (open versus closed circles) are: *(1)* random draws of ancestral lineages; *(2)* large numbers of individuals from nearby allopatric populations; *(3)* a small number of individuals (for A) inside the ancestral species' range; *(4)* a small number of individuals (for B) on the range periphery of the ancestral species; *(5)* moderate numbers of individuals from distant allopatric populations; and *(6)* small numbers of individuals from distant allopatric populations.

might be due to introgressive hybridization that moved some mtDNA lineages of brown bears into polar bears (these species can produce fertile offspring in captivity). Another possibility involves historical lineage sorting. Perhaps polar bears arose within the past few tens of thousands of years from coastal populations of brown bears to which their matrilines appear

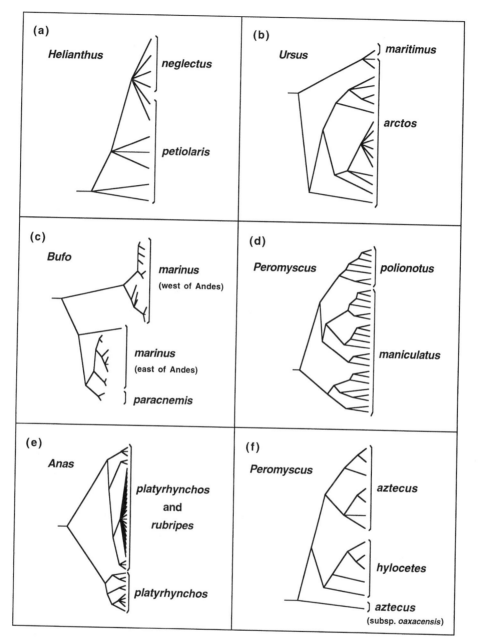

FIGURE 6.5 Examples of gene-tree paraphyly as recorded in data from: *(a)* chloroplast DNA and nuclear rDNA restriction sites in *Helianthus* sunflowers (after Rieseberg and Brouillet, 1994, from data in Rieseberg et al., 1990, and Rieseberg, 1991); *(b and c)* mtDNA sequences in *Ursus* bears (after Talbot and Shields, 1996) and *Bufo* toads (after Slade and Moritz, 1998); *(d and e)* mtDNA restriction sites in *Peromyscus* mice (after Avise et al., 1983) and *Anas* ducks (after Avise et al., 1990); and *(f)* mtDNA sequences in additional *Peromyscus* mice (after Sullivan et al., 1997).

related most closely. If so, polar bears possess a suite of derived morphological characters that arose rapidly and recently. This suggestion has some support from fossil and other evidence (Talbot and Shields, 1996).

Another example involves *Bufo marinus*, a South American toad that is paraphyletic in mtDNA genealogy to *B. paracnemis* (Fig. 6.5c). A major split in the mtDNA gene tree distinguishes *B. marinus* populations east versus west of a regional Andean uplift completed about 2.7 Mya that probably sundered the ancestral toad population. Phylogenetically, *Bufo paracnemis* is nested within the eastern lineage, suggesting that the speciation event postdated the east-west phylogeographic split in *B. marinus*.

The deer mouse, *Peromyscus maniculatus*, occupies most of North America and is paraphyletic in mtDNA phylogeny (Fig. 6.5d) to the old-field mouse, *P. polionotus*, a species confined to the southeastern United States. Similarly, the mallard duck, *Anas platyrhynchos*, with Holarctic distribution appears paraphyletic in mtDNA genealogy (Fig. 6.5e) to the American black duck, *Anas rubripes*, which inhabits eastern North America only. These outcomes might be expected given the geographic distributions of the taxa involved. However, biological species boundaries are somewhat ambiguous in these taxonomic complexes. Some geographic populations of the deer mouse probably warrant sibling species status by several lines of genetic evidence (Hogan et al., 1993), and mallard and black ducks hybridize in nature to an uncertain degree with respect to introgressive consequences (Avise et al., 1990). Thus, it remains arguable as to whether genealogical paraphyly in these named species should be interpreted strictly as a result of incomplete lineage sortings across biological speciation events.

In another group of *Peromyscus* mice, mtDNA lineages in *P. hylocetes* proved to be nested within the broader lineage diversity of *P. aztecus* (Fig. 6.5f). In conjunction with other genetic and biogeographic evidence, this mtDNA pattern led the authors to conclude that *P. aztecus oaxacensis* (the subspecies that produced the paraphyletic outcome) probably should be recognized as a distinct species.

Biological versus Phylogenetic Species Concepts

More than 60 years ago, Dobzhansky (1937) began *Genetics and the Origin of Species* with "an observational fact more or less familiar to everyone . . . . the discontinuity of the organic variation." He extended earlier sentiments by Lamarck, Darwin, Wallace, and others who had identified an important role for reproductive isolation in the origin and maintenance of biotic discontinuities in the living world. This view of speciation as a juncture "at which the once actually or potentially interbreeding array of forms becomes segregated in two or more separate arrays which are physiologically incapable of interbreeding" (Dobzhansky, 1937) became known as the biological species concept (Mayr, 1940). In Dobzhansky's view, "biological classification is simultaneously a man-made system of pigeonholes devised for the pragmatic purpose of recording observations in a convenient manner and an acknowledgment of the fact of organic discontinuity." Throughout this century, the BSC has been the primary conceptual framework orienting research and discussion on the speciation process (Otte and Endler, 1989; Coyne and Orr, 1998).

Several alternative species concepts have been proposed in recent years (see discussions in Avise, 1994; Mayden, 1997; Howard and Berlocher, 1998). One that has risen greatly in popularity (Martin, 1996) deprecates or in the extreme disavows any relevance of reproductive isolation to speciation. Various formulations of a phylogenetic species concept (PSC) have been advanced (Rosen, 1979; Eldredge and Cracraft, 1980; Nelson and Platnick, 1981; Cracraft, 1983, 1987; Donoghue, 1985; Mishler and Brandon, 1987; Nixon and Wheeler, 1990; review in Hull, 1997), but all versions agree that species recognition should emphasize criteria of phylogenetic relationship (descent) and not reproductive relationships (de Queiroz and Donoghue, 1988). Sometimes, even a single synapomorph (shared-derived character) has been deemed sufficient to identify a monophyletic aggregate of individuals worthy of recognition as a phylogenetic species (McKitrick and Zink, 1988; Vogler et al., 1993).

A widespread perception exists of overt conflict between the PSC and the BSC. Proponents of the PSC often express reservations about the BSC that go beyond operational taxonomic issues to doubts about the biological relevance of RIBs to organic discontinuities (McKitrick and Zink, 1988; Cracraft, 1989; Frost and Hillis, 1990; Wheeler and Nixon, 1990). Others counter that the BSC remains useful and that no new species concepts are needed (Coyne et al., 1988; Coyne, 1992). Brief critiques (after Avise and Wollenberg, 1997) follow of three major criticisms leveled against the BSC in a recent review by Zink and McKitrick (1995).

(a) *Reproductive compatibility among populations is a shared primitive (rather than derived) feature, so it provides no criterion for identifying monophyletic units or clades. A serious potential problem of the BSC is the occurrence of paraphyletic, or nonhistorical groups.* As described above, most modes of biological speciation entail initial paraphyly for the ancestral taxon, both in population trees and gene trees. Nevertheless, a paraphyletic species remains a functionally relevant entity from the perspective of the BSC because "adaptive changes occurring in an individual or population may be extended to all members of the species by natural selection; they cannot, however, be passed on to different species" (Ayala, 1976). Furthermore, the paraphyletic status of a biological species likely will be transient as lineage sorting continues, or if the neophyte daughter species phylogenetically nested within it goes extinct. In any event, the phylogenetic content of species trees and their constituent gene trees are informative with respect to historical population demographic events associated with particular biological speciations. In this important sense, paraphyly cannot be equated with "nonhistory."

(b) *A focus on reproductive compatibilities and patterns of interbreeding can cause a misrepresentation of the significance of hybridization among differentiated taxa.* Varying levels of RIB development distinguish currently recognized taxonomic species. It is true that the Linnaean system of binomial nomenclature lends itself poorly to situations with intermediate levels of hybridization and introgression. However, it does so likewise for populations with intermediate levels of phylogenetic separation. Under the genealogical perspectives of coalescent theory, the mosaic histories of gene phylogenies (including introgressed lineages) in an organismal pedigree are of greater empirical content and conceptual import than is any necessarily simplified taxonomic summary of species status.

(c) *A long-recognized drawback of the BSC is its difficulty in ranking al-lopatric populations.* In the absence of sympatry, it is often difficult or im-possible to gauge the effectiveness of suspected intrinsic RIBs and hence to assess biological species status. However, similar grading difficulties can compromise species recognition under the PSC. For example, most al-lopatric populations in low-gene-flow organisms differ from others by at least some recently arisen character states, and a challenge remains of how to rank the differences. Also, demographic factors such as gene flow and effective population size exert overriding influence on phylogeographic pattern. Thus, a strict application of PSC criteria often will bias toward species-level recognition of smaller populations because these will be monophyletic with respect to close relatives more often than will large populations, all else being equal.

Diagnosability criteria have plagued previously formulated versions of the PSC. The resolving power already available in molecular assays of rapidly evolving genes (such as mtDNA or nuclear introns) means that re-cently derived mutations often can be found that distinguish local popula-tions, family units, and even individuals. Furthermore, because unlinked loci have independent transmission routes, they often register gene-tree clades that have the appearance of being mutually inconsistent. In the light of coalescent theory for sexually reproducing organisms, any ap-proach that promulgates species diagnosis on the basis of one or a few synapomorphs makes no biological sense.

A BSC/PSC RECONCILIATION

Some of the apparent philosophical distinctions between the BSC and a re-vised PSC can be reconciled within the framework of coalescent theory. This section will illustrate how considerations of multiple gene trees in a species tree can dispel the false notion of inherent conflict between the phylogenetic and reproductive underpinnings of biotic discontinuities.

In principle, any sticklike phylogenetic representation for separated populations or species may be examined under finer focus by reference to the multitudinous genealogical pathways within it. Avise and Wollenberg (1997) introduced an abstract concept for this purpose that involves the idea of gender-defined transmission pathways. An extended organismal pedigree can be viewed heuristically as a huge collection of historical transmission routes available for alleles. Each gender-defined pathway describes a unique, non-anastomose coalescent tree. For example, one such tree within an extended pedigree consists of the collection of transmission histories through alternating genders in successive generations: $F{\rightarrow}M{\rightarrow}F{\rightarrow}M{\rightarrow} \ldots F{\rightarrow}M$ (where F denotes female and M denotes male). Another genealogical tree is described by the collection of transmission pathways through a gender-based scheme $F{\rightarrow}F{\rightarrow}M{\rightarrow}M{\rightarrow} \ldots F{\rightarrow}F$, and so on.

No real autosomal gene exists whose alleles all will have traversed any one such sex-specified transmission route. Nevertheless, the concept of gender-defined transmission histories offers some epistemological advantages. First, such pathways bear direct analogy to the real-life gender-defined pathways for mtDNA (through females) and the mammalian Y-chromosome (through males). Second, they provide a theoretical common denominator with respect to effective population sizes of genomes, such that (for example) the fourfold increase in coalescent times otherwise expected for autosomal as compared to cytoplasmic gene trees is eliminated. Third, these gender-defined pathways permit explicit tallies of the number of different genealogical routes available for alleles within an organismal pedigree. That number is exactly $2^{(G+1)}$, where G is the number of generations considered. Because every individual has a mother and a father, the relationship between G and the number of gender-defined genealogies holds regardless of population size. Finally, if an organismal pedigree is known or specified, the coalescent tree for any gender-defined pathway within it becomes defined explicitly rather than probabilistically.

The concept of gender-defined lineages can be used to dissect an extended pedigree into component genealogical parts. These gender-defined coalescent trees in turn can be reassembled into a composite picture in a way that reconciles the notions of reproductive isolation and phylogeny in accounting for biotic discontinuities in nature. The approach is illustrated in Fig. 6.6, which portrays the complete pedigree underlying the species phylogeny that was displayed in sticklike fashion in Fig. 6.3b. Given such a pedigree (i.e., mating partners and their offspring in each generation), a coalescent tree for each gender-defined transmission route automatically becomes specified.

Consider, for example, the matrilineal pathway for mtDNA (Fig. 6.7, upper left). All extant females in taxon E trace genealogically through female ancestors to a shared progenitress at $t-5$, those in D coalesce at $t-9$, and those in the D + E assemblage stem to a common ancestor at $t-12$. The great-great-great . . . matrilineal grandmother of all extant individuals in the pedigree existed at $t-20$. With respect to the matrilines in the A–C complex (which coalesce at $t-11$), C1 is paraphyletic to A, and C2 is paraphyletic to B. This coalescent tree reflects the realities of allelic-level ancestry through heredity, as distinguished from empirical *estimates* of ancestry based on molecular or other data.

Three other gender-defined coalescent trees are depicted in Fig. 6.7. Comparisons among them illustrate three main points. First, the multitudinous coalescent trees within an extended pedigree can differ from one another in depth and pattern. For example, extant individuals in species E share a common ancestor in the patrilineal phylogeny at $t-19$ (Fig. 6.7, upper right), whereas they do so in an alternating-gender tree at $t-3$ (Fig. 6.7, lower left). Second, the topologies in such transmission genealogies can differ from the species-level or population-level phylogeny for reasons of historical stochastic lineage sorting. For example, sister species D and E do not display reciprocal monophyly with respect to patrilineal genealogy (Fig. 6.7, upper right), nor do they comprise a sister patrilineal clade to the A–C group. Third, lineage clades identified by synapomorphs in different allelic trees almost inevitably group sexually interbreeding individuals into overlapping arrays such that the phylogenetic units recognized by different transmission routes (and, therefore, different pieces of DNA) are neither mutually exclusive nor nested. For example, all extant C1 males

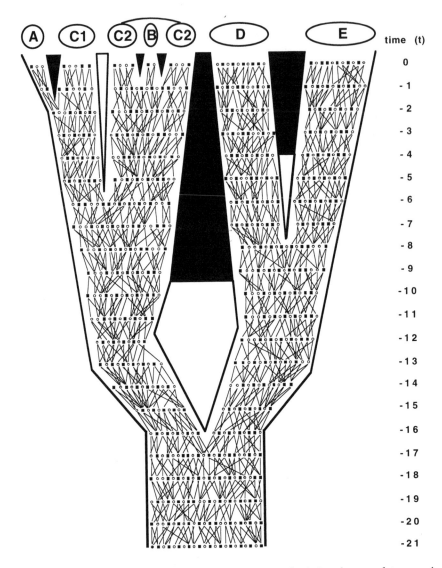

FIGURE 6.6   Same phylogeny as in Fig. 6.3 but here depicting the complete organismal pedigree through 21 discrete generations leading to the present (after Avise and
Wollenberg, 1997). The two lines tracing from each male (filled square) or female (open
circle) in any generation identify parents of that individual. They also describe the geographic dispersal of offspring (assumed to be limited by distance) and the mating
events.

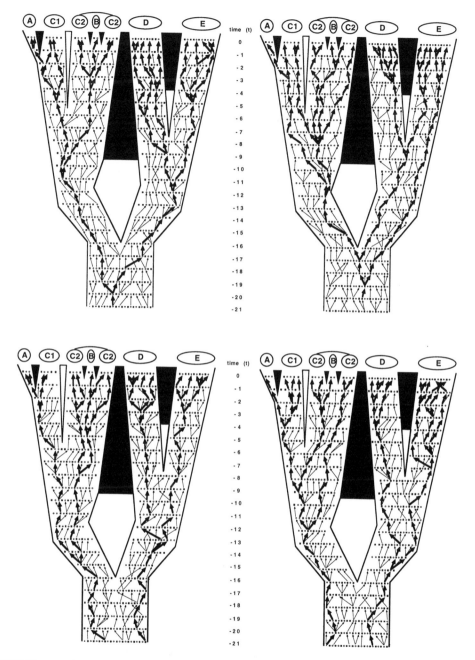

FIGURE 6.7 Identical phylogeny and pedigree to Fig. 6.6, but here with four different allelic transmission pathways (each describing a coalescent tree) highlighted by arrows (after Avise and Wollenberg, 1997). *Upper left,* matrilineal pathway reflecting the $F{\to}F{\to}F{\to}$ . . . transmission route (e.g., of mtDNA); *upper right,* patrilineal pathway reflecting the $M{\to}M{\to}M{\to}$ . . . transmission route (e.g., of the Y chromosome); *lower left,* generation-to-generation pathway through alternating genders $M{\to}F{\to}M{\to}F{\to}$ . . . ; *lower right,* the reciprocal of the latter, $F{\to}M{\to}F{\to}M{\to}$ . . . . Heavy arrows mark transmission routes (and coalescent trees) for lineages leading to individuals alive today, and light arrows mark the same gender-defined transmission routes that terminated before reaching the present.

form a clade in the allelic tree displayed in the lower right of Fig. 6.7, whereas they are allied variously to A or to C2 and B in the patrilineal genealogy (upper right).

The total number of gender-defined coalescent trees in this 21-generation pedigree is $2^{22} = 4,194,304$, and each describes a different genealogical history. However, the degree of coincidence or overlap across pairs of transmission routes varies widely. For example, the $F{\to}F{\to}F{\to} \ldots F{\to}M$ pathway is identical to the matrilineal pathway (Fig. 6.7, upper left) except for the most recent generation, where the transmission was to sons. This pathway describes the history of mtDNA in extant *males*. The reciprocal pathway ($M{\to}M{\to}M{\to} \ldots M{\to}F$) would describe the history of paternally transmitted family surnames as displayed by unmarried *females* in many human societies. All other 4,194,300 genealogical pathways would have been equally available to any piece of autosomal DNA trickling through the organismal pedigree under the rules of Mendelian inheritance.

This genealogical dissection of micro-phylogeny emphasizes why different pieces of DNA can have different histories within and between closely related biotas. This situation is an inevitable outcome of the quasi-independent transmission routes of alleles both within and across loci through the organismal pedigrees of sexual reproducers. Can these dissected gender-defined genealogies be re-amalgamated into a composite picture of organismal phylogeny, and would this consolidation of gene trees bear resemblance to the known species-level tree? Yes, as two approaches indicate (Avise and Wollenberg, 1997).

The first method compares topologies across coalescent genealogies. In general, a consensus tree depicts prevalent phylogenetic patterns across multiple characters or data sets. In the current context, three of the four coalescent trees in Fig. 6.7 portray reciprocal monophyly for D and E and also suggest that these species form a clade distinct from the assemblage A-C. The fourth coalescent tree (upper right) contradicts these patterns but is outvoted in a majority-rule consensus. Avise and Wollenberg (1997) extended this approach to many additional gender-defined coalescent trees through the known pedigree in Fig. 6.6. A comparison of the resulting consensus trees (Fig. 6.8) to the original phylogenetic representations (Fig. 6.3) yielded a salient conclusion: Notwithstanding a considerable variance among gender-defined coalescent trees in the pedigree, a compilation of many such trees properly recaptured the primary topological features of the population-level or species-level phylogeny.

This same conclusion was reached by a consideration of composite indices of genetic relatedness between extant individuals (Avise and Wollenberg, 1997). A coancestry (or kinship) coefficient is the probability that an allele randomly drawn from one individual is identical by descent (autozygous) within the pedigree to an allele drawn from another individual (Hartl and Clark, 1988). Its value is equivalent to an inbreeding coefficient for these individuals' hypothesized offspring. Coancestry coefficients are positive functions of the number of genealogical pathways connecting a pair of individuals to all ancestors in the pedigree, and inverse functions of the lengths of those pathways. A coancestry matrix was generated for all pairs of extant individuals in Fig. 6.6, and the values were clustered (Fig. 6.8c). The topology of the resulting phenogram was similar to that of the underlying population-level phylogeny (Fig. 6.6) upon which, ultimately, it was based. Thus, D and E appear as sister groups separated by a relatively deep node, C1 is paraphyletic to A, C2 is paraphyletic to B, and the A-C assemblage joins the D-E group at the deepest node.

How many genealogical pathways are needed to estimate major features in an organismal phylogeny? The general answer from computer simulations appears to be "rather few" (Wollenberg and Avise, 1998). Sampling properties of allelic transmission routes were examined through extended population pedigrees artificially sundered into reproductively isolated units. For each computer pedigree, a matrix of true coancestry coefficients among all individuals in the final generation was calculated and compared to mean pairwise times to common ancestry as estimated in random samples of allelic transmission pathways. As more genealogies were averaged into the analysis, the statistical recovery of organismal phylogeny improved (Fig. 6.9). The performance curves were asymptotic, showing diminishing returns per unit effort beyond about 5–20 allelic

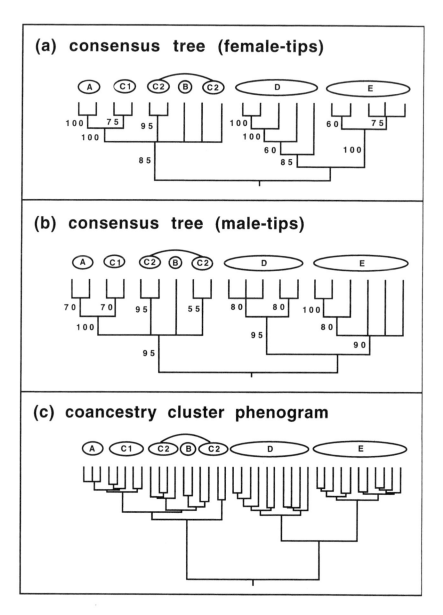

FIGURE 6.8 Composite phylogenetic summaries based on analyses of multiple allelic transmission routes through the organismal pedigree shown in Fig. 6.6 (after Avise and Wollenberg, 1997). *(a and b):* Consensus trees for multiple allelic genealogies whose terminal branches led, respectively, to the 19 extant females or to the 20 extant males. Each consensus tree was generated from 20 gender-defined coalescent trees sampled at random from the pedigree. Numbers indicate percentages of allelic trees in which a clade was identified. *(c):* Phenogram based on cluster analysis (Sneath and Sokal, 1973) of a matrix of coancestry coefficients for the 39 extant individuals in the pedigree. Note the close topological resemblance of these representations to the population- or species-level phylogeny (Figs. 6.3, 6.6).

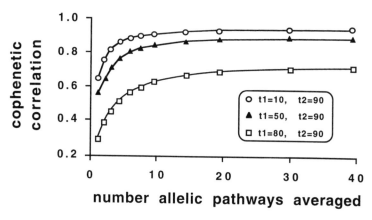

FIGURE 6.9   Sampling curves illustrating how well differing numbers of allelic transmission pathways randomly sampled from a sundered population pedigree perform in
permitting recovery of the true organismal phylogeny (after Wollenberg and Avise,
1998). The organismal phylogeny in these computer simulations involved a 100-
generation pedigree starting at time zero and extending to the present. The pedigree,
consisting of 150 individuals each generation, was sundered by two internal speciation
nodes at generation times $t1$ (oldest) and $t2$ (most recent). As additional allelic genealogies were averaged into the analysis, statistical recovery of the organismal phylogeny
(as measured by cophenetic correlations) improved asymptotically.

trees. In other words, in theory about 5–20 "mtDNA-analogous" (or "Y-
DNA analogous") genealogies for unlinked nuclear loci sufficed to summarize most of the information content in organismal pedigrees of this
size. The curves also depended, however, on population demographics
and the temporal spacing of speciation events. For example, when the
nodes in a species tree were farther apart and older in relation to effective
population size, smaller numbers of gene genealogies were adequate for
recovering the major features of an organismal phylogeny (Wollenberg
and Avise, 1998).

   These theoretical exercises of reamalgamating a population-level phylogeny from known lineage pathways involve some evident circularities
of reasoning. The phylogenetic information in a composite organismal
pedigree cannot differ fundamentally from a proper statistical compilation
of the multitudinous transmission pathways within it. Nonetheless, the

perspectives stemming from such heuristic exercises in coalescent theory are important because they carry the following ramifications for species concepts and speciation theory:

a. No species concept that results in an overly simplified caricature of organismal phylogeny can hope to capture the rich and varied fabric of genealogical histories in multiple pieces of DNA within an extended pedigree. An important challenge is to describe the statistical distributions of gene genealogies in particular instances, and to interpret properly the demographic and evolutionary processes that have shaped them.

b. As a sundering agent at the level of populations and species, extrinsic and intrinsic barriers to interbreeding remain key evolutionary agents motivating genealogical partitions at the level of composite allelic lineages. In effect, they do so by demarcating extended demographic units that are relevant to the lineage-sorting aspects of coalescent theory. Thus, reproductive barriers are important for species concepts (even within a strict phylogenetic framework) because they generate and promote through time increased genealogical depth and concordance across composite DNA transmission pathways. In other words, RIBs (the hallmark of the BSC) tend through time to demarcate phylogenetic bundles or braids of lineages that register important biotic discontinuities in the living world. It is no mere coincidence, for example, that biological species D and E as defined by reproductive criteria (Fig. 6.3) constitute recognizable assemblages of individuals in a historical or genealogical sense (Figs. 6.7, 6.8).

c. Conversely, phylogenetic considerations are important (even within the philosophical framework of the BSC) because they force explicit attention on historical and demographic aspects of the speciation process. For example, the genealogical paraphyly of C to A (Figs. 6.8a,b) and the high coefficients of coancestry between these two groups of individuals (Fig. 6.8c) jointly imply a recent and perhaps bottlenecked separation of A from C, as indeed was the case (Figs. 6.3, 6.6).

d. Concepts of phylogeny and reproductive isolation in speciation theory cannot be divorced from considerations of population genetics and demography. For example, only after reproductive ties have been severed for times (measured in generations) considerably longer than effective population sizes do deeper topologies in multiple DNA transmission pathways tend to come into congruence with one another and with the coarse-focus topologies of the population-level or species-level phylogenies that they comprise. Indeed, speciation under a strict phylogenetic framework could be viewed as an evolutionary process by which tangled twigs in shallow gene genealogies become resolved into organized, bundled branches with deeper separations and increasingly pronounced topological concordance.

NOMENCLATURE

The BSC and the PSC, like several other species concepts (Claridge et al., 1997), are mostly philosophical constructs that consider the ideals of species formation rather than the practical nuts and bolts of species-level taxonomy. No process-oriented theory of speciation exists that invariably can be translated into black-and-white taxonomic decisions about species status: some gray areas inevitably will be present. Under the BSC, RIB development may be partial rather than complete between populations at intermediate stages of evolutionary divergence. Similarly, phylogenetic separations under process-oriented versions of the PSC are a matter of degree, reflecting the fact that any organismal-level cladogram really is a "cloudogram" of genealogical histories with a variance (Maddison, 1997).

With regard to taxonomic guidelines, a compromise between the BSC and PSC is possible. Avise and Ball (1990) suggest that the category species should continue to refer, in principle, to reproductively isolated units. A retention of the philosophical framework of the BSC is warranted in no small part because RIBs generate and maintain biotic discontinuities ("genotypic clusters"; Mallet, 1995) recognizable as historical partitions in gene genealogies and organismal phylogenies. Unfortunately, allopatric populations will remain problematic because the acid test for biological species status in nature is reproductive isolation in sympatry. Also, some biological species (particularly those that arose under sudden-speciation scenarios) will have separated so recently that they may go unrecognized by phylogenetic partitions in most genes other than those underlying the RIBs themselves.

Within a suspected biological species, taxonomic subspecies should be demarcated by any pronounced and concordant phylogeographic partitions observed across multiple genetic traits. Thus, subspecies should conform to evolutionarily significant units as described in Chapter 5. Molecular information such as from gene sequences can contribute to the recognition of subspecies, but in practice concordant differences in genetically based morphological, behavioral, or other phenotypic attributes (Wilson and Brown, 1953) will remain important as readily scored taxonomic features for provisional subspecies assignments. However, levels of phylogenetic separation among isolated populations (as is also true for RIBs) are matters of degree often reflecting gradual accumulation with time. In intermediate situations (and also in hybrid settings), educated nomenclatural judgments will remain necessary at species and subspecies levels.

SPECIES REALITIES

A somewhat different issue is whether existing species-level taxonomies bear facsimile to real biotic discontinuities in the living world. One traditional approach asks whether different human societies perceive biotic units in similar fashion. For example, Mayr (1963; see also Diamond, 1966) found that preliterate peoples of New Guinea had vernacular names for 136 of the 137 native birds recognized as separate species by academic zoologists. Similar conclusions were drawn regarding the recognition of Amazonian trees by native peoples and Western botanists (Pires et al.,

1953). Such examples suggest that species perceptions are not overtly culture-dependent, at least for some taxonomic groups (Coyne, 1994).

Sexual Taxa. In a similar spirit, a new question can be asked: Do the distinctive phylogenetic units identified in molecular analyses bear close resemblance to the biotic discontinuities registered as taxonomic species in traditional morphological appraisals? By the evidence of mtDNA, many animal species (all taxa in phylogeographic Category I) are subdivided into multiple distinctive phylogeographic units (Chapter 4). On the other hand, the number of major (deep) intraspecific units typically is small: only 2–7 in available surveys of most recognized species (Fig. 6.10). Furthermore, even sister species usually appear highly distinct in mtDNA composition (e.g., Avise and Saunders, 1984; Avise and Zink, 1988; review in Johns and Avise, 1998a). These observations suggest that currently named species (by whatever operational criteria they were recognized) often agree rather well in number and composition with biotic entities that might be catalogued by scrutiny of mtDNA genealogies alone. This compatibility of outcomes probably reflects an underlying historical reality to many of the biotic discontinuities traditionally recognized as species.

This conclusion currently applies with greatest force to well-studied vertebrates in temperate regions, where molecular phylogeographic efforts have been concentrated. Outcomes may prove to differ elsewhere, such as in some invertebrates or in tropical biotas. Wake (1997) suggested that repeated patterns of phylogeographic differentiation in several faunas of the eastern United States (Chapter 5) might be due to Pleistocene-mediated range restrictions and population extinctions that "sharpened borders between groups of populations and heightened the genetic cohesion of units." He also suggested that such patterns might be less apparent in areas with more complex geological and climatic histories, such as parts of western North America. In the future, phylogeographic studies of additional taxonomic groups and regions should be conducted to test more broadly whether pronounced gaps in molecular genealogies match those

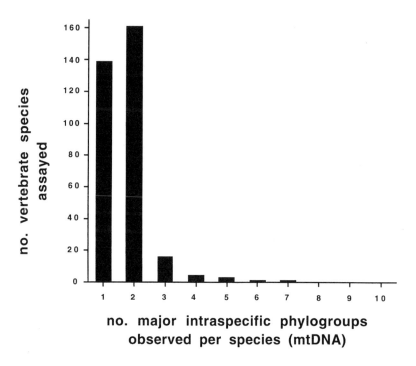

FIGURE 6.10   Histogram of the number of distinctive phylogeographic units per species in mtDNA surveys of 325 named vertebrate species (updated and modified from Avise and Walker, 1999).

presumed to be registered in existing species-level (and subspecies-level) taxonomies.

In conducting such tests and interpreting current evidence, opposing biases arise. Limited geographic and genomic sampling may cause serious underestimates of the numbers of principal intraspecific phylogroups within a taxonomic species. On the other hand, most species surveyed to date were chosen for phylogeographic study because they have large ranges whose histories of occupancy are of special interest. If broadly distributed species are especially prone to consist of multiple historical units (as seems likely), then the estimated numbers of phylogroups per taxonomic species may be biased upward in the available literature. Also, some major mtDNA phylogroups (particularly in species with male-biased dispersal) may not register genome-wide population subdivisions. For this reason also, the number of salient intraspecific units as estimated by mtDNA data could be too high.

Asexual Taxa.   A long-standing conundrum in systematics is how to account for discreet biotic units (i.e., taxonomic species) often suspected in predominantly clonal organisms such as many microbes (Embley and Stackebrandt, 1997; Goodfellow et al., 1997). If truly asexual, these species by definition are not bonded genetically by interbreeding. The issues become complicated by the fact that many such organisms possess mechanisms of genetic exchange (e.g., conjugation, transformation, and transduction in bacteria) that may contribute, probably in conjunction with strong ecological selection pressures, to the genetic unities within and divergences between recognized taxa (Maynard Smith and Szathmáry, 1995; Coyne and Orr, 1998; Cohan, 1999). Nevertheless, historical demographic perspectives prompted by genealogical studies of mtDNA in higher animals raise another intriguing possibility for the perceived taxonomic and biotic patterns in many predominantly asexual forms. Perhaps the unity of a clonal "species" stems from tight vertical connections in phylogeny, due to a recent population expansion from relatively few founders, for example. And, perhaps a species' apparent discontinuity from other clonal taxa may be a byproduct of the extinction of intermediates.

Ironically, this suggestion is prompted by mtDNA patterns observed in sexual species (or evolutionarily significant units within them). There, mtDNA often displays a genetic "cohesiveness" despite the fact that clonal matrilines are connected to one another by historical genealogical ties alone. Recall that mtDNA gene trees within even the most abundant of high-gene-flow species (such as marine copepods) display coalescent depths that are shallower by orders of magnitude than might be expected given current census population sizes. This suggests that evolutionary histories in these species have included severe population contractions (or mtDNA-specific selective sweeps) that squeezed surviving molecular lineages through relatively small numbers of females prior to the more recent population expansions. A related notion is that pronounced genealogical gaps among extant species (and intraspecific phylogroups) also reflect such demographic fluctuations.

A recent case in point involved molecular-genealogical assessment of an abundant crop plant (Eyre-Walker et al., 1998). A 1,400 bp region of the *Adh-1* gene was sequenced in domestic maize (*Zea mays mays*), its presumed progenitor (*Z.m. parviglumis*), and a distant wild relative (*Z. luxurians*). Statistical analyses of the data based on coalescent theory indicated that the genealogical diversity in cultivated maize is consistent with a ten-generation population bottleneck of about 20 founding individuals during a domestication process that took place roughly 7,500 years ago in southern or central Mexico.

A similar molecular genealogical appraisal of an important human microbial parasite, the malarial agent *Plasmodium falciparum*, illustrates another extreme version of the "demographic sweep" scenario (Rich et al., 1998). Sequence analyses of several genes from populations around the world revealed an acute paucity of (presumably neutral) synonymous nucleotide substitutions, suggesting that this parasite spread globally within the last few thousand years from a single ancestral strain probably in tropical Africa. This demographic sweep may have been facilitated by increased association of the mosquito vectors with humans, the spread of agriculture, and worldwide climatic changes following the last glaciation. Furthermore, the recent global expansion of *P. falciparum* and the lack of time for coevolution with human hosts may account in part for the organism's exceptional virulence (Rich et al., 1998). This microbe has an obligate sexual stage in the life cycle, but similar kinds of demographic sweeps may apply to many clonal or semi-clonal microbial forms as well.

The deduced demographic fluctuations in *Zea mays* and *Plasmodium falciparum* might be unusually pronounced, but the broader point is that apparent patterns of genetic cohesion within and discontinuity among many species of multicellular animals and plants may have been sculpted by demographic impacts on historical genealogy as much or more so than by contemporary reproductive associations *per se.* If this interpretation is correct, then by extension similar kinds of historical demographic factors also may account for perceived biotic discontinuities in many taxa that are predominantly asexual. For example, disease agents such as many viruses and bacteria often spread explosively from founder sources, and these nonequilibrium population demographies should leave genealogical signatures of heightened genetic cohesion within extant strains and sharpened borders between them.

In microbes with short generation lengths, population reductions need not be as drastic in absolute numbers as in most higher animals to produce comparable genealogical effects through neutral lineage sorting. Compare, for example, a bacterium and a mammal whose generation lengths are 20 minutes and 10 years, respectively. The bacterial population undergoes about 260,000 rounds of lineage sorting for each such round in the mammalian population. Hence, all else being equal, the microbial population can be that much bigger and still yield a similar sidereal time for genealogical coalescence. This statement holds in theory. An empirical molecular pattern would depend on additional considerations such as whether DNA sequences diverge in step with sidereal time, generational time, or other factors.

The generality of demographic-sweep scenarios for genealogical patterns in asexual (and sexual) taxa is a subject for further evaluation using phylogenetic appraisals of appropriate molecular markers. Indeed, a current challenge in molecular epidemiology is to reconstruct genealogical connections among strains of particular biological disease agents and to deduce the relative impacts of potential impinging evolutionary forces. These forces include reproductive mode and agents of natural selection in addition to historical population demography itself (e.g., Arbeit et al., 1990; Tibayrenc et al., 1990, 1991; Carpenter et al., 1996; Oliveira et al., 1998; Zhu et al., 1998). The genealogical task should be simpler in strictly clonal taxa (if such exist) because all genes in an asexual pedigree are cotransmitted through one pathway. Thus, any selective sweep on a gene would be a population demographic sweep, and a population demographic sweep from any cause would impact all gene trees together.

CONCLUSIONS ABOUT THE BSC/PSC DEBATE

For sexual organisms, biological speciation lies at a pivotal boundary where a partially braided collection of allelic pathways of interbreeding individuals bifurcates into two such collections. This boundary between reticulate and divergent relationships (Hennig, 1966) also demarcates areas of inquiry normally associated with population genetics and phylogenetic biology. The PSC has roots in the field of systematics, but as applied at microevolutionary levels has ignored principles of population genetics (at its peril). Conversely, the BSC has roots in population genetics, but now might profit from an infusion of appropriate phylogenetic considerations to illuminate previously neglected elements of demographic and genealogical history over microevolutionary timescales.

Coalescent theory has shown that to cleanse from species concepts all reference to reproductive isolation would be to leave an unduly sterile epistemological foundation for the origin and maintenance of biotic discontinuities in sexually reproducing organisms. In a sense, reproductive boundaries are important because they demarcate extended populations whose particular demographic histories influence phylogenetic patterns. If concepts resembling the BSC had not existed throughout this century, they now would demand invention. Conversely, to cleanse from species concepts all reference to demographic and genealogical history would be to leave an unduly sterile framework for interpreting the phylogeographic origins of biotic discontinuities (this sentiment applies also to asexual taxa). Thus, the reproductive, demographic, and genealogical aspects of the speciation process are intertwined and mutually illuminating.

Speciation Durations

An overwhelming conclusion from the mtDNA literature is that major matrilineal phylogroups often observed within extant species typically show a strong geographic alignment. Furthermore, the spatial distributions as well as the genealogical depths separating intraspecific phylogroups usually appear to be consistent with historical biogeographic influences of the late Tertiary and Quaternary. The issue to be addressed now concerns temporal durations of the speciation process as estimated from molecular genealogical analyses of extant species.

The Pleistocene Epoch, beginning about 2 Mya, was a time of extraordinary oscillations in climate (Berger, 1984). Pronounced global cooling on a 100,000 year cycle spawned continental glaciers that extended far into northern Europe (to 52°N) and North America (to 40°N). Climatic warming with conditions more like those of the Recent Epoch (last 10,000 years) periodically interrupted the Ice Ages. Climatic sub-oscillations were nested within the major cycles, and similar lesser fluctuations probably occurred in the Tertiary as well. The effects of such climatic changes on the geographic ranges of species were profound (Webb and Bartlein, 1992; Pitelka et al., 1997) as were, apparently, Pleistocene influences on the genesis and distributions of phylogeographic units within many currently recognized species.

If speciation normally proceeds through an intermediate stage of population divergence into distinctive phylogeographic units, then the evolutionary times of phylogroup separation place lower bounds on the temporal duration of the typical speciation process (Fig. 6.11). Upper bounds can be estimated from separation times for extant pairs of sister species (taxonomic species thought to be one another's closest living relatives). Thus, central issues for speciation durations are the evolutionary separation dates for intraspecific phylogroups and for extant sister species.

BIRDS

Conventional wisdom states that Pleistocene climatic cycles precipitated a large proportion of speciation events between extant sister species of birds, particularly in mid- and high latitudes (Rand, 1948; Mengel, 1964; Selander, 1971; Gill, 1995). This paradigm recently was challenged by Klicka and Zink (1997), who in a review of mtDNA sequence divergences

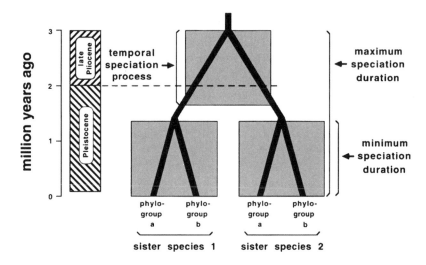

FIGURE 6.11   Phylogeographic approach for assessing temporal durations of the speciation process from genealogical data on extant taxa. Separation times (as inferred for example from net mtDNA sequence differences) for intraspecific phylogroups and for sister species provide minimum and maximum estimates, respectively, of speciation durations. In the available literature, these intra- and interspecific estimates often come from molecular studies on different taxa (rather than from the same organisms as implied in this representation). Under a view of speciation as an extended temporal process rather than a point event in time, many vertebrate speciations initiated in the Pliocene probably were not completed before the Quaternary (see text).

among North American songbirds found that only 11 of 35 assayed pairs of sister species (31 percent) dated to Quaternary separations under a conventional mtDNA clock (Fig. 6.12). The remaining species displayed large genetic distances indicative of a protracted history of various speciations over the past 5 My. The authors concluded: "The most recent glaciations were not, it seems, the force driving songbird diversification so much as they functioned as an ecological obstacle course through which only some species were able to persist. The entrenched paradigm proclaiming that many North American songbird species originated as a consequence of these glaciations is flawed."

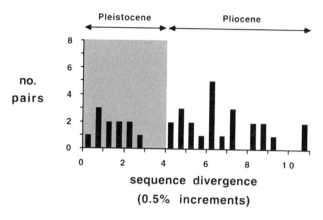

FIGURE 6.12   Histogram of estimated mtDNA sequence divergences (and inferred separation times) between sister-species pairs of North American songbirds (after Klicka and Zink, 1997).

Avise and Walker (1998) reexamined these conclusions in the light of similarly inferred separation times for major phylogroups within extant avian species. Recall that 76 percent of these intraspecific phylogroups in birds dated to the Pleistocene by mtDNA evidence (Fig. 5.24). This percentage is significantly higher than that for avian sister species, most of which date to the Pliocene (Fig. 6.12). This difference makes biological sense: if species separations tend to be older than population separations, then species-level divergences often must have been initiated prior to the Quaternary to accommodate (timewise) suspected Pleistocene effects on the phylogeographic architectures of conspecific avian populations.

Avian speciations probably often proceed through intermediate stages of phylogeographic differentiation, so inferred separation times between intraspecific phylogroups can be thought of as a mean "correction factor" (Fig. 6.11) to be subtracted from estimates of sister-species separation times. When this correction is applied to mtDNA distance estimates in extant birds, about 14 sister species pairs are "bumped" sufficiently to the left in Fig. 6.12 such that their inferred separations fall within the Pleistocene, and 10 other sister-species pairs with previously inferred mid- or late-Pleistocene origins now appear to have separation times indistinguishable from zero years bp. This does not mean that speciations were initiated at these later times, but it does suggest that the speciations were not completed much before then.

These results can be interpreted as follows. If Pliocene vicariant events on avian populations were at least as effective as those operative during the Pleistocene (as suggested by Klicka and Zink, 1997), then many species entering the Pleistocene Epoch already would have been separated into distinctive intraspecific phylogroups, as are many extant bird species today. Such units would be likely candidates for subsequent evolutionary divergence during the Quaternary, eventually achieving the status of sister species in today's taxonomies. Thus, when avian speciation is appreciated as a gradual process rather than a point event in time, then Quaternary biogeographic factors can be seen as important in promoting extensions of phylogeographic differences that often had been inaugurated earlier.

These data also permit quantitative estimates of the temporal durations of avian speciations. Median evolutionary times associated with sequence divergences between major intraspecific phylogroups (1.1 My; Fig. 5.24) and between sister species (2.8 My; Fig. 6.12) can be interpreted as minimum and maximum durations, respectively, of the speciation process. The midpoint between these values, about 2.0 My, supports the notion that population divergence leading to species-level taxonomic recognition in birds often entails substantial evolutionary time.

An observer 2 My hence might view the Pleistocene as a time of active population differentiation leading to many sister species, but this will depend primarily on whether environmental conditions over the next 2 My are conducive to fostering the continuance and further evolutionary divergence of separated intraspecific phylogroups so evident in many of today's avifauna. Such sliding-window perspectives on the temporal framework of biological differentiation apply with equal force to the past.

When viewed from contemporary time, the Pleistocene can be understood to have played a primary role in sponsoring phylogeographic differentiation in many avian species and also in further sculpting incipient, Pliocene-origin phylogeographic variety into extant forms recognized now as sister species.

MAMMALS

Similar assessments of speciation durations from mtDNA have been conducted on mammals (Avise et al., 1998). Recall that based on a standard mtDNA clock, more than 70 percent of the major intraspecific phylogroups within assayed mammals date to Pleistocene separations (Fig. 5.26). By this same criterion, a significantly smaller fraction of mammalian sister-species pairs (25 percent) also date to the Pleistocene (Fig. 6.13a). Median evolutionary times associated with sequence divergences between the phylogroups and between sister species of mammals are about 1.2 and 3.2 My, respectively. The midpoint between these two estimates, 2.2 My, gives a sense of the duration of a typical mammalian speciation as gauged by examination of extant taxa. This estimate is close to that for birds (2.0 My).

OTHER VERTEBRATE GROUPS

Similar exercises have been applied to the poikilothermic vertebrates (Avise et al., 1998), with an added caveat that mtDNA rates are known less securely in many of these groups. If we assume the standard mtDNA clock for homeotherms and follow the same analysis procedures as above (using data summarized in Figs. 5.27, 6.13b, and 6.13c), then typical speciation durations for assayed herpetofauna and for fishes are about 2.3 My and 1.7 My, respectively. These estimates are close to those for the mammals and birds.

On the other hand, lower rates in mtDNA evolution are suspected for some poikilothermic taxa (Avise et al., 1992c; Martin et al., 1992b; Martin and Palumbi, 1993; Rand, 1994; Mindell and Thacker, 1996). The use of suggested slower clock calibrations extends the estimates of speciation durations greatly. For example, under a fourfold slower mtDNA

sister species

(a) mammals

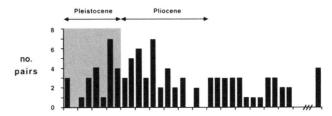

(b) amphibians and reptiles

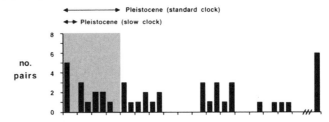

(c) fishes

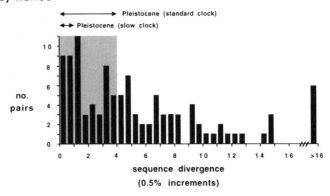

FIGURE 6.13   Histogram of estimated mtDNA sequence divergences (and inferred separation times) between sister-species pairs of nonavian vertebrates (after Johns and Avise, 1998a).

evolutionary rate, speciation durations become about 9.2 and 6.8 My for herpetofauna and fishes, respectively. The broad range of speciation-duration estimates highlights the need for further critical study of molecular evolutionary rates within and among the vertebrates (and other groups).

Comparative compilations of this sort are intended to identify central tendencies, trends, and research challenges that otherwise might be less apparent. They do not supersede focused appraisals of separation times and speciation durations for particular taxa. The primary literature documents considerable heterogeneity in tempo and mode of evolutionary divergence and speciation across vertebrate lineages. Two well-studied examples with radical departures from the phylogeographic norm for vertebrates can suffice here to illustrate this large variance (Fig. 6.14). These studies are directly comparable because both entailed analyses of several hundred base pairs from the cytochrome *b* gene and adjoining mtDNA sequences.

First, the dramatic proliferation of cichlid fishes in Africa's Lake Victoria was examined from the perspective of mtDNA sequences (Meyer et al., 1990). Nearly 200 endemic species are recognized within Victoria, yet the lake system is less than 1 My old. The 14 species genetically assayed (in 9 genera) proved to be nearly identical in mtDNA sequence. This finding in conjunction with other evidence was interpreted to indicate that the entire species flock probably arose within the last few thousand years, a time-frame far shorter than characteristic for most phylogroup separations *within* other vertebrate species. Evolutionary radiations in several other piscine lineages in closed lacustrine environments also imply speciation durations much less than 0.3 My (McCune, 1997; McCune and Lovejoy, 1998). In general, organismal groups that spawn species rapidly (sometimes under sympatric conditions; Schliewen et al., 1994) may have speciation durations far shorter than the vertebrate mean (Givnish and Sytsma, 1997).

Near the other end of the spectrum, genetic differences between salamanders of the *Ensatina eschscholtzii* complex indicate periods of conspecific

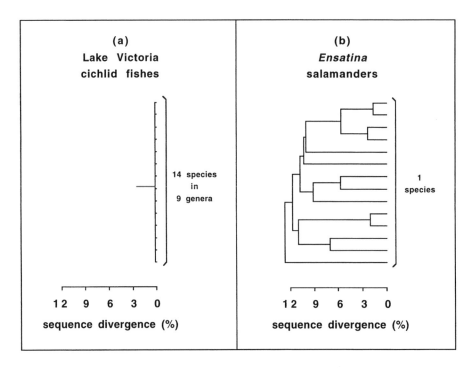

FIGURE 6.14   Contrasting phylogeographic and speciational patterns for: *(a)* cichlid fishes within Lake Victoria (from data in Meyer et al., 1990), and *(b)* a complex of *Ensatina eschscholtzii* salamanders in California (representative of the broader data set in Moritz et al., 1992). Both studies involved comparable molecular assays of mtDNA, and genetic distances here are plotted on the same scale.

population isolation far longer than most vertebrate speciation durations. This "ring species" encircles the Central Valley of California and its populations have been thought to illustrate various stages in the speciation process. Moritz et al. (1992) and Wake (1997) documented huge genetic distances in mtDNA sequences as well as allozymes, and concluded that these register periods of population isolation and differentiation often in excess of 5 My.

   Thus, the contrast between phylogeographic patterns in the cichlid fishes of Lake Victoria and the *Ensatina* salamanders of California could hardly be more dramatic (Fig. 6.14). The biological world is richly diverse in phylogeographic patterns and speciational modes.

DEEP PHYLOGEOGRAPHY

Gene Trees in Ancient Phylogenies

Earlier, it was shown that gene trees in recently separated taxa can differ topologically from one another and from the composite population-level or species-level phylogeny for reasons (among others) of stochastic sorting of ancestral lineages. At first thought, such discordances might seem not to apply to species that separated in ancient time because lineage sorting to reciprocal monophyly in the descendant taxa certainly would have occurred by now. However, the salient issue is whether lineage coalescence took place within the timeframe of the relevant internodes in the species phylogeny. Especially when nodes in a species tree were temporally close relative to effective population sizes at that time, lineage polymorphisms may traverse multiple nodes only to sort idiosyncratically to fixation later on. In such cases, the topologies of gene trees can differ from a species tree (as well as from one another in sexual lineages) at any evolutionary depth. In other words, lineages from an ancestral gene pool that reach fixation in descendant taxa may by chance be those that produce topological discordance between a gene tree and a species tree (Takahata, 1989; Tateno et al., 1989; Wu, 1991).

This concept is illustrated in Fig. 6.15. There, ancient taxa H and I are phylogenetically allied to G in the species tree, but they are genealogically closer to J in the gene tree pictured. This outcome reflects the retention of lineage polymorphisms by ancestral population B across the internodal span t_1 to t_2, followed by the illustrated fixation of these two gene lineages in the descendants G, H, and I. Note that this discordant pattern is locked into place indefinitely provided that the relevant taxa (or their posterity such as K, L, and M at any later time) remain extant. The probability of such a gene-tree discordance due to stochastic lineage sorting across adjacent nodes is $P = 2/3e^{-T/2N_e}$ (Nei, 1987), where e is

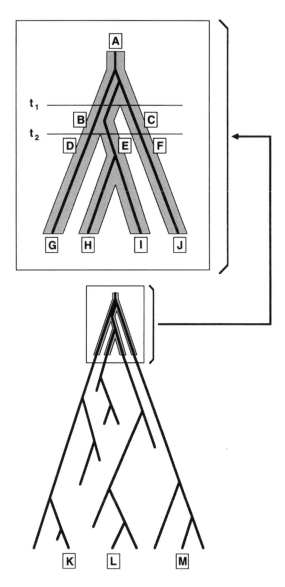

FIGURE 6.15   Diagrammatic representation of how a gene tree and species tree can differ topologically in deep as well as shallow phylogenies (see text).

the base of the natural logarithms, N_e is the effective population size and $T = t_1 - t_2$. For example, suppose that in a complex of copepod species, $t_1 = 15{,}000{,}000$, $t_2 = 5{,}000{,}000$, and $N_e = 5{,}000{,}000$. Then, the probability of a topological conflict between a species phylogeny and a gene tree that traversed the relevant node is about $P = 0.25$. P can be interpreted also as the expected proportion of neutral gene trees that differ in topology from a species tree in sexual reproducers. In general, P is large when $2N_e > T$.

A common practice in molecular systematics has been to estimate a species tree from DNA sequences of a single gene. For such exercises, perspectives from coalescent theory carry cautionary notes that go beyond the conventional concern of how reliably a gene tree itself is recovered from available data. Even a perfectly correct gene tree is likely to conflict with a species tree when nodes in a species phylogeny are close in time.

Furthermore, the likelihood that a species tree is mirrored faithfully in genealogical (or other) data is influenced greatly by the population *demographic* histories of the species involved (an important consideration seldom appreciated in traditional phylogenetic discussions). Molecular phylogeneticists usually have assumed that particular orthologous gene sequences, properly analyzed, should recover higher-level species trees accurately. However, this can be true (at best) only when $2N_e < T$. Given the large census sizes of many species (marine copepods, for example), this state of affairs should not be accepted universally on faith.

These points underscore an irony in the history of systematics. Whereas vast effort has been expended on developing preferred phylogenetic algorithms, historical population demographic factors seldom have been addressed in phylogenetic reconstructions at higher taxonomic levels. Thus, the important connections suggested between demography and phylogeny via studies of mtDNA gene trees in microevolution have gone virtually unnoticed in the broader field of phylogenetic biology.

On the other hand, it seems to be an empirical generality that topologies of gene trees often agree rather well with suspected topologies of higher-level species trees as deduced by independent evidence. Perspectives stemming from intraspecific phylogeography may help to explain this perhaps surprising outcome. As emphasized earlier, effective population sizes of species and intraspecific phylogroups often are vastly smaller than contemporary census sizes (particularly in now-abundant species). Thus, large fluctuations in population size probably have characterized most extant species in recent evolution. It is reasonable to assume that species in the more distant past also fluctuated dramatically due to occasional disease outbreaks, climatic or other environmental challenges, and so on. If so, N_e values along branches of macro-phylogenetic trees may be small in relation to the internodal times (T) between most speciations. Thus, perhaps empirical gene trees from appropriate-rate loci should (after all) tend to agree in topology with one another and with true species trees. Yet, given our considerable ignorance on historical demography and multi-locus genealogy for most species, such statements are conjectural at present.

Consider, for example, the ongoing debate about early branches in the tree of life. For more than two decades, sequence differences in rRNA genes have been interpreted to mark the existence of three principal phylogenetic domains: Eukarya including all plants and animals, and Archaea and Bacteria as two distinct microbial lineages (Fox et al., 1977; Woese and Fox, 1977; Woese et al., 1990). However, recent molecular studies have challenged simple versions of this picture by suggesting that the topologies of different gene trees sometimes are in conflict in these major groups (Pennisi, 1998; Ribeiro and Golding, 1998).

One general possibility is that the surveyed genes evolved at different rates or patterns such that some of them simply provide poor, misleading, or uncertain phylogenetic resolution of ancient history (see Hillis et al., 1994; Schluter, 1995). Another view is that the discordances reflect early horizontal gene movements from branch to branch of the organismal tree, as for example via genetic transformation following food ingestion (Doolittle, 1998). A seldom-considered third possibility, not incompatible with the others, involves population-demographic influences on the coalescent. Could it be that some fraction of the apparent genealogical discordance (in taxa with genetic recombination) is due to idiosyncratic, gene-specific outcomes of stochastic lineage sorting across ancient nodes? This might be especially likely if some microbial populations were extremely large and internodal times in the organismal tree were relatively small.

In principle, such demographic possibilities would only complicate further the already difficult task of recovering the exact topology of phylogenetic trees whose ancient nodes are relatively close in time when viewed from the present. The traditionally suspected Cambrian radiation of animal phyla provides another difficult case in point (Bromham et al., 1998). Thus far, molecular data have been less than definitive in revealing exact phylogenetic relationships of several extant groups that separated near the base of the Metazoan tree (Raff et al., 1994). Much work remains to address biological as well as methodological possibilities for why such empirical problems have seemed so refractory.

Phylogeography Beyond the Species Level

Higher taxa too are linked by phylogenetic ties and have geographic distributions. Thus, phylogeographic analyses by definition can be extended to deeper evolutionary timeframes. Advantages and disadvantages apply to higher-level molecular phylogeographic appraisals. On the plus side, older nodal dates give greater latitude in choice of informative genes and appropriate methods of molecular assay (Avise, 1994). On the other hand, longer timespans also mean more opportunity for complicating factors in particular instances. For example, molecular or other evidence for an ancient vicariant separation might be blurred by multiple instances of secondary dispersal and speciation in the derivative clades.

Phylogenetic perspectives long have been central to biogeographic appraisals above the species level. Thus, the introduction of molecular gene-tree methods has not revolutionized the conceptual orientation of higher-level systematics to the same extent that it has population-level studies. Nonetheless, molecular assays provide a bonanza of empirical data for phylogeographic assessments of higher taxa. Many extended treatments of historical biogeography are available (Brown and Gibson, 1983; Taylor, 1984; Hengeveld, 1990; Cox and Moore, 1993; Ricklefs, 1993;

Tivy, 1993; Morone and Crisci, 1995; Rosenzweig, 1995; Briggs, 1996; Brown and Lomolino, 1998). This section will illustrate typical goals of such studies, and will note some of the strengths and weaknesses of these approaches.

VICARIANCE, DISPERSAL, AND ARRIVAL TIMES

Vicariance and dispersal often are dichotomized as explanations for biotic distributions. However, dispersal barriers arise and disappear repeatedly in evolution and may have different effects depending in part on species' ecologies and natural histories. In the simplest case under vicariance, an area cladogram (the history of subdivision of a geographic region) accurately predicts the phylogenies of all endemic taxa. Departures from this expectation are routine, in part because the biota of a geographic archipelago often consists of multiple components with separate histories, some of which are branching and some reticulate (Enghoff, 1995; Ronquist, 1997). Molecular data can help weigh vicariance and dispersal hypotheses for a given biota by contributing to knowledge on both the branching orders and nodal times in phylogenetic trees superimposed on geography.

The vertebrate fauna of the West Indies is comprised of more than 1,200 species, many of which are confined to particular islands. Taxonomic alliances are to other island endemics or to mainland counterparts in North, Central, or South America. Much debate has centered on the time, mode, and site of origin for various island biotas, with two opposing views prevalent. Under vicariance scenarios, many extant species are descendants of early colonizers who arrived shortly after the islands separated from continental landmasses in the late Cretaceous, about 80 Mya. Alternatively, overwater dispersal might account for current faunal distributions. The vicariance hypothesis predicts ancient lineage separations between present-day island and mainland taxa, whereas dispersalist scenarios predict more recent, perhaps variable dates of lineage separation. Molecular data for 38 pairs of terrestrial vertebrates proved mostly consistent with dispersalist explanations (Hedges et al., 1992b): Genetic distances (in the albumin molecule) showed a large variance across cognate pairs of mainland and island forms, and inferred dates of lineage separation greatly postdated the relevant vicariant events otherwise hypothesized to have separated most of the respective taxa.

Three general insights on the origins of West Indian vertebrates came from a synthesis of traditional biogeographic data and further molecular evidence (Hedges, 1996). First, nearly all independent taxonomic lineages currently on the islands arrived via dispersal during the Cenozoic. Second, South America was the source for most lineages of nonvolant organisms (notably amphibians and reptiles). This outcome can be explained by the nearly unidirectional flow of ocean currents from southeast to northwest, perhaps transporting flotsam (with occasional animal hitchhikers) from the mouths of South American rivers to the West Indies. Third, North America or Central America was the source for most lineages in volant groups (birds and bats), as might be expected based on historical proximity of the islands to these continental regions.

Another example of deep phylogeography addressed by molecular data involved characiform fishes, consisting of about 1,200 extant species in 16 taxonomic families restricted to Africa and South America. Species vary greatly in size and display a great array of trophic specializations. Despite their current confinement to freshwater habitats (making marine dispersal unlikely), morphological evidence suggests that African and Neotropical forms are not reciprocally monophyletic. This suspicion was confirmed in recent mtDNA sequence analyses (Ortí and Meyer, 1997). In assays of slowly evolving rRNA genes, three trans-Atlantic clades were identified (Fig. 6.16), one of which involved predatory forms with ambush lifestyles (pike-like *Hepsetus* in Africa and bowfin-like *Hoplias* in South America). Nevertheless, all intercontinental genetic distances were large by rDNA standards and interpreted as consistent with the ancient (90 Mya) vicariant separation of Africa from South America by continental drift. Two broader messages also emerged from appraisals of the 52 total species included in the survey: Appropriate DNA sequence data (more so than morphological characters) can reveal the temporal framework as well as cladistic relationships within a phylogeny, and recovery of the exact

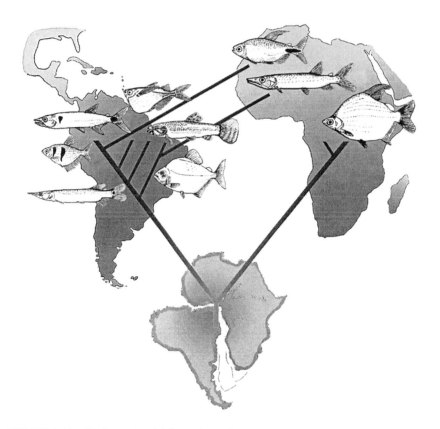

FIGURE 6.16   Phylogeographic hypotheses for characiform fishes in Africa and South America (after Ortí and Meyer, 1997). Three trans-Atlantic sister-group relationships are depicted, one of which involves predatory pike-like forms in Africa and bowfin-like forms in South America. Phylogenetic appraisals of mtDNA sequences support the branching order depicted, but also are consistent with the idea that all phylogenetic connections between the continents predated the opening of the South Atlantic Ocean by continental drift some 90 Mya (see text).

branching order in deep phylogenies can be compromised by homoplasy in DNA sequences due to superimposed nucleotide substitutions and saturation effects.

At least two other speciose groups of circumtropical freshwater fishes have been studied from the perspective of mitochondrial rDNA sequences: Cichlidae (Farias et al., 1999; see also Zardoya et al., 1996; Streel-man et al., 1998), and the Aplocheiloidei complex of killifishes (Murphy and Collier, 1997). Molecular phylogenies in both groups showed similari-ties in phylogeographic topology that appear to be consistent with ancient vicariant separations associated with the breakup of Gondwanaland (Fig. 6.17). Thus, for both the cichlids and killifish, the basal position of an Indian-Madagascar molecular clade probably evidences the detachment of this subcontinent from Africa–South America beginning some 150 Mya, and African–South American dichotomies were interpreted as reflective of the separation of these continents starting nearly 100 Mya.

Molecular studies of other circumtropical groups likewise have ad-dressed the vicariant role of Gondwanaland fission versus secondary dis-persal in accounting for the presence of related biotas throughout the Southern Hemisphere. Parrots (Psittaciformes) occur in Australasia, South America, and Africa. Analyses of DNA hybridization data and mtDNA se-quences suggested that lineages were allocated by ancient vicariance to these continents and subsequently diversified *in situ* (Sibley and Ahlquist, 1990; Miyaki et al., 1998). In the ratite birds, large genetic distances in mito-chondrial (Cooper et al., 1992) and nuclear genomes (Sibley and Ahlquist, 1990) among African ostriches (*Struthio*), South American rheas (*Rhea*), and Australian/New Zealand emus (*Dromaius*), cassowaries (*Casuarius*), and kiwis (*Apteryx*) also were interpreted to support a model of ancient vicariance (Cooper, 1997; Lee et al., 1997; but see Feduccia, 1995, and Här-lid et al., 1998 for dissenting views).

These studies illustrate two broader points. First, DNA hybridization assays offer a composite quantitative index of genetic divergence across a vast number of nuclear genes. Although these methods have not proved useful for microevolutionary studies because of lack of resolving power, they (along with several other classes of molecular assay; Avise, 1994) can be brought to bear on phylogeographic issues for higher taxa. Second, molecular data can be instructive about general separation times in an-cient trees even when they are less than definitive in resolving lineage branching orders (Lee et al., 1997). Resolution of clades is expected to be

(a) killifishes   (b) cichlids

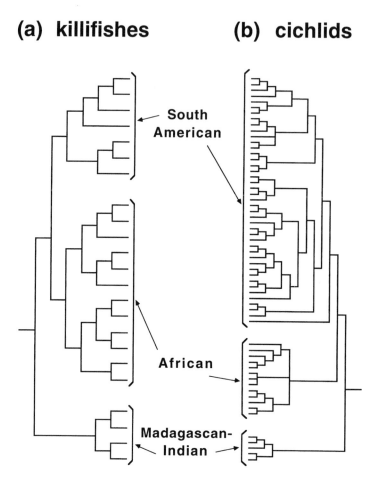

South
American

African

Madagascan-
Indian

FIGURE 6.17 Phylogenetic trees (maximum parsimony) from mitochondrial gene sequences for two speciose assemblages of circumtropical freshwater fishes: *(a)* Cichlids
within the Perciformes (after Farias et al., 1999); and *(b)* Aplocheiloid killifishes within
the Cyprinodontiformes (after Murphy and Collier, 1997). Each external node is a different assayed species. In both cases, continent-specific clades were interpreted to trace to
the ancient breakup of Gondwanaland by plate tectonic movements.

poor when nodes are relatively close in distant time, as probably was true of lineages sundered by the breakup of Gondwanaland, for example.

Muroid rodents in the subfamily Sigmodontinae are a speciose group found primarily in South America. Insecure knowledge of phylogenetic relationships and the absence of a strong fossil record has fueled debate over the entry time of sigmodontines onto the continent from their uncontested ancestral homeland in North America. An early-arrival hypothesis posits that sigmodontines arrived in South America no later than the Miocene (20 Mya), whereas a late-arrival hypothesis views the current diversity of species as tracing to one or more ancestral genera that invaded the continent within the last 3–4 My as a part of the "Great American Interchange" following the rise of the Panamanian Land Bridge.

Phylogenetic analyses of mtDNA sequences support neither hypothesis, but instead indicate an invasion date of about 5–9 Mya, probably from waif dispersal across a narrow sea channel (the Bolivar Trough) present at that time (Engel et al., 1998; see also Smith and Patton, 1993). This conclusion, based on provisional mtDNA clock calibrations, led to a new model for the distributions of the two major phylogeographic assemblages of rodents in the Americas (the other lineage being a speciose neotomine-peromyscine clade mostly in North America). In this model, chance historical events set the stage for the adaptive success of the sigmodontines in South America: ". . . the ancestors to what are now the South American sigmodontines fortuitously reached South America prior to the Great American Interchange and, once there, never relinquished the advantage afforded to them by their early arrival" (Engel et al., 1998).

PHYLOGENETIC CHARACTER MAPPING ON GEOGRAPHY

A popular exercise in recent years has been to map morphological, behavioral, or other species traits on a cladogram, the purpose usually being to reveal the phylogenetic origins of particular adaptations (Brooks and McLennan, 1991; Harvey and Pagel, 1991; Martins, 1996; Sheldon and Whittingham, 1997). The cladograms themselves often are estimated from molecular data presumed to be functionally independent of the traits to be fitted onto the tree. When conducted in a geographic context, the entire exercise could be referred to as phylogeographic character mapping. Such efforts can be conducted at intraspecific levels, especially when the traits of interest differ between isolated populations (Foster and Cameron, 1996) or between ecoforms specialized to different microhabitat types (Stanhope et al., 1993). However, phylogeographic character mapping more often has been conducted on phylogenies for species or higher taxa.

One recent example involved molecular analyses of a species flock of plants in the genus *Aquilegia* (Hodges and Arnold, 1994). These columbines are renowned for broad variation in floral morphology and ecology, yet sequence data from nuclear and chloroplast DNA implies that the species arose recently. An evolutionary key innovation (Heard and Hauser, 1995)—multiple floral nectar spurs—maps to the base of the *Aquilegia* phylogeny and may have facilitated rapid speciation through specialization to specific pollinators. In contrast to most other recognized species flocks that are confined to small areas such as island chains or lake systems (Echelle and Kornfield, 1984; Givnish and Sytsma, 1997; but see also Johns and Avise, 1998b), the radiation of *Aquilegia* occurred on a broad geographic setting that includes Europe, Asia, and North America.

An example on a finer geographic scale involved nine species of land crab endemic to Jamaica. They are unique among all crabs in providing parental care for larvae and juveniles, but otherwise the Jamaican crabs display ecological and morphological specializations that have complicated phylogenetic reconstructions. One hypothesis held that the Jamaican group is monophyletic, having adaptively radiated following a single colonization event of unknown age. A competing view posited independent colonization events by multiple unrelated ancestors. A recent global phylogeographic survey of the crab family Grapsidae appears to have settled the issue in favor of the *in situ* radiation hypothesis (Schubart et al., 1998). In a mtDNA gene tree, all Jamaican species formed a clade whose basal node dates to about 4 Mya. Similarly, molecular data suggest that the *Eleutherodactylus* frogs and *Anolis* lizards of Jamaica each radiated *in situ* from a single colonizer taxon following emergence of the island from a Mid-Tertiary inundation by the Caribbean Sea (Hedges, 1989; Hedges and Burnell, 1990).

Overlays of phylogenetic trees on geography also can be powerful in identifying cases of convergent evolution. For example, several striking morphotypic adaptations are shared by various cichlids in different African lakes, and an unanswered issue was whether these reflect ancestral inter-lake migrations or convergent evolution. Six pairs of cichlids with similar morphological features from Lakes Malawi and Tanganyika (Fig. 6.18) were assayed for mtDNA sequence (Kocher et al., 1993). The resulting molecular phylogeny showed that these phenotypic features (and associated behaviors) evolved independently in fishes in the two lakes. Similar conclusions were reached for *Anolis* lizards in the Caribbean region (Jackman et al., 1997; Losos et al., 1998). Repeated morphological and ecological features, when mapped onto a mtDNA-based phylogeny for 55 species, indicated that similar suites of sympatric "ecomorphs" had evolved independently on each of four islands.

In general, phylogeographic character mapping entails the same promises and cautions that apply to conventional phylogenetic character mapping (Frumhoff and Reeve, 1994; Ryan, 1996). Reservations include the following. First, proper mapping requires knowledge of the species tree, and how best to estimate this phylogeny (with respect to data bases and tree-building algorithms) is a matter of considerable debate (Swofford et al., 1996). Once available, however, a phylogeny provides a necessary but not sufficient framework for deducing the number of independent origins for a trait displayed by multiple species (Felsenstein, 1985b; Maddison, 1994; Nee et al., 1996b; Martins and Hansen, 1997).

Second, even for a species tree known with certainty, assumptions of the evolutionary model can alter interpretations. Suppose, for example, that face-value parsimony suggests a single acquisition for a character near one tip of a three-species tree. A nonexcluded alternative is that the trait originated at the base of the tree and later was lost independently by

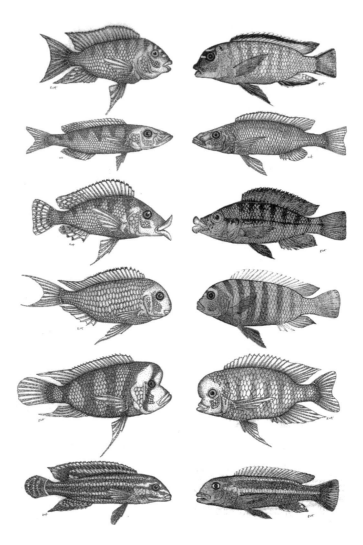

FIGURE 6.18   Six pairs of morphologically similar African cichlid fishes from *(left)* Lake Tanganyika and *(right)* Lake Malawi (redrawn by R. Craig Albertson after Kocher et al., 1993). Placement of these species on a mtDNA-based phylogeny indicates that the phenotypic similarities in each case probably arose via convergent evolution.

two species. One gain and two losses of a character provide a more complicated explanation than a single gain, but there is no guarantee that evolution proceeds in a most parsimonious fashion (Schluter et al., 1997).

A third caveat about phylogenetic character mapping involves issues of trait homology. A morphological or behavioral trait may be mapped correctly onto a species tree, but nonetheless might have different genetic etiologies in different phylogenetic branches (Avise, 1994, p. 320). In other words, for quantitative characters that by definition are influenced by multiple interacting loci, different suites of alleles (or, perhaps, gene-regulatory mechanisms) may be responsible for a given level or pattern of trait expression (Shubin, 1998). The rise of coalescent theory merely has given new weight to old concerns about homology in phylogenetic character analysis. Because different loci (and alleles) have different genealogical histories within an extended sexual pedigree, it is even more evident that a shared complex character might be forged from a mixture of homologous and nonhomologous genetic factors. Whether this matters in a given study depends upon the particular goal of the analysis.

SUMMARY

1. A species phylogeny in sexually reproducing organisms can be conceived as a central tendency for multitudinous genealogical pathways through extended organismal pedigrees. Many evolutionary and demographic factors can affect the variance across gene genealogies within a species tree. Also, gene trees can differ topologically from one another and from a species phylogeny at any evolutionary depth, particularly when internodal times are short in relation to effective population size. These realizations have ramifications for species concepts, for phylogenetic character-state mapping, and for virtually all interpretations in phylogeography.

2. Population demography and geography have played key roles in most speciation scenarios. Each demo-geographic model of speciation entails logical predictions about the temporal transition in

phylogenetic patterns among related taxa. At the level of both species trees and gene trees, paraphyly rather than reciprocal monophyly is an anticipated (and observed) outcome for many biological species that separated recently.

3. No species concept that results in an overly simplified caricature of organismal phylogeny can hope to capture the rich and varied fabric of genealogical histories at multiple loci. Perspectives from a neophyte multilocus coalescent theory may reconcile perceived incompatibilities between phylogenetic and biological species concepts. Reproductive barriers are important for species concepts in part because they demarcate extended demographic units that are relevant to lineage sorting processes generating phylogenetic discontinuities in the biological world. Conversely, phylogenetic considerations are important also, because they force explicit attention on historical-demographic correlates of reproductive isolation.

4. In the vertebrates, current species-level taxonomies often bear close facsimile to presumptive biotic discontinuities in the living world as gauged by molecular genealogical evidence. This agreement probably reflects an underlying historical reality to many of the biological units traditionally recognized as species. Dynamic population-demographic processes (suspected fluctuations in population size, relatively low N_e values, and frequent population extinctions) often deduced from empirical genealogical patterns in animal mtDNA also raise an intriguing possibility regarding "species realities" in predominantly asexual taxa. Perhaps a suspected genetic cohesiveness within and distinctiveness between some facultatively asexual microbes reflect historical demographic factors more so than contemporary patterns of reproductive connections *per se.*

5. Approximate temporal durations of the speciation process can be estimated from genealogical patterns in extant organisms through appeals to molecular clock calibrations. Inferred separation times for intraspecific phylogroups and sister species provide minimum

and maximum estimates, respectively, of speciation durations. Available data from mtDNA suggest that a typical vertebrate speciation extends across approximately 2 My, although there is a large variance about this value in studied taxa.

6. Higher taxa too are linked by phylogenetic ties and have geographic distributions, so by definition phylogeographic analyses can be extended to problems in deeper evolutionary time. Available literature documents the utility of molecular appraisals in weighing the roles of ancient vicariance and dispersal in species' distributions, in estimating arrival times for colonizing taxa, and in mapping the origins and distributions of morphological or other organismal traits in a phylogeographic context.

SYNOPSIS, AND THE FUTURE OF PHYLOGEOGRAPHY

Phylogeography as a formal discipline was born little more than a decade ago. Its gestation in many respects began in the mid-1970s with the introduction of mtDNA analyses to population genetics, and with the profound shift toward genealogical thought at the intraspecific level (now formalized as coalescent theory) that these methods prompted. In broad terms, phylogeography's most important conceptual and empirical contributions have been to emphasize the historical, nonequilibrium aspects of microevolution; clarify the tight connections between population demography and genealogy; and build bridges between the nominally separate fields of population genetics and phylogenetic biology.

What does the future hold for this adolescent discipline? The field is now growing vigorously. There will be further empirical studies on diverse organisms, with results to be interpreted against a backdrop of comparative information from ecology, paleontology, population demography, ethology, natural history, and related fields. On the practical side, conservation efforts in particular will benefit from this emerging synthesis. There will also be a further expansion of interest in the utility of coalescent theory for interpreting genealogical data under the nonequilibrium population demographic conditions that are a sine qua non of historical biology.

Specifically, ample room remains for the expansion of phylogeography in at least three main areas, each tied to a distinct aspect of genealogical concordance. First, much more empirical effort is needed to develop usable methods for the rapid recovery and interpretation of genealogical data from nuclear loci. Associated with this effort should be further cultivation of a multi-locus coalescent theory that deals with both the means and the variances in phylogeographic patterns across genes as a function of varied population demographic histories. Second, many more empirical studies are needed of comparative phylogeography on a regional scale, using multiple codistributed species including those with diverse natural histories. Apart from the evolutionary lessons to be learned from such analyses, the results also carry ramifications for species-specific and regional conservation efforts. Third, additional attention should be devoted to integrating molecular genealogical data with traditional classes of biogeographic information such as species' distributional patterns and the paleontological record. The result will be a mutual enrichment of phylogeography and the other biodiversity disciplines to which it is allied.

Phylogeography is off to an auspicious start. The greatest benefits and research opportunities will continue to arise from the field's central, integrative position within the evolutionary and ecological sciences. I hope that this book may contribute to an awareness of this growing discipline, and serve to place phylogeography on a solid epistemological foundation that will facilitate further achievements.

WORKS CITED / INDEX

WORKS CITED

Abell, R., S. Olson, E. Dinerstein, S. Walters, P. Hurley, W. Wettengel, C. Loucks, and P. Hedao. 1998. *A Conservation Assessment of the Freshwater Ecoregions of North America.* World Wildlife Fund, Washington, D.C.

Adachi, J., Y. Cao, and M. Hasegawa. 1993. Tempo and mode of mitochondrial DNA evolution in vertebrates at the amino acid level: Rapid evolution in warm blooded vertebrates. *J. Mol. Evol.* 36:270–281.

Adams, C. C. 1901. Baseleveling and its faunal significance, with illustrations from southeastern United States. *Amer. Nat.* 35:839–852.

Afonso, J. M., A. Volz, M. Hernandez, H. Ruttkay, M. Gonzalez, J. M. Larruga, V. M. Cabrera, and D. Sperlich. 1990. Mitochondrial DNA variation and genetic structure in Old-World populations of *Drosophila subobscura. Mol. Biol. Evol.* 7:123–142.

Agnese, J. F., B. Adepo-Gourene, E. K. Abban, and Y. Fermon. 1997. Genetic differentiation among natural populations of the Nile tilapia *Oreochromis niloticus* (Teleostei, Cichlidae). *Heredity* 79:88–96.

Aguadé, M. and C. Langley. 1994. Polymorphism and divergence in regions of low recombination in *Drosophila.* In *Non-Neutral Evolution: Theories and Molecular Data,* B. Golding (ed.). New York: Chapman & Hall, pp. 67–76.

Allard, M. W., M. M. Miyamoto, K. A. Bjorndal, A. B. Bolten, and B. W. Bowen. 1994. Support for natal homing in green turtles from mitochondrial DNA sequences. *Copeia* 1994:34–41.

Amato, G., D. Wharton, Z. Z. Zainuddin, and J. R. Powell. 1995. Assessment of conservation units for the Sumatran rhinoceros (*Dicerorhinus sumatrensis*). *Zoo Biol.* 14:395–402.

Amos, B., C. Schlotterer, and D. Tautz. 1993. Social structure of pilot whales revealed by analytical DNA fingerprinting. *Science* 30:670–672.

Anderson, S., A. T. Bankier, B. G. Barrell, M. H. L. De Bruijn, A. R. Coulson, J. Drouin, I. C. Eperon, D. P. Nierlich, B. A. Roe, F. Sanger, P. H. Schreier, A. J. H. Smith, R. Staden, and I. G. Young. 1981. Sequence and organization of the human mitochondrial genome. *Nature* 290:457–465.

Anderson, S., M. H. L. De Bruijn, A. R. Coulson, I. C. Eperon, F. Sanger, and I. G. Young. 1982. The complete sequence of bovine mitochondrial DNA: Conserved features of the mammalian mitochondrial genome. *J. Mol. Biol.* 156:683–717.

Anderson, T. J. C., R. Komuniecki, P. R. Komuniecki, and J. Jaenike. 1995. Are mitochondria paternally inherited in *Ascaris? Int. J. Parasitol.* 25:1001–1004.

Angers, B. and L. Bernatchez. 1998. Combined use of SMM and non-SMM methods to infer fine structure and evolutionary history of closely related brook charr (*Salvelinus fontinalis*, Salmonidae) populations from microsatellites. *Mol. Biol. Evol.* 15:143–159.

AOU (American Ornithologists' Union). 1995. Fortieth supplement to the American Ornithologists' Union Check-list of North American Birds. *Auk* 112: 819–830.

Apostolidis, A. P., C. Triantaphyllidis, A. Kouvatsi, and P. S. Economidis. 1997. Mitochondrial DNA sequence variation and phylogeography among *Salmo trutta* L. (Greek brown trout) populations. *Mol. Ecol.* 6:531–542.

Aquadro, C. F. 1992. Why is the genome variable? Insights from *Drosophila. Trends Genet.* 8:355–362.

———1993. Molecular population genetics of *Drosophila.* In *Molecular Approaches to Fundamental and Applied Entomology,* J. Oakeshott and M. J. Whitten (eds.). New York: Springer-Verlag, pp. 222–266.

Aquadro, C. F. and J. C. Avise. 1982. An assessment of "hidden" heterogeneity within electromorphs at three enzyme loci in deer mice. *Genetics* 102:269–284.

Aquadro, C. F. and D. J. Begun. 1993. Evidence for and implications of genetic hitchhiking in the *Drosophila* genome. In *Mechanisms of Molecular Evolution,* N. Takahata and A. Clark (eds.). Sunderland, Mass: Sinauer, pp. 159–178.

Aquadro, C. F., D. J. Begun, and E. C. Kindahl. 1994. Selection, recombination, and DNA polymorphism in *Drosophila.* In *Non-Neutral Evolution: Theories and Molecular Data.* New York: Chapman & Hall, pp. 46–56.

Aquadro, C. F., S. F. Deese, M. M. Bland, C. H. Langley, and C. C. Laurie-Ahlberg. 1986. Molecular population genetics of the alcohol dehydrogenase gene region of *Drosophila melanogaster*. *Genetics* 114:1165–1190.

Aquadro, C. F. and B. D. Greenberg. 1983. Human mitochondrial DNA variation and evolution: Analysis of nucleotide sequences from seven individuals. *Genetics* 103:287–312.

Arbeit, R. D., M. Arthur, R. Dunn, C. Kim, R. K. Selander, and R. Goldstein. 1990. Resolution of recent evolutionary divergence among *Escherichia coli* from related lineages: The application of pulsed field electrophoresis to molecular epidemiology. *J. Infect. Dis.* 161:230–235.

Arctander, P., P. W. Kat, R. A. Aman, and H. R. Siegismund. 1996. Extreme genetic differences among populations of *Gazella granti*, Grant's gazelle, in Kenya. *Heredity* 76:465–475.

Armour, J. A. L., and 9 others. 1996. Minisatellite diversity supports a recent African origin for modern humans. *Nature Genet.* 13:154–160.

Arnason, E. and S. Pálsson. 1996. Mitochondrial cytochrome *b* DNA sequence variation of Atlantic cod *Gadus morhua* from Norway. *Mol. Ecol.* 5:715–724.

Arndt, A. and M. J. Smith. 1998. Genetic diversity and population structure in two species of sea cucumber: Differing patterns according to mode of development. *Mol. Ecol.* 7:1053–1064.

Arnold, J. 1993. Cytonuclear disequilibria in hybrid zones. *Annu. Rev. Ecol. Syst.* 24:521–554.

Arnold, M. 1997. *Natural Hybridization and Evolution.* New York: Oxford University Press.

Ashley, M. V., D. J. Melnick, and D. Western. 1990. Conservation genetics of the black rhinoceros (*Diceros bicornis*), I: Evidence from the mitochondrial DNA of three populations. *Conserv. Biol.* 4:71–77.

Attardi, G. 1985. Animal mitochondrial DNA: An extreme example of genetic economy. *Int. Rev. Cytol.* 93:93–145.

Auffray, J. C., F. Vanlerberghe, and J. Britton-Davidson. 1990. The house mouse progression in Eurasia: A paleontological and archaeozoological approach. *Biol. J. Linn. Soc.* 41:13–25.

Austin, J. J., R. W. G. White, and J. R. Ovenden. 1994. Population-genetic structure of a philopatric, colonially nesting seabird, the short-tailed shearwater (*Puffinus tenuirostris*). *Auk* 111:70–79.

Avise, J. C. 1974. Systematic value of electrophoretic data. *Syst. Zool.* 23:465–481.

———1983. Commentary. In *Perspectives in Ornithology*, A. H. Brush and G. A. Clark (eds.). New York: Cambridge University Press, pp. 262–270.

————1986. Mitochondrial DNA and the evolutionary genetics of higher animals. *Phil. Trans. Roy. Soc. Lond. B* 312:325–342.

————1987. Identification and interpretation of mitochondrial DNA stocks in marine species. In *Proc. Stock Identification Workshop*, H. Kumpf and E. L. Nakamura (eds.). Panama City, Fla.: Publ. Natl. Oceanographic and Atmospheric Administration, pp. 105–136.

————1989a. A role for molecular genetics in the recognition and conservation of endangered species. *Trends Ecol. Evol.* 4:279–281.

————1989b. Gene trees and organismal histories: A phylogenetic approach to population biology. *Evolution* 43:1192–1208.

————1989c. Nature's family archives. *Natur. Hist.* 3:24–27.

————1991. Ten unorthodox perspectives on evolution prompted by comparative population genetic findings on mitochondrial DNA. *Annu. Rev. Genet.* 25:45–69.

————1992. Molecular population structure and the biogeographic history of a regional fauna: A case history with lessons for conservation biology. *Oikos* 63:62–76.

————1993. The evolutionary biology of aging, sexual reproduction, and DNA repair. *Evolution* 47:1293–1301.

————1994. *Molecular Markers, Natural History and Evolution*. New York: Chapman & Hall.

————1995. Mitochondrial DNA polymorphism and a connection between genetics and demography of relevance to conservation. *Conserv. Biol.* 9:686–690.

————1996a. Space and time as axes in intraspecific phylogeography. In *Past and Future Rapid Environmental Changes: The Spatial and Evolutionary Responses of Terrestrial Biota*, B. Huntley, W. Cramer, A. V. Morgan, H. C. Prentice, and J. R. M. Allen (eds.). New York: Springer-Verlag, pp. 381–388.

————1996b. Toward a regional conservation genetics perspective: Phylogeography of faunas in the southeastern United States. In *Conservation Genetics: Case Histories from Nature*, J. C. Avise and J. L. Hamrick (eds.). New York: Chapman & Hall, pp. 431–470.

————1996c. Three fundamental contributions of molecular genetics to avian ecology and evolution. *Ibis* 138:16–25.

————1998a. The history and purview of phylogeography: A personal reflection. *Mol. Ecol.* 7:371–379.

————1998b. Conservation genetics in the marine realm. *J. Heredity* 89:377–382.

Avise, J. C., R. T. Alisauskas, W. S. Nelson, and C. D. Ankney. 1992b. Matriarchal population genetic structure in an avian species with female natal philopatry. *Evolution* 46:1084–1096.

Avise, J. C., C. D. Ankney, and W. S. Nelson. 1990. Mitochondrial gene trees and the evolutionary relationship of mallard and black ducks. *Evolution* 44:1109–1119.

Avise, J. C., J. Arnold, R. M. Ball, Jr, E. Bermingham, T. Lamb, J. E. Neigel, C. A. Reeb, and N. C. Saunders. 1987a. Intraspecific phylogeography: The mitochondrial DNA bridge between population genetics and systematics. *Annu. Rev. Ecol. Syst.* 18:489–522.

Avise, J. C. and R. M. Ball, Jr. 1990. Principles of genealogical concordance in species concepts and biological taxonomy. *Oxford Surv. Evol. Biol.* 7:45–67.

———1991. Mitochondrial DNA and avian microevolution. *Proc. Int. Orn. Congr.* 20:514–524.

Avise, J. C., R. M. Ball, Jr., and J. Arnold. 1988. Current versus historical population sizes in vertebrate species with high gene flow: A comparison based on mitochondrial DNA lineages and inbreeding theory for neutral mutations. *Mol. Biol. Evol.* 5:331–344.

Avise, J. C., E. Bermingham, L. G. Kessler, and N. C. Saunders. 1984b. Characterization of mitochondrial DNA variability in a hybrid swarm between subpecies of bluegill sunfish (*Lepomis macrochirus*). *Evolution* 38:931–941.

Avise, J. C. and B. W. Bowen. 1994. Investigating sea turtle migration using DNA markers. *Curr. Opin. Genet. Develop.* 4:882–886.

Avise, J. C., B. W. Bowen, and T. Lamb. 1989. DNA fingerprints from hypervariable mitochondrial genotypes. *Mol. Biol. Evol.* 6:258–269.

Avise, J. C., B. W. Bowen, T. A. Lamb, A. B. Meylan, and E. Bermingham. 1992c. Mitochondrial DNA evolution at a turtle's pace: Evidence for low genetic variability and reduced microevolutionary rate in the Testudines. *Mol. Biol. Evol.* 9:457–473.

Avise, J. C., C. Giblin-Davidson, J. Laerm, J. C. Patton, and R. A. Lansman. 1979b. Mitochondrial DNA clones and matriarchal phylogeny within and among geographic populations of the pocket gopher, *Geomys pinetis. Proc. Natl. Acad. Sci. USA* 76:6694–6698.

Avise, J. C. and J. L. Hamrick (eds.). 1996. *Conservation Genetics: Case Histories from Nature.* New York: Chapman & Hall.

Avise, J. C., G. S. Helfman, N. C. Saunders, and L. S. Hales. 1986. Mitochondrial DNA differentiation in North Atlantic eels: Population genetic consequences of an unusual life history pattern. *Proc. Natl. Acad. Sci. USA* 83:4350–4354.

Avise, J. C. and R. A. Lansman. 1983. Polymorphism of mitochondrial DNA in populations of higher animals. In *Evolution of Genes and Proteins*, M. Nei and R. K. Koehn (eds.). Sunderland, Mass.: Sinauer, pp. 147–164.

Avise, J. C., R. A. Lansman, and R. O. Shade. 1979a. The use of restriction endonucleases to measure mitochondrial DNA sequence relatedness in natural populations. I. Population structure and evolution in the genus *Peromyscus*. *Genetics* 92:279–295.

Avise, J. C., J. E. Neigel, and J. Arnold. 1984a. Demographic influences on mitochondrial DNA lineage survivorship in animal populations. *J. Mol. Evol.* 20:99–105.

Avise, J. C. and W. S. Nelson. 1989. Molecular genetic relationships of the extinct dusky seaside sparrow. *Science* 243:646–648.

Avise, J. C., W. S. Nelson, and H. Sugita. 1994. A speciational history of "living fossils": Molecular evolutionary patterns in horseshoe crabs. *Evolution* 48: 1986–2001.

Avise, J. C., P. C. Pierce, M. J. Van Den Avyle, M. H. Smith, W. S. Nelson, and M. A. Asmussen. 1997. Cytonuclear introgressive swamping and species turnover of bass after an introduction. *J. Heredity* 88:14–20.

Avise, J. C., J. M. Quattro, and R. C. Vrijenhoek. 1992a. Molecular clones within organismal clones: Mitochondrial DNA phylogenies and the evolutionary histories of unisexual vertebrates. *Evol. Biol.* 26:225–246.

Avise, J. C., C. A. Reeb, and N. C. Saunders. 1987b. Geographic population structure and species differences in mitochondrial DNA of mouthbrooding marine catfishes (Ariidae) and demersal spawning toadfishes (Batrachoididae). *Evolution* 41:991–1002.

Avise, J. C. and N. C. Saunders. 1984. Hybridization and introgression among species of sunfish (*Lepomis*): Analysis by mitochondrial DNA and allozyme markers. *Genetics* 108:237–255.

Avise, J. C., J. F. Shapira, S. W. Daniel, C. F. Aquadro, and R. A. Lansman. 1983. Mitochondrial DNA differentiation during the speciation process in *Peromyscus*. *Mol. Biol. Evol.* 1:38–56.

Avise, J. C. and M. H. Smith. 1974. Biochemical genetics of sunfish: I. Geographic variation and subspecific intergradation in the bluegill, *Lepomis macrochirus*. *Evolution* 28:42–56.

Avise, J. C., M. H. Smith, and R. K. Selander. 1979c. Biochemical polymorphism and systematics in the genus *Peromyscus*. VII. Geographic differentiation in members of the *truei* and *maniculatus* species groups. *J. Mammal.* 60:177–192.

Avise, J. C. and R. J. Vrijenhoek. 1987. Mode of inheritance and variation of mitochondrial DNA in hybridogenetic fishes of the genus *Poeciliopsis*. *Mol. Biol. Evol.* 4:514–525.

Avise, J. C. and D. Walker. 1998. Pleistocene phylogeographic effects on avian populations and the speciation process. *Proc. Roy. Soc. Lond. B* 265:457–463.

———1999. Species realities and numbers in sexual vertebrates: Perspectives from an asexually transmitted genome. *Proc. Natl. Acad. Sci. USA* 96:992–995.

Avise, J. C., D. Walker, and G. C. Johns. 1998. Speciation durations and Pleistocene effects on vertebrate phylogeography. *Proc. Roy. Soc. Lond. B* 265: 1707–1712.

Avise, J. C. and K. Wollenberg. 1997. Phylogenetics and the origin of species. *Proc. Natl. Acad. Sci. USA* 94:7748–7755.

Avise, J. C. and R. M. Zink. 1988. Molecular genetic divergence between avian sibling species: King and clapper rails, long-billed and short-billed dowitchers, boat-tailed and great-tailed grackles, and tufted and black-crested titmice. *Auk* 105:516–528.

Ayala, F. J. 1976. Molecular genetics and evolution. In *Molecular Evolution*, F. J. Ayala (ed.). Sunderland, Mass.: Sinauer, pp. 1–20.

———1995a. The myth of Eve: Molecular biology and human origins. *Science* 270:1930–1936.

———1995b. Adam, Eve, and other ancestors: A story of human origins told by genes. *Hist. Phil. Life Sci.* 17:303–313.

———1996. HLA sequence polymorphism and the origin of humans. *Science* 274:1554.

Ayala, F. J., A. Escalante, C. O'Huigin, and J. Klein. 1994. Molecular genetics of speciation and human origins. *Proc. Natl. Acad. Sci. USA* 91:6787–6794.

Ayres, J. M. and T. H. Clutton-Brock. 1992. River boundaries and species range size in Amazonian primates. *Amer. Nat.* 140:531–537.

Bagley, M. J. and G. A. E. Gall. 1998. Mitochondrial and nuclear DNA sequence variability among populations of rainbow trout (*Oncorhynchus mykiss*). *Mol. Ecol.* 7:945–961.

Baker, A. J., C. H. Daugherty, R. Colbourne, and J.L. McLennan. 1995. Flightless brown kiwis of New Zealand possess extremely subdivided population structure and cryptic species like small mammals. *Proc. Natl. Acad. Sci. USA* 92:8254–8258.

Baker, A. J. and H. D. Marshall. 1997. Mitochondrial control region sequences as tools for understanding evolution. In *Avian Molecular Evolution and Systematics*, D. P. Mindell (ed.). New York: Academic Press, pp. 51–82.

Baker, A. J., T. Piersma, and L. Rosenmeier. 1994. Unraveling the intraspecific phylogeography of knots *Calidris canutus:* A progress report on the search for genetic markers. *J. Ornithol.* 135:599–608.

Baker, C. S. and S. R. Palumbi. 1996. Population structure, molecular systematics, and forensic identification of whales and dolphins. In *Conservation Genetics: Case Histories from Nature*, J. C. Avise and J. L. Hamrick (eds.). New York: Chapman & Hall, pp. 10–49.

Baker, C. S., L. Medrano-Gonzalez, J. Calmabokidis, A. Perry, F. Pichler, H. Rosen-
    baum, J. M. Straley, J. Urbán-Ramirez, M. Yamaguchi, and O. Von Ziegesar.
    1998. Population structure of nuclear and mitochondrial DNA variation
    among humpback whales in the North Pacific. *Mol. Ecol.* 7:695–707.

Baker, C. S., S. R. Palumbi, R. H. Lambertsen, M. T. Weinrich, J. Calmabokidis,
    and S. J. O'Brien. 1990. Influence of seasonal migration on geographic dis-
    tribution of mitochondrial DNA haplotypes in humpback whales. *Nature*
    34:238–240.

Baker, C. S., A. Perry, J. L. Bannister, M. T. Weinrich, R. B. Abernethy, J.
    Calmabokidis, J. Lien, R. H. Lambertsen, J. Urbán-Ramírez, O. Vasquez, P. J.
    Clapham, A. Alling, S. J. O'Brien, and S.R. Palumbi. 1993. Abundant mito-
    chondrial DNA variation and world-wide population structure in humpback
    whales. *Proc. Natl. Acad. Sci. USA* 90:8239–8243.

Baker, C. S., A. Perry, G. K. Chambers, and P. J. Smith. 1995. Population variation in
    the mitochondrial cytochrome *b* gene of the orange roughy *Hoplostethus at-
    lanticus* and the hoki *Macruronus novaezelandiae*. *Marine Biol.* 122:503–509.

Baker, C. S., R. W. Slade, J. L. Bannister, R. B. Abernethy, M. T. Weinrich, J. Lien, J.
    Urban, P. Corkeron, J. Calmabokidis, O. Vasquez, and S. R. Palumbi. 1994. Hi-
    erarchical structure of mitochondrial DNA gene flow among humpback
    whales *Megaptera novaeangliae*, world-wide. *Mol. Ecol.* 3:313–327.

Ball, R. M., Jr. and J. C. Avise. 1992. Mitochondrial DNA phylogeographic differen-
    tiation among avian populations and the evolutionary significance of sub-
    species. *Auk* 109:626–636.

Ball, R. M., Jr., F. C. James, S. Freeman, E. Bermingham, and J. C. Avise. 1988. Phylo-
    geographic population structure of red-winged blackbirds assessed by mito-
    chondrial DNA. *Proc. Natl. Acad. Sci. USA* 85:1558–1562.

Ball, R. M., Jr., J. E. Neigel, and J. C. Avise. 1990. Gene genealogies within the organ-
    ismal pedigrees of random-mating populations. *Evolution* 44:360–370.

Ballinger, S. W., T. G. Schurr, A. Torroni, Y. Y. Gan, J. A. Hodge, K. Hassan, K.-H.
    Chen and D. C. Wallace. 1992. Southeast Asian mitochondrial DNA analysis
    reveals genetic continuity of ancient Mongoloid migrations. *Genetics* 130:
    139–152.

Bamshad, M., W. S. Watkins, M. E. Dixon, L. B. Jorde, B. B. Rao, J. M. Naidu, B. V. R.
    Prasad, A. Rasanayagam, and M. F. Hammer. 1998. Female gene flow stratifies
    Hindu castes. *Nature* 395:651–652.

Barber, P. H. 1996. Phylogeography and evolutionary history of the canyon
    treefrog, *Hyla arenicolor*. *Amer. Zool.* 36:122 (abstract).

Barbujani, G., G. Bertorelle, and L. Chikhi. 1998. Evidence for Paleolithic and Neo-
    lithic gene flow in Europe. *Am. J. Human Genet.* 62:488–491.

Barraclough, T. G., A. P. Vogler, and P. H. Harvey. 1998. Revealing the factors that promote speciation. *Phil. Trans. Roy. Soc. Lond. B* 353:241–249.

Barratt, E. M., R. Deaville, T. M. Burland, M. W. Bruford, G. Jones, P. A. Racey, and R. K. Wayne. 1997. DNA answers the call of pipistrelle bat species. *Nature* 387:138–139.

Barrowclough, G. F. 1983. Biochemical studies of microevolutionary processes. In *Perspectives in Ornithology*, A. H. Brush and G. A. Clark, Jr. (eds.). New York: Cambridge University Press, pp. 223–261.

Barrowclough, G. F. 1992. Systematics, biodiversity, and conservation biology. In *Systematics, Ecology, and the Biodiversity Crisis*, N. Eldredge (ed.). New York: Columbia University Press, pp. 121–143.

Barton, N. H. and G. M. Hewitt. 1985. Analysis of hybrid zones. *Annu. Rev. Ecol. Syst.* 16:113–148.

Barton, N. H. and M. Slatkin. 1986. A quasi-equilibrium theory of the distribution of rare alleles in a subdivided population. *Heredity* 56:409–415.

Barton, N. H. and I. Wilson. 1995. Genealogies and geography. *Phil. Trans. Roy. Soc. Lond B* 349:49–59.

———1996. Genealogies and geography. In *New Uses for New Phylogenies*, P. H. Harvey, A. J. Leigh Brown, J. Maynard Smith, and S. Nee (eds.). New York: Oxford University Press, pp. 23–56.

Bass, A. L., D. A. Good, K. A. Bjorndal, J. I. Richardson, Z.-M. Hillis, J. A. Horrocks, and B. W. Bowen. 1996. Testing models of female reproductive migratory behavior and population structure in the Caribbean hawksbill turtle, *Eretmochelys imbricata*, with mtDNA sequences. *Mol. Ecol.* 5:321–328.

Baum, D. A. and K. L. Shaw. 1995. Genealogical perspectives on the species problem. In *Experimental and Molecular Approaches to Plant Biosystematics*, P. C. Hoch and A. G. Stephenson (eds.). St. Louis: Missouri Botanical Garden, Monogr. Syst. Bot. Missouri Bot. Gard. 53, pp. 289–303.

Becker, I. I., W. S. Grant, R. Kirby, and F. T. Robb. 1988. Evolutionary divergence between sympatric species of southern African hakes, *Merluccius capensis* and *M. paradoxus*. II. Restriction enzyme analysis of mitochondrial DNA. *Heredity* 61:21–30.

Begun, D. J. and C. F. Aquadro. 1992. Levels of naturally occurring DNA polymorphisms correlate with recombination rates in *D. melanogaster*. *Nature* 356: 519–520.

Bell, G. 1997. *Selection: The Mechanism of Evolution*. New York: Chapman & Hall.

Bentzen, P., G. C. Brown, and W. C. Leggett. 1989. Mitochondrial DNA polymorphism, population structure, and life history variation in American shad (*Alosa sapidissima*). *Can. J. Fish. Aquat. Sci.* 46:1446–1454.

Berger, A. 1984. Accuracy and frequency stability of the Earth's orbital elements during the Quaternary. In *Milankovitch and Climate, Part 1*, A. Berger, J. Imbrie, J. Hays, G. Kukla, and B. Saltzmann (eds.). Dordrecht: Reidel, pp. 527–537.

Bergstrom, D. E., Jr., 1997. The phylogeny and historical biogeography of Missouri's *Amblyopsis rosae* (Ozark cavefish) and *Typhlichthys subterraneus* (southern cavefish). Master's thesis, University of Missouri, Columbia.

Bergström, T. F., A. Josefsson, H. A. Erlich, and U. Gyllensten. 1998. Recent origin of *HLA-DRB1* alleles and implications for human evolution. *Nature Genet.* 18:237–242.

Bermingham, E. and J. C. Avise. 1986. Molecular zoogeography of freshwater fishes in the southeastern United States. *Genetics* 113:939–965.

Bermingham, E., T. Lamb, and J. C. Avise. 1986. Size polymorphism and heteroplasmy in the mitochondrial DNA of lower vertebrates. *J. Heredity* 77:249–252.

Bermingham, E. and H. Lessios. 1993. Rate variation of protein and mtDNA evolution as revealed by sea urchins separated by the Isthmus of Panama. *Proc. Natl. Acad. Sci. USA* 90:2734–2738.

Bermingham, E. and A. P. Martin. 1998. Comparative mtDNA phylogeography of neotropical freshwater fish: Testing shared history to infer the evolutionary landscape of lower Central America. *Mol. Ecol.* 7:499–517.

Bermingham, E., S. S. McCafferty, and A. P. Martin. 1997. Fish biogeography and molecular clocks: Perspectives from the Panamanian Isthmus. In *Molecular Systematics of Fishes*, T. D. Kocher and C. A. Stepien (eds.). San Diego, Calif.: Academic Press, pp. 113–128.

Bermingham, E. and C. Moritz. 1998. Comparative phylogeography: Concepts and applications. *Mol. Ecol.* 7:367–369.

Bermingham, E., S. Rohwer, S. Freeman, and C. Wood. 1992. Vicariance biogeography in the Pleistocene and speciation in North American wood warblers: A test of Mengel's model. *Proc. Natl. Acad. Sci. USA* 89:6624–6628.

Bermingham, E., G. Seutin, and R. E. Ricklefs. 1996. Regional approaches to conservation biology: RFLPs, DNA sequence and Caribbean birds. In *Molecular Genetic Approaches in Conservation*, T. B. Smith and R. K. Wayne (eds.). New York: Oxford University Press, pp. 104–124.

Bernardi, G., P. Sordino, and D. A. Powers. 1993. Concordant mitochondrial and nuclear DNA phylogenies for populations of the teleost fish *Fundulus heteroclitus. Proc. Natl. Acad. Sci. USA* 90:9271–9274.

Bernatchez, L. 1995. A role for molecular systematics in defining evolutionarily significant units (ESU) in fishes. *Amer. Fish. Soc. Symp.* 17:114–132.

————1997. Mitochondrial DNA analysis confirms the existence of two glacial races of rainbow smelt *Osmerus mordax* and their reproductive isolation in the St. Lawrence River estuary (Québec, Canada). *Mol. Ecol.* 6:73–83.

Bernatchez, L. and R. G. Danzmann. 1993. Congruence in control-region sequence and restriction-site variation in mitochondrial DNA of brook charr (*Salvelinus fontinalis* Mitchell). *Mol. Biol. Evol.* 10:1002–1014.

Bernatchez, L. and J. J. Dodson. 1990. Allopatric origin of sympatric populations of lake whitefish (*Coregonus clupeaformis*) as revealed by mitochondrial–DNA restriction analysis. *Evolution* 44:1263–1271.

————1991. Phylogeographic structure in mitochondrial DNA of the lake whitefish (*Coregonus clupeaformis*) and its relation to Pleistocene glaciations. *Evolution* 45:1016–1035.

————1994. Phylogenetic relationships among Palearctic and Nearctic whitefish (*Coregonus* sp.) populations as revealed by mitochondrial DNA variation. *Can. J. Fish. Aquat. Sci.* 51:240–251.

Bernatchez, L., R. Guyomard, and F. Bonhomme. 1992. DNA sequence variation of the mitochondrial control region among geographically and morphologically remote European brown trout *Salmo trutta* populations. *Mol. Ecol.* 1:161–173.

Bernatchez, L. and A. Osinov. 1995. Genetic diversity of trout (genus *Salmo*) from its most eastern native range based on mitochondrial DNA and nuclear gene variation. *Mol. Ecol.* 4:285–297.

Bernatchez, L., J. A. Vuorinen, R. A. Bodaly, and J. J. Dodson. 1996. Genetic evidence for reproductive isolation and multiple origins of sympatric trophic ecotypes of whitefish (*Coregonus*). *Evolution* 50:624–635.

Bernatchez, L. and C. C. Wilson. 1998. Comparative phylogeography of Nearctic and Palearctic fishes. *Mol. Ecol.* 7:431–452.

Bert, T. 1986. Speciation in western Atlantic stone crabs (genus *Menippe*): The role of geological processes and climatic events in the formation and distribution of species. *Marine Biol.* 93:157–170.

Bertorelle, G. and G. Barbujani. 1995. Analysis of DNA diversity by spatial autocorrelation. *Genetics* 140:811–819.

Bérubé, M., A. Aguilar, D. Dendanto, F. Larsen, G. Notarbartolo di Sciara, R. Sears, J. Sigurjónsson, J. Urban-R., and P. J. Palsbøll. 1998. Population genetic structure of North Atlantic, Mediterranean Sea and Sea of Cortez fin whales, *Balaenoptera physalus* (Linnaeus 1758): Analysis of mitochondrial and nuclear loci. *Mol. Ecol.* 7:585–599.

Bibb, M. J., R. A. Van Etten, C. T. Wright, M. W. Walberg, and D. A. Clayton. 1981. Sequence and gene organization of mouse mitochondrial DNA. *Cell* 26:167–180.

Bibby, C. 1994. A global view of priorities for bird conservation: A summary. *Ibis* 137:S247–S248.

Bickham, J. W., J. C. Patton, and T. R. Loughlin. 1996. High variability for control-region sequences in a marine mammal—implications for conservation and biogeography of Stellar sea lions (*Eumetopias jubatus*). *J. Mammal.* 77:95–108.

Bickham, J. W., C. C. Wood, and J. C. Patton. 1995. Biogeographic implications of cytochrome *b* sequences and allozymes in sockeye (*Oncorhynchus nerka*). *J. Heredity* 86:140–144.

Biju-Duval, C., H. Ennafaa, N. Dennebouy, M. Monnerot, F. Mignotte, R. C. Soriguer, A. E. Gaaïed, A. E. Hili, and J.-C. Mounolou. 1991. Mitochondrial DNA evolution in Lagomorphs: Origin of systematic heteroplasmy and organization of diversity in European rabbits. *J. Mol. Evol.* 33:92–102.

Billington, N. and R. M. Strange. 1995. Mitochondrial DNA analysis confirms the existence of a genetically divergent walleye population in northeastern Mississippi. *Trans. Amer. Fish. Soc.* 124:770–776.

Bilton, D. T. 1994. Phylogeography and recent historical biogeography of *Hydroporus glabriusculus* Aubé (Coleoptera: Dytiscidae) in the British Isles and Scandinavia. *Biol. J. Linn. Soc.* 51:293–307.

Birky, C. W., Jr. 1978. Transmission genetics of mitochondria and chloroplasts. *Annu. Rev. Genet.* 12:471–512.

———1983. Relaxed cellular controls and organelle heredity. *Science* 222:468–475.

———1991. Evolution and population genetics of organelle genes: Mechanisms and models. In *Evolution at the Molecular Level*, R. K. Selander, A. G. Clark and T. S. Whittam (eds.). Sunderland, Mass: Sinauer, pp. 112–134.

Birky, C. W., Jr., A. R. Acton, R. Dietrich, and M. Carver. 1982. Mitochondrial transmission genetics: Replication, recombination, and segregation of mitochondrial DNA and its inheritance in crosses. In *Mitochondrial Genes*, P. Slonimski, P. Borst, and G. Attardi (eds.). Cold Spring Harbor, New York: Cold Spring Harbor Laboratory, pp. 333–348.

Birky, C. W., Jr., P. Fuerst, and T. Maruyama. 1989. Organelle gene diversity under migration, mutation and drift: Equilibrium expectations, approach to equilibrium, effects of heteroplasmic cells, and comparison to nuclear genes. *Genetics* 121:613–627.

Birky, C. W., Jr., T. Maruyama, and P. A. Fuerst. 1983. An approach to population and evolutionary genetic theory for genes in mitochondria and chloroplasts and some results. *Genetics* 103:513–527.

Birky, C. W., Jr. and R. V. Skavaril. 1976. Maintenance of genetic homogeneity in systems with multiple genomes. *Genet. Res.* 27:249–265.

Birt, T. P., V. L. Friesen, R. D. Birt, J. M. Green, and W. S. Davidson. 1995. Mitochondrial DNA variation in Atlantic capelin, *Mallotus villosus:* A comparison of restriction and sequence analyses. *Mol. Ecol.* 4:771–776.

Birt-Friesen, V. L., W. A. Montevecchi, A. J. Gaston, and W. S. Davidson. 1992. Genetic structure of thick-billed murre (*Uria lomvia*) populations examined using direct sequence analysis of amplified DNA. *Evolution* 46:267–272.

Blair, W. F. 1940. A study of prairie deer-mouse populations in southern Michigan. *Amer. Midland Nat.* 24:273–305.

Bogenhagen, D. and D. A. Clayton. 1977. Mouse L cell mitochondrial DNA molecules are selected randomly for replication throughout the cell cycle. *Cell* 11:719–727.

Bohlmeyer, D. A. and J. R. Gold. 1991. Genetic studies of marine fishes II. A protein electrophoretic analysis of population structure in the red drum *Sciaenops ocellatus. Marine Biol.* 108:197–206.

Boissinot, S. and P. Boursot. 1997. Discordant phylogeographic patterns between the Y chromosome and mitochondrial DNA in the house mouse: Selection on the Y chromosome? *Genetics* 146:1019–1034.

Bolten, A. B., K. A. Bjorndal, H. R. Martins, T. Dellinger, M. J. Biscoito, S. E. Encalada, and B. W. Bowen. 1998. Transatlantic developmental migrations of loggerhead sea turtles demonstrated by mtDNA sequence analysis. *Ecol. Appl.* 8:1–7.

Boore, J. L., T. M. Collins, D. Stanton, L. L. Daehler, and W. M. Brown. 1995. Deducing the pattern of arthropod phylogeny from mitochondrial DNA rearrangements. *Nature* 376:163–165.

Boskovic, R., K. M. Kovacs, M. O. Hammill, and B. N. White. 1996. Geographic distribution of mitochondrial DNA haplotypes in grey seals (*Halichoerus grypus*). *Can. J. Zool.* 74:1787–1796.

Bossart, J. L. and D. P. Prowell. 1998. Genetic estimates of population structure and gene flow: Limitations, lessons and new directions. *Trends Ecol. Evol.* 13:202–206.

Boursot, P., W. Din, R. Anand, D. Darviche, B. Dod, F. Von Deimling, G. P. Talwar, and F. Bonhomme. 1996. Origin and radiation of the house mouse: Mitochondrial DNA phylogeny. *J. Evol. Biol.* 9:391–415.

Bowcock, A. M., A. Ruiz-Linares, J. Tomfohrde, E. Minch, J. R. Kidd, and L. L. Cavalli-Sforza. 1994. High resolution of human evolutionary trees with polymorphic microsatellites. *Nature* 368:455–457.

Bowen, B. W. 1995. Tracking marine turtles with genetic markers: Voyages of the ancient mariners. *BioScience* 45:528–534.

———1996a. Literature on marine turtle population structure, molecular evolution, conservation genetics, and related topics. In *Proceedings of the International Symposium on Sea Turtle Conservation Genetics,* B. W. Bowen and W. N. Witzell (eds.). NOAA Tech. Memo. NMFS-SEFSC-396, pp. 9–16.

———1996b. Tracking marine turtles with genetic markers. In *Proceedings of the International Symposium on Sea Turtle Conservation Genetics,* B. W. Bowen and W. N. Witzell (eds.). NOAA Tech. Memo. NMFS-SEFSC–396, pp. 109–117.

Bowen, B. W., F. A. Abreu-Grobois, G. H. Balazs, N. Kamezaki, C. J. Limpus, and R. J. Ferl. 1995. Trans-Pacific migrations of the loggerhead sea turtle demonstrated with mitochondrial DNA markers. *Proc. Natl. Acad. Sci. USA* 92:3731–3734.

Bowen, B. W. and J. C. Avise. 1990. Genetic structure of Atlantic and Gulf of Mexico populations of sea bass, menhaden, and sturgeon: Influence of zoogeographic factors and life-history patterns. *Marine Biol.* 107:371–381.

———1996. Conservation genetics of marine turtles. In *Conservation Genetics: Case Histories from Nature,* J. C. Avise and J. L. Hamrick (eds.). New York: Chapman & Hall, pp. 190–237.

Bowen, B. W., J. C. Avise, J. I. Richardson, A. B. Meylan, D. Margaritoulis, and S. R. Hopkins-Murphy. 1993b. Population structure of loggerhead turtles (*Caretta caretta*) in the northwestern Atlantic Ocean and Mediterranean Sea. *Conserv. Biol.* 7:834–844.

Bowen, B. W., A. L. Bass, A. Garcia, C. E. Diez, R. van Dam, A. Bolten, K. A. Bjorndal, M. M. Miyamoto, and R. J. Ferl. 1996. The origin of hawksbill turtles in a Caribbean feeding area as indicated by genetic markers. *Ecol. Appl.* 6:566–572.

Bowen, B. W., A. M. Clark, F. A. Abreu-Grobois, A. Chaves, H. A. Reichart, and R. J. Ferl. 1998. Global phylogeography of the ridley sea turtles (*Lepidochelys* spp.) as inferred from mitochondrial DNA sequences. *Genetica* 101:179–189.

Bowen, B. W. and W. S. Grant. 1997. Phylogeography of the sardines (*Sardinops* spp.): Assessing biogeographic models and population histories in temperate upwelling zones. *Evolution* 51:1601–1610.

Bowen, B. W., N. Kamezaki, C. J. Limpus, G. R. Hughes, A. B. Meylan, and J. C. Avise. 1994. Global phylogeography of the loggerhead turtle (*Caretta caretta*) as indicated by mitochondrial DNA genotypes. *Evolution* 48:1820–1828.

Bowen, B. W. and S. A. Karl. 1997. Population genetics, phylogeography, and molecular evolution. In *The Biology of Sea Turtles,* P. L. Lutz and J. A. Musick (eds.). Boca Raton, Fla: CRC Press, pp. 29–50.

Bowen, B. W., A. B. Meylan, and J. C. Avise. 1989. An odyssey of the green sea turtle: Ascension Island revisited. *Proc. Natl. Acad. Sci. USA* 86:573–576.

———1991. Evolutionary distinctiveness of the endangered Kemp's ridley sea turtle. *Nature* 352:709–711.

Bowen, B. W., A. B. Meylan, J. Perran Ross, C. J. Limpus, G. H. Balazs, and J. C.
Avise. 1992. Global population structure and natural history of the green turtle
(*Chelonia mydas*) in terms of matriarchal phylogeny. *Evolution* 46:865–881.

Bowen, B. W., W. S. Nelson, and J. C. Avise. 1993a. A molecular phylogeny for marine turtles: Trait mapping, rate assessment, and conservation relevance. *Proc.
Natl. Acad. Sci. USA* 90:5574–5577.

Bowen, B. W. and W. N. Witzell (eds.). 1996. Proceedings of the International Symposium on Sea Turtle Conservation Genetics. NOAA Tech. Memo. NMFS-
SEFSC-396.

Bowers, N., J. R. Stauffer, and T. D. Kocher. 1994. Intra- and interspecific mitochondrial DNA sequence variation within two species of rock-dwelling cichlids
(Teleostei: Cichlidae) from Lake Malawi, Africa. *Mol. Phylogen. Evol.* 3:75–82.

Boyce, A. J. and C. G. N. Mascie-Taylor (eds.). 1996. *Molecular Biology and Human
Diversity.* Cambridge: Cambridge University Press.

Boyce, T. M., M. E. Zwick, and C. F. Aquadro. 1994. Mitochondrial DNA in the bark
weevils: Phylogeny and evolution in the *Pissodes strobi* species group
(Coleoptera: Curculionidae). *Mol. Biol. Evol.* 11:183–194.

Bradley, R. D. and D. M. Hillis. 1997. Recombinant DNA sequences generated by
PCR amplification. *Mol. Biol. Evol.* 14:592–593.

Bräuer, G. and F. H. Smith (eds.). 1992. *Continuity or Replacement: Controversies in*
Homo sapiens *Evolution.* Rotterdam: Balkema.

Brawn, J. D., T. M. Collins, M. Medina, and E. Bermingham. 1996. Associations between physical isolation and geographical variation within three species of
Neotropical birds. *Mol. Ecol.* 5:33–46.

Bremer, J. R. A., A. J. Baker, and J. Mejuto. 1995. Mitochondrial-DNA control region
sequences indicate extensive mixing of swordfish (*Xiphias gladius*) populations in the Atlantic Ocean. *Can. J. Fish. Aquat. Sci.* 52:1720–1732.

Briggs, J. C. 1958. A list of Florida fishes and their distributions. *Bull. Fla. State Mus.
Biol. Sci.* 2:223–318.

———1974. *Marine Zoogeography.* New York: McGraw-Hill.

———1996. *Global Biogeography.* Amsterdam: Elsevier.

Britten, R. J. 1986. Rates of DNA sequence evolution differ between taxonomic
groups. *Science* 231:1393–1398.

Broderick, D. and C. Moritz. 1996. Hawksbill breeding and foraging populations in the Indo-Pacific region. In *Proceedings of the International Symposium on Sea Turtle Conservation Genetics*, B. W. Bowen and W. N. Witzell (eds.). NOAA Tech. Memo. NMFS-SEFSC-396, pp. 119–128.

Broderick, D., C. Moritz, J. D. Miller, M. Guinea, R. J. Prince, and C.J. Limpus. 1994. Genetic studies of the hawksbill turtle: Evidence for multiple stocks and mixed feeding grounds in Australian waters. *Pacific Conserv. Biol.* 1:123–131.

Bromham, L., A. Rambaut, R. Fortey, A. Cooper, and D. Penny. 1998. Testing the Cambrian explosion hypothesis by using a molecular dating technique. *Proc. Natl. Acad. Sci. USA* 95:12386–12389.

Brooks, D. R. 1985. Historical ecology: A new approach to studying the evolution of ecological associations. *Ann. Missouri Bot. Gard.* 72:660–680.

———1990. Parsimony analysis and historical biogeography: Methodological and theoretical update. *Syst. Zool.* 39:14–30.

Brooks, D. R. and D. A. McLennen. 1991. *Phylogeny, Ecology, and Behavior.* Chicago: University of Chicago Press.

Brower, A.V. Z. 1994. Rapid morphological radiation and convergence among races of the butterfly *Heliconius erato* inferred from patterns of mitochondrial DNA evolution. *Proc. Natl. Acad. Sci. USA* 91:6491–6495.

Brower, A.V. Z. and T. M. Boyce. 1991. Mitochondrial DNA variation in monarch butterflies. *Evolution* 45:1281–1286.

Brown, J. H. and A. C. Gibson. 1983. *Biogeography.* St. Louis: Mosby.

Brown, J. H. and M. V. Lomolino. 1998. *Biogeography,* 2d ed. Sunderland, Mass.: Sinauer.

Brown, J. M., W. G. Abrahamson, and P. A. Way. 1996. Mitochondrial DNA phylogeography of host races of the goldenrod ball gallmaker, *Eurosta solidaginis* (Diptera: Tephritidae). *Evolution* 50:777–786.

Brown, J. M., J. H. Leebens-Mack, J. N. Thompson, O. Pellmyr, and R. G. Harrison. 1997. Phylogeography and host association in a pollinating seed parasite *Greya politella* (Lepidoptera: Prodoxidae). *Mol. Ecol.* 6:215–224.

Brown, J. R., A. T. Beckenbach, and M. J. Smith. 1993. Intraspecific DNA sequence variation of the mitochondrial control region of white sturgeon (*Acipenser transmontanus*). *Mol. Biol. Evol.* 10:326–341.

Brown, R. P. and J. Pestano. 1998. Phylogeography of skinks (*Chalcides*) in the Canary Islands inferred from mitochondrial DNA sequences. *Mol. Ecol.* 7:1183–1191.

Brown, W. M. 1980. Polymorphism in mitochondrial DNA of humans as revealed by restriction endonuclease analysis. *Proc. Natl. Acad. Sci. USA* 77:3605–3609.

————1981. Mechanisms of evolution in animal mitochondrial DNA. *Ann. N. Y. Acad. Sci.* 361:119–134.

————1983. Evolution of animal mitochondrial DNA. In *Evolution of Genes and Proteins,* M. Nei and R. K. Koehn (eds.). Sunderland, Mass.: Sinauer, pp. 62–88.

Brown, W. M., M. George, Jr., and A. C. Wilson. 1979. Rapid evolution of animal mitochondrial DNA. *Proc. Natl. Acad. Sci. USA* 76:1967–1971.

Brown, W. M. and M. H. Goodman. 1979. Quantitation of intrapopulation variation by restriction endonuclease analysis of human mitochondrial DNA. In *Extrachromosomal DNA,* D. J. Cummings, P. Borst, I. B. Dawid, S. M. Weissman, and C. F. Fox (eds.). New York: Academic Press, pp. 485–499.

Brown, W. M., E. M. Prager, A. Wang, and A. C. Wilson. 1982. Mitochondrial DNA sequences of primates: Tempo and mode of evolution. *J. Mol. Evol.* 18:225–239.

Brown, W. M. and J. Vinograd. 1974. Restriction endonuclease cleavage maps of animal mitochondrial DNAs. *Proc. Natl. Acad. Sci. USA* 71:4617–4621.

Brown, W. M. and J. W. Wright. 1975. Mitochondrial DNA and the origin of parthenogenesis in whiptail lizards (*Cnemidophorus*). *Herpetol. Rev.* 6:70–71.

————1979. Mitochondrial DNA analysis and the origin and relative age of parthenogenetic lizards (genus *Cnemidophorus*). *Science* 203:1247–1249.

Brown Gladden, J. G., M. M. Ferguson, and J. W. Clayton. 1997. Matriarchal genetic population structure of North American beluga whales *Delphinapterus leucas* (Cetacea: Monodontidae). *Mol. Ecol.* 6:1033–1046.

Bucklin, A., C. Caudill, and M. Guarnieri. 1998. Population genetics and phylogeny of marine planktonic copepods. In *Molecular Approaches to the Study of the Ocean,* K. Cooksey (ed.). London: Chapman & Hall, pp. 303–317.

Bucklin, A. and T. D. Kocher. 1996. Source regions for recruitment of *Calanus finmarchicus* to Georges Bank: Evidence from molecular population genetic analysis of mtDNA. *Deep-Sea Res.* 43:1665–1681.

Bucklin, A., T. C. LaJeunesse, E. Curry, J. Wallinga, and K. Garrison. 1996a. Molecular diversity of the copepod, *Nannocalanus minor:* Genetic evidence of species and population structure in the North Atlantic Ocean. *J. Marine Res.* 54:285–310.

Bucklin, A., S. B. Smolenack, and A. M. Bentley. 1997. Gene flow patterns of the euphausiid, *Meganyctiphanes norvegica,* in the NW Atlantic based on mtDNA sequences for cytochrome *b* and cytochrome oxidase I. *J. Plankton Res.* 19:1763–1781.

Bucklin, A., R. C. Sundt, and G. Dahle. 1996b. The population genetics of *Calanus finmarchicus* in the North Atlantic. *Ophelia* 44:29–45.

Bucklin, A. and P. H. Wiebe. 1998. Low mitochondrial diversity and small effective population sizes of the copepods *Calanus finmarchicus* and *Nannocalanus minor:* Possible impact of climatic variation during recent glaciation. *J. Heredity* 89:383–392.

Bull, J. J., J. P. Huelsenbeck, C. W. Cunningham, D. L. Swofford, and P. J. Waddell. 1993. Partitioning and combining data in phylogenetic analysis. *Syst. Biol.* 42:384–397.

Buroker, N. E. 1983. Population genetics of the American oyster *Crassostrea virginica* along the Atlantic coast and the Gulf of Mexico. *Marine Biol.* 75:99–112.

Burrows, P. M. and C. C. Cockerham. 1974. Distributions of time to fixation of neutral genes. *Theor. Pop. Biol.* 5:192–207.

Burrows, W. and O. A. Ryder. 1997. Y-chromosome variation in great apes. *Nature* 385:125–126.

Burton, R. S. 1983. Protein polymorphisms and genetic differentiation of marine invertebrate populations. *Marine Biol. Letters* 4:193–206.

———1998. Intraspecific phylogeography across the Point Conception biogeographic boundary. *Evolution* 52:734–745.

Burton, R. S. and B.-N. Lee. 1994. Nuclear and mitochondrial gene genealogies and allozyme polymorphism across a major phylogeographic break in the copepod *Tigriopus californicus*. *Proc. Natl. Acad. Sci. USA* 91:5197–5201.

Bush, G. L. 1975. Modes of Animal Speciation. *Annu. Rev. Ecol. Syst.* 6:339–364.

———1994. Sympatric speciation in animals: New wine in old bottles. *Trends Ecol. Evol.* 9:285–288.

Bush, M. B. 1994. Amazonian speciation: A necessarily complex model. *J. Biogeogr.* 21:5–17.

Butlin, R. K. 1989. Reinforcement of premating isolation. In *Speciation and Its Consequences,* D. Otte and J. A. Endler (eds.). Sunderland, Mass.: Sinauer, pp. 158–179.

Butlin, R. K. and G. M. Hewitt. 1985. A hybrid zone between *Chorthippus parallelus parallelus* and *Chorthippus parallelus erythropus* (Orthoptera: Acrididae). I. Morphological and electrophoretic characters. *Biol. J. Linn. Soc.* 26:269–285.

Byun, S. A., B. F. Koop, and T. E. Reimchen. 1997. North American black bear mtDNA phylogeography: Implications for morphology and the Haida Gwaii glacial refugium controversy. *Evolution* 51:1647–1653.

Caccone, A., M. C. Milinkovitch, V. Sbordoni, and J. R. Powell. 1997. Mitochondrial DNA rates and biogeography of European newts (genus *Euproctus*). *Syst. Biol.* 46:126–144.

Camper, J. D., R. C. Barber, L. R. Richardson, and J. R. Gold. 1993. Mitochondrial DNA variation among red snapper (*Lutjanus campechanus*) from the Gulf of Mexico. *Mol. Mar. Biol. Biotech.* 2:154–161.

Canatore, R., M. Roberti, G. Pesole, A. Ludovico, F. Milella, M. N. Gadaleta, and C. Saccone. 1994. Evolutionary analysis of cytochrome b sequences in some Perciformes: Evidence for a slower rate of evolution than in mammals. *J. Mol. Evol.* 39:589–597.

Cann, R. L. and A. C. Wilson. 1983. Length mutations in human mitochondrial DNA. *Genetics* 104:699–711.

Cann, R. L., W. M. Brown, and A. C. Wilson. 1982. Evolution of human mitochondrial DNA: molecular, genetic, and anthropological implications. In *Proc. 6th Intl. Congr. Human Genetics, Part A*, B. Bonné-Tamir (ed.). New York: Alan R. Liss, pp. 157–165.

————1984. Polymorphic sites and the mechanism of evolution in human mitochondrial DNA. *Genetics* 106:479–499.

Cann, R. L., M. Stoneking, and A. C. Wilson. 1987. Mitochondrial DNA and human evolution. *Nature* 325:31–36.

Carlton, J. T. and J. B. Geller. 1993. Ecological roulette: The global transport of nonindigenous marine organisms. *Science* 261:78–82.

Carpenter, M. A., E. W. Brown, M. Culver, W. E. Johnson, J. Pecon-Slattery, D. Brousset, and S. J. O'Brien. 1996. Genetic and phylogenetic divergence of feline immunodeficiency virus in the puma (*Puma concolor*). *J. Virol.* 70:6682–6693.

Carr, A. and P. J. Coleman. 1974. Seafloor spreading theory and the odyssey of the green turtle from Brazil to Ascension Island, Central Atlantic. *Nature* 249:128–130.

Carr, S. M., S. W. Ballinger, J. N. Derr, L. H. Blackenship, and J. W. Bickham. 1986. Mitochondrial DNA analysis of hybridization between sympatric white-tailed deer and mule deer in west Texas. *Proc. Natl. Acad. Sci. USA* 83:9576–9580.

Carr, S. M. and H. D. Marshall. 1991. Detection of intraspecific DNA sequence variation in the mitochondrial cytochrome *b* gene of Atlantic cod (*Gadus morhua*) by the polymerase chain reaction. *Can. J. Fish. Aquat. Sci.* 48:48–52.

Castelloe, J. and A. R. Templeton. 1994. Root probabilities for intraspecific gene trees under neutral coalescent theory. *Mol. Phylogen. Evol.* 3:102–113.

Cavalli-Sforza, L. L. 1997. Genes, peoples, and langauges. *Proc. Natl. Acad. Sci. USA* 94:7719–7724.

Cavalli-Sforza, L. L., P. Menozzi, and A. Piazza. 1994. *The History and Geography of Human Genes.* Princeton, N.J.: Princeton University Press.

Chapco, W., R. A. Kelln, and D. A. McFayden. 1992. Intraspecific mitochondrial DNA variation in the migratory grasshopper, *Melanoplus sanguinipes*. *Heredity* 69:547–557.

———1994. Mitochondrial DNA variation in North American melanopline grasshoppers. *Heredity* 72:1–9.

Chapman, R. W., J. C. Stephens, R. A. Lansman, and J. C. Avise. 1982. Models of mitochondrial DNA transmission genetics and evolution in higher eukaryotes. *Genet. Res.* 40:41–57.

Charlesworth, B., M. T. Morgan, and D. Charlesworth. 1993. The effect of deleterious mutations on neutral molecular variation. *Genetics* 134:1289–1303.

Chenoweth, S. F. and J. M. Hughes. 1997. Genetic population structure of the catadromous perciform: *Macquaria novemaculeata* (Percichthyidae). *J. Fish Biol.* 50:721–733.

Chow, S. and S. Inoue. 1993. Intra- and interspecific restriction fragment length polymorphism in mitochondrial genes of *Thunnus* tuna species. *Bull. Nat. Res. Inst. Far Seas Fish.* 30:207–225.

Chow, S. and S. Ushiama. 1995. Global population structure of albacore (*Thunnus alalunga*) inferred by RFLP analysis of the mitochondrial ATPase gene. *Marine Biol.* 123:39–45.

Chu, J. Y. and 13 others. 1998. Genetic relationships of populations in China. *Proc. Natl. Acad. Sci. USA* 95:11763–11768.

Chubb, A. L., R. M. Zink, and J. M. Fitzsimons. 1998. Patterns of mtDNA variation in Hawaiian freshwater fishes: the phylogeographic consequences of amphidromy. *J. Heredity* 89:8–16.

Claridge, M. F., H. A. Dawah, and M. R. Wilson (eds.). 1997. *Species: The Units of Biodiversity*. New York: Chapman & Hall.

Clark, A. G. 1993. Evolutionary inferences from molecular characterization of self-incompatibility alleles. In *Mechanisms of Molecular Evolution*, N. Takahata and A. G. Clark (eds.). Sunderland, Mass.: Sinauer, pp. 79–108.

———1997. Neutral behavior of shared polymorphism. *Proc. Natl. Acad. Sci. USA* 94:7730–7734.

Clegg, S. M., P. Hale, and C. Moritz. 1998. Molecular population genetics of the red kangaroo (*Macropus rufus*): mtDNA variation. *Mol. Ecol.* 7:679–686.

Cockerham, C. C. and B. S. Weir. 1993. Estimation of gene flow from *F*-statistics. *Evolution* 47:855–863.

Cohan, F. M. 1999. Genetic structure of bacterial populations. In *Evolutionary Genetics from Molecules to Morphology*, R. Singh and C. Krimbas (eds.). Cambridge: Cambridge University Press, in press.

Collins, T. M., K. Frazer, A. R. Palmer, G. J. Vermeij, and W. M. Brown. 1996. Evolutionary history of Northern Hemisphere *Nucella* (Gastropoda, Muricidae): Molecular, morphological, ecological, and paleontological evidence. *Evolution* 50:2287–2304.

Comas, D., S. Pääbo, and J. Bertranpetit. 1995. Heteroplasmy in the control region of human mitochondrial DNA. *Genome Res.* 5:89–90.

Cooke, F., C. D. MacInnes, and J. P. Prevett. 1975. Gene flow between breeding populations of lesser snow geese. *Auk* 92:493–510.

Cooper, A. 1997. Studies of avian ancient DNA: From Jurassic Park to modern island extinctions. In *Avian Molecular Evolution and Systematics,* D. P. Mindell (ed.). New York: Academic Press, pp. 345–373.

Cooper, A., C. Mourer-Chauviré, G. K. Chambers, A. von Haeseler, A. C. Wilson, and S. Pääbo. 1992. Independent origins of the New Zealand moas and kiwis. *Proc. Natl. Acad. Sci. USA* 89:8741–8744.

Cooper, A., H. N. Poinar, S. Pääbo, J. Radovcic, A. Debénath, M. Caparros, C. Barroso-Ruiz, J. Bertranpetit, C. Nielsen-Marsh, R. E. M. Hedges, and B. Sykes. 1997. Neandertal genetics. *Science* 277:1021–1024.

Cooper, S. J. B. and G. M. Hewitt. 1993. Nuclear DNA sequence divergence between parapatric subspecies of the grasshopper *Chorthippus parallelus. Insect Mol. Biol.* 2:1–10.

Cooper, S.J.B., K.M. Ibrahim, and G.M. Hewitt. 1995. Postglacial expansion and genome subdivision in the European grasshopper *Chorthippus parallelus. Mol. Ecol.* 4:49–60.

Cornuet, J. M. and L. Garnery. 1991. Mitochondrial DNA variability in honeybees and its phylogeographic implications. *Apidologie* 22:627–642.

Cox, C. B. and P. D. Moore. 1993. *Biogeography: An Ecological and Evolutionary Approach,* 5th ed. Oxford: Blackwell.

Coyne, J. A. 1992. Much ado about species. *Nature* 357:289–290.

———1994. Ernst Mayr and the origin of species. *Evolution* 48:19–30.

Coyne, J. A. and H. A. Orr. 1998. The evolutionary genetics of speciation. *Phil. Trans. Roy. Soc. Lond. B* 353:287–305.

Coyne, J. A., H. A. Orr, and D. J. Futuyma. 1988. Do we need a new species concept? *Syst. Zool.* 37:190–200.

Cracraft, J. 1983. Species concepts and speciation analysis. *Curr. Ornithol.* 1:159–187.

———1987. Species concepts and the ontology of evolution. *Biol. Philos.* 2:397–414.

———1988. Deep-history biogeography: retrieving the historical pattern of evolving continental biotas. *Syst. Zool.* 37:221–236.

————1989. Speciation and its ontology: The empirical consequences of alternative species concepts for understanding patterns and processes of differentiation. In *Speciation and Its Consequences,* D. Otte and J. A. Endler (eds.). Sunderland, Mass.: Sinauer, pp. 28–59.

Cracraft, J. and R. O. Prum. 1988. Patterns and processes of diversification: Speciation and historical congruence in some neotropical birds. *Evolution* 42:603–620.

Craddock, C., W. R. Hoeh, R. A. Lutz, and R. C. Vrijenhoek. 1995. Extensive gene flow among mytilid (*Bathymodiolus thermophilus*) populations from hydrothermal vents of the eastern Pacific. *Marine Biol.* 124:137–146.

Crandall, K. A. and A. R. Templeton. 1993. Empirical tests of some predictions from coalescent theory with applications to intraspecific phylogeny reconstruction. *Genetics* 134:959–969.

————1996. Applications of intraspecific phylogenetics. In *New Uses for New Phylogenies,* P. H. Harvey, A. J. Leigh Brown, J. Maynard Smith, and S. Nee (eds.). New York: Oxford University Press, pp. 81–99.

Crandall, K. A., A. R. Templeton, and C. F. Sing. 1994. Intraspecific phylogenetics: Problems and solutions. In *Phylogeny Reconstruction,* R. W. Scotland, D. J. Siebert, and D. M. Williams (eds.). Oxford: Clarendon Press, pp. 273–297.

Crews, S., D. Ojala, J. Posakony, J. Nishiguchi, and G. Attardi. 1979. Nucleotide sequence of a region of human mitochondrial DNA containing the precisely defined origin of replication. *Nature* 277:192–198.

Croizat, L., G. Nelson, and D. E. Rosen. 1974. Centers of origin and related concepts. *Syst. Zool.* 23:265–287.

Cronin, M. A. 1992. Intraspecific variation in mitochondrial DNA of North American cervids. *J. Mammal.* 73:70–82.

————1993. Mitochondrial DNA in wildlife taxonomy and conservation biology: cautionary notes. *Wildl. Soc. Bull.* 21:339–348.

Cronin, M. A., S. C. Amstrup, G. W. Garner, and E. R. Vyse. 1991b. Interspecific and intraspecific mitochondrial DNA variation in North American bears (*Ursus*). *Can. J. Zool.* 69:2985–2992.

Cronin, M. A., J. Bodkin, B. Ballachey, J. Estes, and J. C. Patton. 1996a. Mitochondrial-DNA variation among subspecies and populations of sea otters (*Enhydra lutris*). *J. Mammal.* 77:546–557.

Cronin, M. A., J. B. Grand, D. Esler, D. V. Derksen, and K. T. Scribner. 1996b. Breeding populations of northern pintails have similar mitochondrial DNA. *Can. J. Zool.* 74:992–999.

Cronin, M. A., M. E. Nelson, and D. F. Pac. 1991a. Spatial heterogeneity of mito-chondrial DNA and allozymes among populations of white-tailed deer and mule deer. *J. Heredity* 82:118–127.

Cronin, T. M. 1988. Evolution of marine climates of the U.S. Atlantic coast during the last four million years. In *The Past Three Million Years: Evolution of Climatic Variability in the North Atlantic Region*, N. J. Shackleton, R. G. West, and D. Q. Bowen (eds.). London: The Royal Society, pp. 327–356.

Crosetti, D., W. S. Nelson, and J. C. Avise. 1993. Pronounced genetic structure of mi-tochondrial DNA among populations of the circumglobally distributed grey mullet (*Mugil cephalus*). *J. Fish Biol.* 44:47–58.

Crow, J. F. and K. Aoki. 1984. Group selection for a polygenic behavioral trait: Estimating the degree of population subdivision. *Proc. Natl. Acad. Sci. USA* 81: 6073–6077.

Crow, J. F. and M. Kimura. 1970. *An Introduction to Population Genetics Theory.* New York: Harper & Row.

Crozier, R. 1992. Genetic diversity and the agony of choice. *Biol. Cons.* 61:11–15.

———1997. Preserving the information content of species: Genetic diversity, phy-logeny, and conservation worth. *Annu. Rev. Ecol. Syst.* 28:243–268.

Cummings, D. J., P. Borst, I. B. Dawid, S. M. Weissman, and C. F. Fox (eds.). 1979. *Extrachromosomal DNA.* New York: Academic Press.

Cunningham, C. W., N. W. Blackstone, and L. W. Buss. 1992. Evolution of king crabs from hermit crab ancestors. *Nature* 355:539–542.

Cunningham, C.W., L.W. Buss, and C. Anderson. 1991. Molecular and geologic ev-idence of shared history between hermit crabs and the symbiotic genus *Hy-dractinia*. *Evolution* 45:1301–1316.

Cunningham, C.W. and T.M. Collins. 1998. Beyond area relationships: Extinction and recolonization in molecular marine biogeography. In *Molecular Approaches to Ecology and Evolution*, R. DeSalle and B. Schierwater (eds.). Basel: Birk-häuser, pp. 297–321.

Cunningham, M. and C. Moritz. 1998. Genetic effects of forest fragmentation on a rainforest restricted lizard (Scincidae: *Gnypetoscincus queenslandiae*). *Biol. Cons.* 83:19–30.

da Silva, M. N. F. and J. L. Patton. 1993. Amazonian phylogeography: mtDNA se-quence variation in arboreal echimyid rodents (Caviomorpha). *Mol. Phylogen. Evol.* 2:243–255.

———1998. Molecular phylogeography and the evolution and conservation of Amazonian mammals. *Mol. Ecol.* 7:475–486.

Daugherty, C. H., A. Cree, J. M. Hay, and M. B. Thompson. 1990. Neglected taxonomy and continuing extinctions of tuatara (*Sphenodon*). *Nature* 347:177–179.

Dawid, I. B. and A. W. Blackler. 1972. Maternal and cytoplasmic inheritance of mtDNA in *Xenopus*. *Develop. Biol.* 29:152–161.

Dawkins, R. 1995. *River Out of Eden*. New York: Basic Books.

Dawley, R. M. and J. P. Bogart (eds.). 1989. *Evolution and Ecology of Unisexual Vertebrates*. Albany: New York State Museum.

Day, W. H. 1983. Properties of the nearest neighbor interchange metric for trees of small size. *J. Theor. Biol.* 101:275–288.

DeBry, R. W. 1992. The consistency of several phylogeny-inference methods under varying evolutionary rates. *Mol. Biol. Evol.* 9:537–551.

Degnan, S. M. 1993. The perils of single gene trees—mitochondrial versus single-copy nuclear DNA variation in white-eyes (Aves: Zosteropidae). *Mol. Ecol.* 2:219–225.

Degnan, S. M. and C. Moritz. 1992. Phylogeography of mitochondrial DNA in two species of white-eyes in Australia. *Auk* 109:800–811.

Demastes, J. W., M. S. Hafner, and D. J. Hafner. 1996. Phylogeographic variation in two central American pocket gophers (*Orthogeomys*). *J. Mammal.* 77:917–927.

Demesure, B., B. Comps, and R. J. Petit. 1996. Chloroplast DNA phylogeography of the common beech (*Fagus sylvatica* L.) in Europe. *Evolution* 50:2515–2520.

Denaro, M., H. Blanc, M. J. Johnson, K. H. Chen, E. Wilmsen, L. L. Cavalli-Sforza, and D. C. Wallace. 1981. Ethnic variation in *Hpa*I endonuclease cleavage patterns of human mitochondrial DNA. *Proc. Natl. Acad. Sci. USA* 78:5768–5772.

Densmore, L. D., III, C. C. Moritz, J. W. Wright, and W. M. Brown. 1989. Mitochondrial DNA analyses and the origin and relative age of parthenogenetic lizards (genus *Cnemidophorus*). IV. Nine *sexlineatus*-group unisexuals. *Evolution* 43:969–983.

Densmore, L. D., J. W. Wright, and W. M. Brown. 1985. Length variation and heteroplasmy are frequent in mitochondrial DNA from parthenogenetic and bisexual lizards (genus *Cnemidophorus*). *Genetics* 110:689–707.

de Queiroz, K. and M. J. Donoghue. 1988. Phylogenetic systematics and the species problem. *Cladistics* 4:317–338.

DeSalle, R., A. V. Z. Brower, R. Baker, and J. Remsen. 1997. A hierarchical view of the Hawaiian Drosophilidae. *Pacific Sci.* 51:462–474.

DeSalle, R., T. Freedman, E. M. Prager, and A. C. Wilson. 1987. Tempo and mode of evolution in mitochondrial DNA of Hawaiian *Drosophila*. *J. Mol. Evol.* 26:157–164.

DeSalle, R. and L. V. Giddings. 1986. Discordance of nuclear and mitochon-drial DNA phylogenies in Hawaiian *Drosophila*. *Proc. Natl. Acad. Sci. USA* 83: 6902–6906.

de Stordeur, E. 1997. Nonrandom partition of mitochondria in heteroplasmic *Drosophila*. *Heredity* 79:615–623.

Diamond, J. D. 1966. Zoological classification system of a primitive people. *Science* 151:1102–1104.

Dice, L. R. and W. E. Howard. 1951. Distances of dispersal by prairie deer mice from birthplaces to breeding sites. *Cont. Lab. Vert. Biol. Univ. Mich.* 50:1–15.

Din, W., R. Anand, P. Boursot, D. Darviche, B. Dod, E. Jouvinmarche, A. Orth, G. P. Talwar, P. A. Cazenave, and F. Bonhomme. 1996. Origin and radiation of the house mouse—clues from nuclear genes. *J. Evol. Biol.* 9:519–539.

Dinerstein, E. and E. Wikramanyake. 1993. Beyond "hot spots": How to prioritize investments in biodiversity in the Indo-Pacific region. *Conserv. Biol.* 7:53–65.

Di Rienzo, A. and A. C. Wilson. 1991. Branching pattern in the evolutionary tree for human mitochondrial DNA. *Proc. Natl. Acad. Sci. USA* 88:1597–1601.

Dizon, A. E., C. Lockyer, W. F. Perrin, D. P. Demaster, and J. Sisson. 1992. Rethink-ing the stock concept: A phylogeographic approach. *Conserv. Biol.* 6:24–36.

Dizon, A. E., S. O. Southern, and W. F. Perrin. 1991. Molecular analysis of mtDNA types in exploited populations of spinner dolphins (*Stenella longirostris*). *Rep. Int. Whaling Comm.* (special issue) 13:183–202.

Dobzhansky, T. 1937. *Genetics and the Origin of Species*. New York: Columbia Uni-versity Press.

Dodson, J. J., J. E. Carscadden, L. Bernatchez, and F. Colombani. 1991. Relation-ship between spawning mode and phylogeographic structure in mitochon-drial DNA of North Atlantic capelin *Mallotus villosus*. *Mar. Ecol. Progr. Ser.* 76:103–113.

Dodson, J. J., F. Colombani, and P. K. L. Ng. 1995. Phylogeographic structure in mi-tochodrial DNA of a South-east Asian freshwater fish, *Hemibagrus nemurus* (Siluroidei: Bagridae) and Pleistocene sea-level changes on the Sunda shelf. *Mol. Ecol.* 4:331–346.

Dong, J. and D. B. Wagner. 1994. Paternally inherited chloroplast polymorphism in *Pinus*: Estimation of diversity and population subdivision, and tests of dis-equilibrium with a maternally inherited mitochondrial polymorphism. *Genet-ics* 136:1187–1194.

Donnelly, P. and S. Tavaré. 1986. The ages of alleles and a coalescent. *Adv. Appl. Prob.* 18:1–19.

———1995. Coalescents and genealogical structure under neutrality. *Annu. Rev. Genet.* 29:401–421.

Donnelly, P. and S. Tavaré (eds.). 1997. *Progress in Population Genetics and Human Evolution.* New York: Springer-Verlag.

Donnelly, P., S. Tavaré, D. J. Balding, and R. C. Griffiths. 1996. Estimating the age of the common ancestor of men from the ZFY intron. *Science* 272:1357–1359.

Donoghue, M. J. 1985. A critique of the biological species concept and recommendations for a phylogenetic alternative. *Bryologist* 88:172–181.

Doolittle, W. F. 1998. You are what you eat: A gene transfer ratchet could account for bacterial genes in eukaryotic nuclear genomes. *Trends in Genet.* 14:307–311.

Dopazo, H., L. Boto, and P. Alberch. 1998. Mitochondrial DNA variability in viviparous and ovoviparous populations of the urodele *Salamandra salamandra. J. Evol. Biol.* 11:365–378.

Dorit, R. L., H. Akashi, and W. Gilbert. 1995. Absence of polymorphism at the ZFY locus on the human Y chromosome. *Science* 268:1183–1185.

———1996 Estimating the age of the common ancestor of men from the ZFY intron. *Science* 272:1361–1362.

Douris, V., R. A. D. Cameron, G. C. Rodakis, and R. Lecanidou. 1998. Mitochondrial phylogeography of the land snail *Albinaria* in Crete: Long-term geological and short-term vicariance effects. *Evolution* 52:116–125.

Dowling, T. E., R. E. Broughton, and B. D. DeMarais. 1997. Significant role for historical effects in the evolution of reproductive isolation: Evidence from patterns of introgression between the cyprinid fishes, *Luxilus cornutus* and *Luxilus chrysocephalus. Evolution* 51:1574–1583.

Dowling, T. E. and W. B. Brown. 1993. Population structure of the bottlenose dolphin (*Tursiops truncatus*) as determined by restriction endonuclease analysis of mitochondrial DNA. *Marine Mammal Sci.* 9:138–155.

Dowling, T. E. and B. D. DeMarais. 1993. Evolutionary significance of introgressive hybridization in cyprinid fishes. *Nature* 362:444–446.

Dowling, T. E. and A. L. Secor. 1997. The role of hybridization and introgression in the diversification of animals. *Annu. Rev. Ecol. Syst.* 28:593–619.

Doyle, J. J. 1997. Trees within trees: Genes and species, molecules and morphology. *Syst. Biol.* 46:537–553.

Doyle, J. J., J. L. Doyle, and A. H. D. Brown. 1990. Chloroplast DNA polymorphism and phylogeny in the B genome of *Glycine* subgenus *Glycine* (Leguminosae). *Amer. J. Bot.* 77:772–782.

Duggins, C. F., Jr., A. A. Karlin, T. A. Mousseau, and K. G. Relyea. 1995. Analysis of a hybrid zone in *Fundulus majalis* in a northeastern Florida ecotone. *Heredity* 74:117–128.

Dumolin-Lapegue, S., B. Demesure, S. Fineschi, V. Le Corre, and R. J. Petit. 1997. Phylogeographic structure of white oaks throughout the European continent. *Genetics* 146:1475–1487.

Echelle, A. A. and T. E. Dowling. 1992. Mitochondrial DNA variation and evolution of the Death Valley pupfishes (*Cyprinodon,* Cyprinodontidae). *Evolution* 46:193–206.

Echelle, A. A., T. E. Dowling, C. C. Moritz, and W. M. Brown. 1989. Mitochondrial-DNA diversity and the origin of the *Menidia clarkhubbsi* complex of unisexual fishes (Atherinidae). *Evolution* 43:984–993.

Echelle, A. A. and I. Kornfield (eds.). 1984. *Evolution of Fish Species Flocks.* Orono: University of Maine Press.

Edmands, S., P. E. Moberg, and R. S. Burton. 1996. Allozyme and mitochondrial DNA evidence of population subdivision in the purple sea urchin *Strongylocentrotus purpuratus. Marine Biol.* 126:443–450.

Edwards, S. V. 1993a. Long-distance gene flow in a cooperative breeder suggested by genealogies of mitochondrial DNA sequences. *Proc. Roy. Soc. Lond. B* 252:177–185.

———1993b. Mitochondrial gene genealogy and gene flow among island and mainland populations of a sedentary songbird, the grey-crowned babbler. *Evolution* 47:1118–1137.

———1997. Relevance of microevolutionary processes to higher level molecular systematics. In *Avian Molecular Evolution and Systematics,* D. P. Mindell (ed.). New York: Academic Press, pp. 251–278.

Edwards, S. V. and A. C. Wilson. 1990. Phylogenetically informative length polymorphism and sequence variability in mitochondrial DNA of Australian songbirds (*Pomatostomus*). *Genetics* 126:695–711.

Ehrlich, P. R. and E. O. Wilson. 1991. Biodiversity studies: Science and policy. *Science* 253:758–761.

Eizirik, E., S. L. Bonatto, W. E. Johnson, P. G. Crawshaw, Jr., J. C. Vié, D.M. Brousset, S. J. O'Brien, and F. M. Salzano. 1998. Phylogeographic patterns and evolution of the mitochondrial DNA control region in two neotropical cats (Mammalia, Felidae). *J. Mol. Evol.* 47:613–624.

Eldredge, N. and J. Cracraft. 1980. *Phylogenetic Patterns and the Evolutionary Process.* New York: Columbia University Press.

Eller, E. and H. C. Harpending. 1996. Simulations show that neither population expansion nor population stationarity in a West African population can be rejected. *Mol. Biol. Evol.* 13:1155–1157.

Ellsworth, D. L., R. L. Honeycutt, N. J. Silvy, J. H. Bickham, and W. D. Klimstra. 1994a. Historical biogeography and contemporary patterns of mitochondrial DNA variation in white-tailed deer from the southeastern United States. *Evolution* 48:122–136.

Ellsworth, D. L., R. L. Honeycutt, N. J. Silvy, K. D. Rittenhouse, and M. H. Smith. 1994b. Mitochondrial-DNA and nuclear-gene differentiation in North American prairie grouse (genus *Tympanuchus*). *Auk* 111:661–671.

El Mousadik, A. and R. J. Petit. 1996. Chloroplast DNA phylogeography of the argan tree of Morocco. *Mol. Ecol.* 5:547–555.

Embley, T. M. and E. Stackebrandt. 1997. Species in practice: Exploring uncultured prokaryote diversity in natural samples. In *Species: The Units of Biodiversity,* M. F. Claridge, H. A. Dawah, and M. R. Wilson (eds.). New York: Chapman & Hall, pp. 61–81.

Encalada, S. E. 1996. Conservation genetics of Atlantic and Mediterranean green turtles: Inferences from mtDNA sequences. In *Proceedings of the International Symposium on Sea Turtle Conservation Genetics,* B. W. Bowen and W. N. Witzell (eds.). NOAA Tech. Memo. NMFS-SEFSC-396, pp. 33–40.

Encalada, S. E., K. A. Bjorndal, A. B. Bolten, J. C. Zurita, B. Schroeder, E. Possardt, C. J. Sears, and B. W. Bowen. 1997. Population structure of loggerhead turtle (*Caretta caretta*) nesting colonies in the Atlantic and Mediterranean as inferred from mitochondrial DNA control region sequences. *Marine Biol.* 130:567–575.

Encalada, S. E., P. N. Lahanas, K. A. Bjorndal, A. B. Bolton. M. M. Miyamoto, and B. W. Bowen. 1996. Phylogeography and population structure of the Atlantic and Mediterranean green turtle *Chelonia mydas:* A mitochondrial DNA control region sequence assessment. *Mol. Ecol.* 5:473–483.

Endler, J. A. 1977. *Geographic Variation, Speciation, and Clines.* Princeton, N.J.: Princeton University Press.

———1982. Pleistocene forest refuges: Fact or fancy? In *Biological Diversification in the Tropics,* G. T. Prince (ed.). New York: Columbia University Press, pp. 641–657.

———1986. *Natural Selection in the Wild.* Princeton, N.J.: Princeton University Press.

Engel, S. R., K. M. Hogan, J. F. Taylor, and S. K. Davis. 1998. Molecular systematics and paleobiogeography of the South American sigmodontine rodents. *Mol. Biol. Evol.* 15:35–49.

Engels, W. R. 1981. Estimating genetic divergence and genetic variability with re-striction endonucleases. *Proc. Natl. Acad. Sci. USA* 78:6329–6333.

Enghoff, H. 1995. Historical biogeography of the Holarctic: Area relationships, an-cestral areas, and dispersal of non-marine animals. *Cladistics* 11:223–263.

Epifanio, J. M., J. B. Koppelman, M. A. Nedbal, and D. P. Philipp. 1996. Geographic variation of paddlefish allozymes and mitochondrial DNA. *Trans. Amer. Fish. Soc.* 125:546–561.

Epifanio, J. M., P. E. Smouse, C. J. Kobak, and B. L. Brown. 1995. Mitochondrial DNA divergence among populations of American shad (*Alosa sapidissima*): How much variation is enough for mixed-stock analysis? *Can. J. Fish. Aquat. Sci.* 52:1688–1702.

Epperson, B. K. 1993. Recent advances in correlation studies of spatial patterns of genetic variation. *Evol. Biol.* 27:95–155.

Erlich, H. A. (ed.). 1989. *PCR Technology: Principles and Applications for DNA Amplifi-cation.* New York: Stockton Press.

Erlich, H. A., T. F. Bergström, M. Stoneking, and U. Gyllensten. 1996. HLA se-quence polymorphism and the origin of humans. *Science* 274:1552–1554.

Erwin, T. L. 1991. An evolutionary basis for conservation strategies. *Science* 253:750–752.

Esposti, M., S. DeVries, M. Crimi, A. Ghelli, T. Patarnello, and A. Meyer. 1993. Mito-chondrial cytochrome *b*: Evolution and structure of the protein. *Biochem. Bio-phys. Acta* 1143:243–271.

Estoup, A., M. Solignac, J.-M. Cornuet, J. Goudet, and A. Scholl. 1996. Genetic dif-ferentiation of continental and island populations of *Bombus terrestris* (Hy-menoptera: Apidae) in Europe. *Mol. Ecol.* 5:19–31.

Evans, B. J., J. C. Morales, M. D. Picker, D. B. Kelley, and D. J. Melnick. 1997. Comparative molecular phylogeography of two *Xenopus* species, *X. gilli* and *X. laevis*, in the south-western Cape Province, South Africa. *Mol. Ecol.* 6:333–343.

Excoffier, L. 1990. Evolution of human mitochondrial DNA: Evidence for depar-ture from a pure neutral model of populations at equilibrium. *J. Mol. Evol.* 30:125–139.

Excoffier, L. and P. E. Smouse. 1994. Using allele frequencies and geographic subdi-vision to reconstruct gene trees within a species: Molecular variance parsi-mony. *Genetics* 136:343–359.

Excoffier, L., P. E. Smouse, and J. M. Quattro. 1992. Analysis of molecular variance inferred from metric distances among DNA haplotypes: Applications to hu-man mitochondrial DNA restriction data. *Genetics* 131:479–491.

Eyre-Walker, A., R. L. Gaut, H. Hilton, D. L. Feldman, and B. S. Gaut. 1998. Investigation of the bottleneck leading to the domestication of maize. *Proc. Natl. Acad. Sci. USA* 95:4441–4446.

Faber, J. E. and C. A. Stepien. 1997. The utility of mitochondrial DNA control region sequences for analyzing phylogenetic relationships among populations, species, and genera of the Percidae. In *Molecular Systematics of Fishes,* T. D. Kocher and C. A. Stepien (eds.). San Diego, Calif.: Academic Press, pp. 129–143.

Faith, D. P. 1992. Conservation evaluation and phylogenetic diversity. *Biol. Cons.* 61:1–10.

———1994. Genetic diversity and taxonomic priorities for conservation. *Biol. Cons.* 68:69–74.

Faith, D. P. and P. A. Walker. 1996. How do indicator groups provide information about the relative biodiversity of different sets of areas? On hotspots, complementarity and pattern-based approaches. *Biodiversity Letters* 3:18–25.

Fajen, A. and F. Breden. 1992. Mitochondrial DNA sequence variation among natural populations of the Trinidad guppy, *Poecilia reticulata. Evolution* 46:1457–1465.

Farias, I. P., G. Ortí, I. Sampaio, H. Schneider, and A. Meyer. 1999. Mitochondrial DNA phylogeny of the family Cichlidae: Monophyly and high genetic divergence of the neotropical assemblage. *J. Mol. Evol.,* in press.

Farris, J. S., M. Källersjö, A. G. Kluge, and C. Bult. 1994. Testing significance of incongruence. *Cladistics* 10:315–319.

Faulkes, C. G., D. H. Abbott, H. P. O'Brien, L. Lau, M. R. Roy, R. K. Wayne, and M. W. Bruford. 1997. Micro- and macrogeographical genetic structure of colonies of naked mole-rats *Heterocephalus glaber. Mol. Ecol.* 6:615–628.

Fauron, C. M.-R. and D. R. Wolstenholme. 1980a. Extensive diversity among *Drosophila* species with respect to nucleotide sequences within the adenine + thymine-rich region of mitochondrial DNA molecules. *Nucleic Acids Res.* 8:2439–2452.

———1980b. Intraspecific diversity of nucleotide sequences within the adenine + thymine-rich region of mitochondrial DNA molecules of *Drosophila mauritiana, Drosophila melanogaster,* and *Drosophila simulans. Nucleic Acids Res.* 8:5391–5410.

Federov, V., M. Jaarola, and K. Fredga. 1996. Low mitochondrial DNA variation and recent colonization of Scandinavia by the wood lemming *Myopus schisticolor. Mol. Ecol.* 5:577–581.

Feduccia, A. 1995. Explosive evolution in Tertiary birds and mammals. *Science* 267:637–638.

Felder, D. L. and J. L. Staton. 1994. Genetic differentiation in trans-Floridian species complexes of *Sesarma* and *Uca* (Decapoda: Brachyura). *J. Crust. Biol.* 14:191–209.

Felsenstein, J. 1971. The rate of loss of multiple alleles in finite haploid populations. *Theor. Pop. Biol.* 2:391–403.

———1982. How can we infer geography and history from gene frequencies? *J. Theor. Biol.* 96:9–20.

———1985a. Confidence limits on phylogenies: An approach using the bootstrap. *Evolution* 39:783–791.

———1985b. Phylogenies and the comparative method. *Amer. Nat.* 125:1–15.

———1988. Phylogenies from molecular sequences: Inference and reliability. *Annu. Rev. Genet.* 22:521–565.

———1992a. Estimating effective population size from samples of sequences: Inefficiency of pairwise and segregating sites as compared to phylogenetic estimates. *Genet. Res.* 59:139–147.

———1992b. Estimating effective population size from samples of sequences: A bootstrap Monte Carlo integration method. *Genet. Res.* 60:209–220.

———1993. PHYLIP (Phylogeny Inference package), version 3.5c. Seattle: Department of Genetics, University of Washington.

Ferraris, J. D. and S. R. Palumbi (eds.). 1996. *Molecular Zoology.* New York: Wiley-Liss.

Ferris, S. D., W. M. Brown, W. S. Davidson, and A. C. Wilson. 1981a. Extensive polymorphism in the mitochondrial DNA of apes. *Proc. Natl. Acad. Sci. USA* 78:6319–6323.

Ferris, S. D., R. D. Sage, C.-M. Huang, J. T. Nielsen, U. Ritte, and A.C. Wilson. 1983a. Flow of mitochondrial DNA across a species boundary. *Proc. Natl. Acad. Sci. USA* 80:2290–2294.

Ferris, S. D., R. D. Sage, E. M. Prager, U. Ritte, and A. C. Wilson. 1983b. Mitochondrial DNA evolution in mice. *Genetics* 105:681–721.

Ferris, S. D, A. C. Wilson, and W. M. Brown. 1981b. Evolutionary tree for apes and humans based on cleavage maps of mitochondrial DNA. *Proc. Natl. Acad. Sci. USA* 78:2432–2436.

Figueroa, F., E. Günther, and J. Klein. 1988. MHC polymorphisms pre-dating speciation. *Nature* 335:265–271.

Finnerty, J. R. and B. A. Block. 1992. Direct sequencing of mitochondrial DNA detects highly divergent haplotypes in blue marlin (*Makaira nigricans*). *Mol. Mar. Biol. Biotech.* 1:206–214.

Fisher, R. A. 1930. *The Genetical Theory of Natural Selection.* Oxford: Clarendon Press.

FitzSimmons, N. N., C. J. Limpus, J. A. Norman, A. R. Goldizen, J. D. Miller, and C. Moritz. 1997a. Philopatry of male marine turtles inferred from mitochondrial DNA markers. *Proc. Natl. Acad. Sci. USA* 94:8912–8917.

FitzSimmons, N. N., C. Moritz, C. J. Limpus, L. Pope, and R. Prince. 1997b. Geographic structure of mitochondrial and nuclear gene polymorphisms in Australian green turtle populations and male-biased gene flow. *Genetics* 147: 1843–1854.

Fleischer, R. C., C. E. McIntosh, and C. L. Tarr. 1998. Evolution on a volcanic conveyor belt: Using phylogeographic reconstructions and K-Ar-based ages of the Hawaiian Islands to estimate molecular evolutionary rates. *Mol. Ecol.* 7:533–545.

Foster, S. A. and S. A. Cameron. 1996. Geographic variation in behavior: A phylogenetic framework for comparative studies. In *Phylogenies and the Comparative Method in Animal Behavior,* E. P. Martins (ed.). New York: Oxford University Press, pp. 138–165.

Fox, G. E., L. J. Magrum, W. E. Balch, R. S. Wolfe, and C. R. Woese. 1977. Classification of methanogenic bacteria by 16S ribosomal RNA characterization. *Proc. Natl. Acad. Sci. USA* 74:4537–4541.

Francisco, J. F., G. G. Brown, and M. V. Simpson. 1979. Further studies on types A and B rat mtDNA's: Cleavage maps and evidence for cytoplasmic inheritance in mammals. *Plasmid* 2:426–436.

Freitag, S. and T. J. Robinson. 1993. Phylogeographic patterns in mitochondrial DNA of the ostrich (*Struthio camelus*). *Auk* 110:614–622.

Friesen, V. L., B. C. Congdon, H. E. Walsh, and T. P. Birt. 1997. Intron variation in marbled murrelets detected using analyses of single-stranded conformational polymorphisms. *Mol. Ecol.* 6:1047–1058.

Friesen, V. L., W. A. Montevecchi, A. J. Baker, R. T. Barrett, and W. S. Davidson. 1996. Population differentiation and evolution in the common guillemot *Uria aalge. Mol. Ecol.* 5:793–805.

Frost, D. R. and D. M. Hillis. 1990. Species in concept and practice: Herpetological applications. *Herpetologica* 46:87–104.

Frumhoff, P. C. and H. K. Reeve. 1994. Using phylogenies to test hypotheses of adaptation: A critique of some current proposals. *Evolution* 48:172–180.

Fu, Y.-X. 1994a. A phylogenetic estimator of effective population size or mutation rate. *Genetics* 136:685–692.

———1994b. Estimating effective population size or mutation rate using the frequencies of mutations of various classes in a sample of DNA sequences. *Genetics* 138:1375–1386.

Fu, Y.-X. and W.-H. Li. 1993. Statistical tests of neutrality of mutations. *Genetics* 133:693–709.

———1996. Estimating the age of the common ancestor of men from the *ZFY* intron. *Science* 272:1356–1357.

Fujii, N. 1997. Phylogeographic studies in Japanese alpine plants based on intraspecific chloroplast DNA variations. *Bull. Biogeog. Soc. Japan* 52:59–69.

Fullerton, S. M., R. M. Harding, A. J. Boyce, and J. B. Clegg. 1994. Molecular and population genetic analysis of allelic sequence diversity at the human B-globin locus. *Proc. Natl. Acad. Sci. USA* 91:1805–1809.

Funderburg, J. B., Jr. and T. L. Quay. 1983. In *The Seaside Sparrow, Its Biology and Management,* T. L. Quay, J. B. Funderburg, Jr., D. S. Lee, E. F. Potter, and C. S. Robbins (eds.). Raleigh: North Carolina State Museum, pp. 19–27.

Futuyma, D. J. 1998. *Evolutionary Biology,* 3d ed. Sunderland, Mass.: Sinauer.

Gach, M. H. 1996. Geographic variation in mitochondrial DNA and biogeography of *Culaea inconstans* (Gasterosteidae). *Copeia* 1996:563–575.

Galis, F. and J. A. J. Metz. 1998. Why are there so many cichlid species? *Trends Ecol. Evol.* 13:1–2.

García-París, M., M. Alcobendas, and P. Alberch. 1998. Influence of the Guadalquivir River Basin on mitochondrial DNA evolution of *Salamandra salamandra* (Caudata: Salamandridae) from southern Spain. *Copeia* 1998:173–176.

Garcia-Rodriguez, A. I., B. W. Bowen, D. Domning, A. A. Mignucci-Giannoni, M. Marmontel, R. A. Montoya-Ospina, B. Morales-Vela, M. Rudin, R. K. Bonde, and P. M. McGuire. 1998. Phylogeography of the West Indian manatee (*Trichechus manatus*): How many populations and how many taxa? *Mol. Ecol.* 7:1137–1149.

Garner, K. J. and O. A. Ryder. 1996. Mitochondrial DNA diversity in gorillas. *Mol. Phylogen. Evol.* 6:39–48.

Garnery, L., J.-M. Cornuet, and M. Solignac. 1992. Evolutionary history of the honey bee *Apis mellifera* inferred from mitochondrial DNA analysis. *Mol. Ecol.* 1:145–154.

Gavrilets, S. 1997. Evolution and speciation on holey adaptive landscapes. *Trends Ecol. Evol.* 12:307–312.

Geller, J. B., E. D. Walton, E. D. Grosholz, and G. M. Ruiz. 1997. Cryptic invasions of the crab *Carcinus* detected by molecular phylogeography. *Mol. Ecol.* 6:901–906.

George, M., Jr. and O. A. Ryder. 1986. Mitochondrial DNA evolution in the genus *Equus. Mol. Biol. Evol.* 3:535–546.

Georgiadis, N., L. Bischof, A. Templeton, J. Patton, W. Karesh, and D. Western. 1994. Structure and history of African elephant populations: I. Eastern and Southern Africa. *J. Heredity* 85:100–104.

Gibbons, A. 1996. The peopling of the Americas. *Science* 274:31–34.

———1998. Calibrating the mitochondrial clock. *Science* 279:28–29.

Giddings, L. V., K. Y. Kaneshiro, and W. W. Anderson (eds.). 1989. *Genetics, Specia-tion and the Founder Principle.* New York: Oxford University Press.

Giles, R. E., H. Blanc, H. M. Cann, and D. C. Wallace. 1980. Maternal inheritance of human mitochondrial DNA. *Proc. Natl. Acad. Sci. USA* 77:6715–6719.

Gill, F. B. 1995. *Ornithology,* 2d ed. New York: Freeman.

Gill, F. B., A. M. Mostrom, and A. L. Mack. 1993. Speciation in North Amer-ican chickadees. I. Patterns of mtDNA genetic divergence. *Evolution* 47:195–212.

Gill, F. B. and B. Slikas. 1992. Patterns of mitochondrial divergence in North Amer-ican crested titmice. *Condor* 94:20–28.

Gillespie, J. H. 1986. Variability of evolutionary rates of DNA. *Genetics* 113:1077–1091.

Gillham, N. W. (ed.). 1978. *Organelle Heredity.* New York: Raven Press.

Girman, D. J., P. W. Kat, M. G. L. Mills, J. R. Ginsberg, M. Borner, V. Wilson, J. H. Fanshawe, C. Fitzgibbon, L. M. Lau, and R. K. Wayne. 1993. Molecular genetic and morphological analyses of the African wild dog (*Lycaon pictus*). *J. Heredity* 84:450–459.

Giuffra, E., L. Bernatchez, and R. Guyomard. 1994. Mitochondrial control region and protein coding genes sequence variation among phenotypic forms of brown trout *Salmo trutta* from northern Italy. *Mol. Ecol.* 3:161–171.

Giuffra, E., R. Guyomard, and G. Forneris. 1996. Phylogenetic relationships and introgression patterns between incipient parapatric species of Italian brown trout (*Salmo trutta* L. complex). *Mol. Ecol.* 5:207–220.

Givnish, T. J. and K. J. Sytsma (eds.). 1997. *Molecular Evolution and Adaptive Radia-tion.* Cambridge: Cambridge University Press.

Gold, J. R. and L. R. Richardson. 1994. Mitochondrial DNA variation among "red fishes" from the Gulf of Mexico. *Fish. Res.* 20:137–150.

———1998. Mitochondrial DNA diversification and population structure in fishes from the Gulf of Mexico and western Atlantic. *J. Heredity* 89:404–414.

Gold, J. R., L. R. Richardson, C. Furman, and T. L. King. 1993. Mitochondrial DNA differentiation and population structure in red drum (*Sciaenops ocellatus*) from the Gulf of Mexico and Atlantic Ocean. *Marine Biol.* 116:175–185.

Gold, J. R., L. R. Richardson, C. Furman, and F. Sun. 1994. Mitochondrial DNA di-versity and population structure in marine fish species from the Gulf of Mex-ico. *Can. J. Fish. Aquat. Sci. 51* (supplement 1):205–214.

Goldberg, T. L. and M. Ruvolo. 1997. Molecular phylogenetics and historical bio-geography of east African chimpanzees. *Biol. J. Linn. Soc.* 61:301–324.

Golding, G. B. 1997. The effect of purifying selection on genealogies. In *Progress in Population Genetics and Human Evolution,* P. Donnelly and S. Tavaré (eds.). New York: Springer-Verlag, pp. 271–285.

Goldman, N. and N. H. Barton. 1992. Genetics and geography. *Nature* 357:440–441.

Goldstein, D. B., A. R. Linares, L. L. Cavalli-Sforza, and M. W. Feldman. 1995. Genetic absolute dating based on microsatellites and the origin of modern humans. *Proc. Natl. Acad. Sci. USA* 92:6723–6727.

Goldstein, P. Z. 1997. Phyloconservation. *Conserv. Biol.* 11:582–583.

Gonder, M. K., J. F. Oates, T. R. Disotell, M. R. J. Forstner, J. C. Morales, and D. J. Melnick. 1997. A new west African chimpanzee subspecies? *Nature* 388:337.

González, S., J. E. Maldonado, J. A. Leonard, C. Vila, J. M. Barbanti Duarte, M. Merino, N. Brum-Zorrilla, and R. K. Wayne. 1998. Conservation genetics of the endangered Pampas deer (*Ozotoceros bezoarticus*). *Mol. Ecol.* 7:47–56.

González-Villaseñor, L. I. and D. A. Powers. 1990. Mitochondrial-DNA restriction-site polymorphisms in the teleost *Fundulus heteroclitus* support secondary intergradation. *Evolution* 44:27–37.

Good, S. V., D. F. Williams, K. Ralls, and R. C. Fleischer. 1997. Population structure of *Dipodomys ingens* (Heteromyidae): The role of spatial heterogeneity in maintaining genetic diversity. *Evolution* 51:1296–1310.

Goodbred, C. O. and J. E. Graves. 1996. Genetic relationships among geographically isolated populations of bluefish (*Pomatomus saltatrix*). *Mar. Freshwater Res.* 47:347–355.

Goodfellow, M., G. P. Manfio, and J. Chun. 1997. Towards a practical species concept for cultivable bacteria. In *Species: The Units of Biodiversity,* M. F. Claridge, H. A. Dawah, and M. R. Wilson (eds.). New York: Chapman & Hall, pp. 25–59.

Gotoh, O., J.-I. Hayashi, H. Yonekawa, and Y. Tagashira. 1979. An improved method for estimating sequence divergence between related DNAs from changes in restriction endonuclease cleavage sites. *J. Mol. Evol.* 14:301–310.

Grandjean, F., C. Souty-Grosset, R. Raimond, and D. M. Holdich. 1997. Geographical variation of mitochondrial DNA between populations of the white-clawed crayfish *Austropotamobius pallipes. Freshwater Biol.* 37:493–501.

Grant, W. S. 1987. Genetic divergence between congeneric Atlantic and Pacific Ocean fishes. In *Population Genetics and Fisheries Management,* N. Ryman and F. Utter (eds.). Seattle: University of Washington Press, pp. 225–246.

Grant, W. S. and B. W. Bowen. 1998. Shallow population histories in deep evolutionary lineages of marine fishes: Insights from sardines and anchovies and lessons for conservation. *J. Heredity,* 89:415–426.

Grant, W. S., A.-M. Clark, and B. W. Bowen. 1999. Why RFLP analysis of mitochondrial DNA failed to resolve sardine (*Sardinops*) biogeography: Insights from mitochondrial DNA cytochrome *b* sequences. *Can. J. Fish. Aquat. Sci.*, in press.

Grant, W. S. and R. W. Leslie. 1996. Late Pleistocene dispersal of Indian-Pacific sardine populations in an ancient lineage of the genus *Sardinops*. *Marine Biol.* 126:133–142.

Graves, J. E. 1996. Conservation genetics of fishes in the pelagic marine realm. In *Conservation Genetics: Case Histories from Nature*, J. C. Avise and J. L. Hamrick (eds.). New York: Chapman & Hall, pp. 335–366.

———1998. Molecular insights into the population structures of cosmopolitan marine fishes. *J. Heredity* 89:427–437.

Graves, J. E. and A. E. Dizon. 1989. Mitochondrial DNA sequence similarity of Atlantic and Pacific albacore tuna. *Can. J. Fish. Aquat. Sci.* 46:870–873.

Graves, J. E., S. D. Ferris, and A. E. Dizon. 1984. High genetic similarity of Atlantic and Pacific skipjack tuna demonstrated with restriction endonuclease analysis of mitochondrial DNA. *Marine Biol.* 79:315–319.

Graves, J. E. and J. R. McDowell. 1994. Genetic analysis of striped marlin *Tetrapturus audax* population structure in the Pacific Ocean. *Can. J. Fish. Aquat. Sci.* 51:1762–1768.

———1995. Inter-ocean genetic differentiation of istiophorid billfishes. *Marine Biol.* 122:193–203.

Graves, J. E., J. R. McDowell, and M. L. Jones. 1992. A genetic analysis of weakfish *Cynoscion regalis* stock structure along the mid-Atlantic coast. *Fish. Bull.* 90:469–475.

Graybeal, A. 1995. Naming species. *Syst. Biol.* 44:237–250.

Greenberg, B. D., J. E. Newbold, and A. Sugino. 1983. Intraspecific nucleotide sequence variability surrounding the origin of replication in human mitochondrial DNA. *Gene* 21:33–49.

Greenberg, J. H., G. G. Turner III, and S. L. Zegura. 1986. The settlement of the Americas: A comparison of linguistic, dental and genetic evidence. *Curr. Anthrop.* 27:477–497.

Greenberg, R., P. J. Cordero, S. Droege, and R. C. Fleischer. 1998. Morphological adaptation with no mitochondrial DNA differentiation in the coastal plain swamp sparrow. *Auk* 115:706–712.

Greenlaw, J. S. 1993. Behavioral and morphological diversification in Atlantic coastal sharp-tailed sparrows (*Ammodramus caudacutus*). *Auk* 110:286–303.

Greenwood, P. J. 1980. Mating systems, philopatry and dispersal in birds and mammals. *Anim. Behav.* 28:1140–1162.

Greenwood, P. J. and P. H. Harvey. 1982. The natal and breeding dispersal of birds. *Annu. Rev. Ecol. Syst.* 13:1–21.

Griffiths, R. C. 1980. Lines of descent in the diffusion approximation of neutral Wright-Fisher models. *Theor. Pop. Biol.* 17:40–50.

Griffiths, R. C. and S. Tavaré. 1997. Computational methods for the coalescent. In *Progress in Population Genetics and Human Evolution,* P. Donnelly and S. Tavaré (eds.). New York: Springer-Verlag, pp. 165–182.

Grijalva-Chon, J. M., K. Numachi, O. Sosa-Nishizaki, and J. de la Rosa-Velez. 1994. Mitochondrial DNA analysis of north Pacific swordfish (*Xiphias gladius*) population structure. *Mar. Ecol. Progr. Ser.* 115:15–19.

Groves, P. 1997. Intraspecific variation in mitochondrial DNA of muskoxen, based on control-region sequences. *Can. J. Zool.* 75:568–575.

Gutierrez, P. C. 1994. Mitochondrial-DNA polymorphism in the oilbird (*Steatornis caripensis,* Steatornithidae) in Venezuela. *Auk* 111:573–578.

Gyllensten, U. 1985. The genetic structure of fish: Differences in the intraspecific distribution of biochemical genetic variation between marine, anadromous, and freshwater species. *J. Fish. Biol.* 26:691–699.

Gyllensten, U. and H. A. Erlich. 1989. Ancient roots for polymorphism of the HLA-DQα locus in primates. *Proc. Natl. Acad. Sci. USA* 86:9986–9990.

Gyllensten, U., M. Sundvall, and H. A. Erlich. 1991b. Allelic diversity is generated by intraexon sequence exchange at the *DRB1* locus of primates. *Proc. Natl. Acad. Sci. USA* 88:3686–3690.

Gyllensten, U., D. Wharton, A. Joseffson, and A. C. Wilson. 1991a. Paternal inheritance of mitochondrial DNA in mice. *Nature* 352:255–257.

Gyllensten, U., D. Wharton, and A. C. Wilson. 1985. Maternal inheritance of mitochondrial DNA during backcrossing of two species of mice. *J. Heredity* 76:321–324.

Gyllensten, U. and A. C. Wilson. 1987. Interspecific mitochondrial DNA transfer and the colonization of Scandinavia by mice. *Genet. Res.* 49:25–29.

Haffer, J. 1969. Speciation in Amazonian forest birds. *Science* 165:131–137.

Haglund, T. R., D. G. Buth, and R. Lawson. 1992. Allozyme variation and phylogenetic relationships of Asian, North American, and European populations of the threespine stickleback, *Gasterosteus aculeatus. Copeia* 1992:432–443.

Haldane, J. B. S. 1932. *The Causes of Evolution.* London: Longmans and Green.

Hale, L. R. and R. S. Singh. 1987. Mitochondrial DNA variation and genetic structure in populations of *Drosophila melanogaster. Mol. Biol. Evol.* 4:622–637.

Hall, H. G. and K. Muralidharan. 1989. Evidence from mitochondrial DNA that African honey bees spread as continuous maternal lineages. *Nature* 339:211–213.

Hall, H. G. and D. R. Smith. 1991. Distinguishing African and European honeybee matrilines using amplified mitochondrial DNA. *Proc. Natl. Acad. Sci. USA* 88:4548–4552.

Hammer, M. F. 1995. A recent common ancestry for human Y chromosomes. *Nature* 378:376–378.

Hammer, M. F., A. B. Spurdle, T. Karafet, M. R. Bonner, E. T. Wood, A. Novelletto, P. Malaspina, R. J. Mitchell, S. Horai, T. Jenkins, and S. L. Zegura. 1997. The geographic distribution of human Y chromosome variation. *Genetics* 145:787–805.

Hanni, C., V. Laudet, D. Stehelin, and P. Taberlet. 1994. Tracking the origins of the cave bear (*Ursus spelaeus*) by mitochondrial DNA sequencing. *Proc. Natl. Acad. Sci. USA* 91:12336–12340.

Hanski, I. and M. E. Gilpin. 1997. *Metapopulation Biology: Ecology, Genetics, and Evolution.* San Diego: Academic Press.

Harding, R. M. 1996. New phylogenies: An introductory look at the coalescent. In *New Uses for New Phylogenies,* P. H. Harvey, A. J. Leigh Brown, J. Maynard Smith, and S. Nee (eds.). New York: Oxford University Press, pp. 15–22.

———1997. Lines of descent from mitochondrial Eve: An evolutionary look at the coalescence. In *Progress in Population Genetics and Human Evolution,* P. Donnelly and S. Tavaré (eds.). New York: Springer-Verlag, pp. 15–31.

Harding, R. M., S. M. Fullerton, R. C. Griffiths, J. Bond, M. J. Cox, J. A. Schneider, D. S. Moulin, and J. B. Clegg. 1997a. Archaic African and Asian lineages in the genetic ancestry of modern humans. *Amer. J. Hum. Genet.* 60:772–789.

Harding, R. M., S. M. Fullerton, R. C. Griffiths, and J. B. Clegg. 1997b. A gene tree for *ß*-globin sequences from Melanesia. *J. Mol. Evol.* 44 (Suppl. 1):S133–S138.

Hare, M. P. 1998. Using mitochondrial DNA gene trees and nuclear RFLPs to predict genealogical patterns at nuclear loci: Examples from the American oyster. In *Proceedings of the Trinational Workshop on Molecular Evolution,* M. K. Uyenoyama and A. von Haeseler (eds.). Durham, N.C.: Duke Publ. Group, pp. 125–130.

Hare, M. P. and J. C. Avise. 1996. Molecular genetic analysis of a stepped multilocus cline in the American oyster (*Crassostrea virginica*). *Evolution* 50:2305–2315.

———1998. Population structure in the American oyster as inferred by nuclear gene genealogies. *Mol. Biol. Evol.* 15:119–128.

Härlid, A., A. Janke, and U. Arnason. 1998. The complete mitochondrial genome of *Rhea americana* and early avian divergences. *J. Mol. Evol.* 46:669–679.

Harpending, H. C., M. A. Batzer, M. Gurven, L. B. Jorde, A. R. Rogers, and S. T. Sherry. 1998. Genetic traces of ancient demography. *Proc. Natl. Acad. Sci. USA* 95:1961–1967.

Harpending, H. C., S. T. Sherry, A. R. Rogers, and M. Stoneking. 1993. The genetic structure of ancient human populations. *Curr. Anthrop.* 34:483–496.

Harris, H. 1966. Enzyme polymorphism in man. *Proc. Roy. Soc. Lond. B* 164:298–310.

Harrison, R. G. 1989. Animal mitochondrial DNA as a genetic marker in population and evolutionary biology. *Trends Ecol. Evol.* 4:6–11.

———1990. Hybrid zones: Windows on evolutionary process. *Oxford Surv. Evol. Biol.* 7:69–128.

———1991. Molecular changes at speciation. *Annu. Rev. Ecol. Syst.* 22:281–308.

———(ed.). 1993. *Hybrid Zones and the Evolutionary Process.* New York: Oxford University Press.

Harrison, R. G., D. M. Rand, and W. C. Wheeler. 1985. Mitochondrial DNA size variation within individual crickets. *Science* 228:1446–1448.

———1987. Mitochondrial DNA variation in field crickets across a narrow hybrid zone. *Mol. Biol. Evol.* 4:144–158.

Hartl, D. L. and A. G. Clark. 1988. *Principles of Population Genetics,* 2d ed. Sunderland, Mass.: Sinauer.

Hartl, G. B., F. Kurt, R. Tiedemann, C. Gmeiner, K. Nadlinger, K. Mar, and A. Rübel. 1996. Population genetics and systematics of Asian elephant (*Elephas maximus*): A study based on sequence variation at the *cyt b* gene of PCR-amplified mitochondrial DNA from hair bulbs. *Int. J. Mammal. Biol.* 61:285–294.

Harvey, P. H. and M. D. Pagel. 1991. *The Comparative Method in Evolutionary Biology.* New York: Oxford University Press.

Hasegawa, M. and S. Horai. 1991. Time of the deepest root for polymorphism in human mitochondrial DNA. *J. Mol. Evol.* 32:37–42.

Hasegawa, M., A. Di Rienzo, T. D. Kocher, and A.C. Wilson. 1993. Toward a more accurate time scale for the human mitochondrial DNA tree. *J. Mol. Evol.* 37:347–354.

Hauswirth, W. W. and P. J. Laipis. 1982. Mitochondrial DNA polymorphism in a maternal lineage of Holstein cows. *Proc. Natl. Acad. Sci. USA* 79:4686–4690.

Hauswirth, W. W., M. J. Van de Walle, P. J. Laipis, and P. D. Olivo. 1984. Heterogeneous mitochondrial DNA D-loop sequences in bovine tissue. *Cell* 37: 1001–1007.

Hayashi, J. I., H. Yonekawa, O. Gotoh, J. Watanabe, and Y. Tagashira. 1978. Strictly maternal inheritance of rat mitochondrial DNA. *Biochem. Biophys. Res. Comm.* 83:1032–1038.

Hayes, J. P. and R. G. Harrison. 1992. Variation in mitochondrial DNA and the biogeographic history of woodrats (*Neotoma*) of the eastern United States. *Syst. Biol.* 41:331–344.

Heard, S. B. and D. L. Hauser. 1995. Key evolutionary innovations and their ecological mechanisms. *Hist. Biol.* 10:151–173.

Hedgecock, D. 1986. Is gene flow from pelagic larval dispersal important in the adaptation and evolution of marine invertebrates? *Bull. Marine Sci.* 39: 550–564.

Hedgecock, D., V. Chow, and W. E. Waples. 1992. Effective population numbers of shellfish broodstocks estimated from temporal variance in allelic frequencies. *Aquaculture* 108:215–232.

Hedgecock, D. and F. Sly. 1990. Genetic drift and effective population sizes of hatchery-propagated stocks of the Pacific oyster, *Crassostrea gigas. Aquaculture* 88:21–38.

Hedgecock, D., M. L. Tracey, and K. Nelson. 1982. Genetics. In *The Biology of Crustacea*, vol. 2, L. G. Abele (ed.). New York: Academic Press, pp. 297–403.

Hedges, S. B. 1989. An island radiation: Allozyme evolution in Jamaican frogs of the genus *Eleutherodactylus* (Leptodactylidae). *Carib. J. Sci.* 25:123–147.

——1996. Historical biogeography of West Indian vertebrates. *Annu. Rev. Ecol. Syst.* 27:163–196.

Hedges, S. B. and K. L. Burnell. 1990. The Jamaican radiation of *Anolis* (Sauria: Iguanidae): An analysis of relationships and biogeography using sequential electrophoresis. *Carib. J. Sci.* 26:31–44.

Hedges, S. B., C. A. Hass, and L. R. Maxon. 1992b. Caribbean biogeography: Molecular evidence for dispersal in West Indian terrestrial vertebrates. *Proc. Natl. Acad. Sci. USA* 89:1909–1913.

Hedges, S. B., S. Kumar, K. Tamura, and M. Stoneking. 1992a. Human origins and the analysis of mitochondrial DNA sequences. *Science* 255:737–739.

Hedin, M. C. 1997. Molecular phylogenetics at the population/species interface in cave spiders of the southern Appalachians (Araneae: Nesticidae: *Nesticus*). *Mol. Biol. Evol.* 14:309–324.

Hein, J. 1990. Reconstructing evolution of sequences subject to recombination using parsimony. *Mathemat. Biosci.* 98:185–200.

——1993. A heuristic method to reconstruct the history of sequences subject to recombination. *J. Mol. Evol.* 36:396–405.

Heist, E. J., J. E. Graves, and J. A. Musick. 1995. Population genetics of the sandbar shark (*Carcharhinus plumbeus*) in the Gulf of Mexico and Mid-Atlantic Bight. *Copeia* 1995:555–562.

Heist, E. J., J. A. Musick, and J. E. Graves. 1996a. Mitochondrial DNA diversity and divergence among sharpnose sharks, *Rhizoprionodon terraenovae*, from the Gulf of Mexico and Mid-Atlantic Bight. *Fish. Bull.* 94:664–668.

————1996b. Genetic population structure of the shortfin mako (*Isurus oxyrinchus*) inferred from restriction fragment length polymorphism analysis of mitochondrial DNA. *Can. J. Fish. Aquat. Sci.* 53:583–588.

Helbig, A. J., I. Seibold, J. Martens, and M. Wink. 1995. Genetic differentiation and phylogenetic relationships of Bonelli's warbler *Phylloscopus bonelli* and green warbler *P. nitidus*. *J. Avian Biol.* 26:138–153.

Henderson, R. W. and S. B. Hedges. 1995. Origin of West Indian populations of the geographically widespread boa *Corallus enydris* inferred from mitochondrial DNA sequences. *Mol. Phylogen. Evol.* 4:88–92.

Hengeveld, R. 1990. *Dynamic Biogeography.* Cambridge: Cambridge University Press.

Hennig, W. 1966. *Phylogenetic Systematics.* Urbana: University of Illinois Press.

Herbots, H. M. 1997. The structured coalescent. In *Progress in Population Genetics and Human Evolution,* P. Donnelly and S. Tavaré (eds.). New York: Springer-Verlag, pp. 231–255.

Hertzberg, M., N. P. Mickelson, S. W. Serjeantson, J. F. Prior, and J. Trent. 1989. An Asian-specific 9-bp deletion of mitochondrial DNA is frequently found in Polynesians. *Amer. J. Hum. Genet.* 44:504–510.

Hewitt, G. M. 1993. After the ice: *parallelus* meets *erythropus* in the Pyrenees. In *Hybrid Zones and the Evolutionary Process,* R. G. Harrison (ed.). New York: Oxford University Press, pp. 140–146.

————1996. Some genetic consequences of ice ages, and their role in divergence and speciation. *Biol. J. Linn. Soc.* 58:247–276.

Hey, J. 1994. Bridging phylogenetics and population genetics with gene tree models. In *Molecular Ecology and Evolution: Approaches and Applications,* B. Schierwater, B. Streit, G. P. Wagner, and R. DeSalle (eds.). Basel: Birkhäuser Verlag, pp. 435–449.

Hey, J. and R. M. Klinman. 1993. Population genetics and phylogenetics of DNA sequence variation at multiple loci within the *Drosophila melanogaster* species complex. *Mol. Biol. Evol.* 10:804–822.

Hickson, R. E. and R. L. Cann. 1997. *Mhc* allelic diversity and modern human origins. *J. Mol. Evol.* 45:589–598.

Hill, A. V. S., C. E. M. Allsopp, D. Kwiatkowski, T. E. Taylor, S. N. R. Yates, N. M. Anstey, J. J. Wirima, D. R. Brewster, A. J. McMichael, M. E. Molyneux, and B. M. Greenwood. 1992. Extensive genetic diversity in the HLA class II region of Africans, with a focally predominant allele, *DRB*1304*. *Proc. Natl. Acad. Sci. USA* 89:2277–2281.

Hillis, D. M. 1987. Molecular versus morphological approaches to systematics. *Annu. Rev. Ecol. Syst.* 18:23–42.

Hillis, D. M., M. W. Allard, and M. M. Miyamoto. 1993. Analysis of DNA sequence data: Phylogenetic inference. *Meth. Enzymol.* 242:456–487.

Hillis, D. M. and J. P. Huelsenbeck. 1995. Assessing molecular phylogenies. *Science* 267:255–256.

Hillis, D. M., J. P. Huelsenbeck, and C. W. Cunningham. 1994. Application and accuracy of molecular phylogenies. *Science* 264:671–677.

Hillis, D. M., C. Moritz, and B. K. Mable (eds.). 1996. *Molecular Systematics,* 2d ed. Sunderland, Mass.: Sinauer.

Hodges, S. A. and M. L. Arnold. 1994. Columbines: A geographically widespread species flock. *Proc. Natl. Acad. Sci. USA* 91:5129–5232.

Hoeh, W. R., K. H. Blakley, and W. M. Brown. 1991. Heteroplasmy suggests limited biparental inheritance of *Mytilus* mitochondrial DNA. *Science* 251:1488–1490.

Hoelzel, A. R. 1994. Genetics and ecology of whales and dolphins. *Annu. Rev. Ecol. Syst.* 25:377–399.

Hoelzel, A. R., M. Dahlheim, and S. J. Stern. 1998a. Low genetic variation among killer whales (*Orcinus orca*) in the eastern North Pacific and genetic differentiation between foraging specialists. *J. Heredity* 89:121–128.

Hoelzel, A. R. and G. A. Dover. 1991. Genetic differentiation between sympatric killer whale populations. *Heredity* 66:191–195.

Hoelzel, A. R., C. W. Potter, and P. B. Best. 1998b. Genetic differentiation between parapatric "nearshore" and "offshore" populations of the bottlenose dolphin. *Proc. Roy. Soc. Lond. B* 265:1177–1183.

Hoelzer, G. A. 1997. Inferring phylogenies from mtDNA variation: Mitochondrial-gene trees versus nuclear-gene trees revisited. *Evolution* 51:622–626.

Hoelzer, G. A., W. P. J. Dittus, M. V. Ashley, and D. J. Melnick. 1994. The local distribution of highly divergent mitochondrial DNA haplotypes in toque macaques *Macaca sinica* at Polonnaruwa, Sri Lanka. *Mol. Ecol.* 3:451–458.

Hoelzer, G. A., J. Wallman, and D. J. Melnick. 1998. The effects of social structure, geographical structure, and population size on the evolution of mitochondrial DNA: II. Molecular clocks and the lineage sorting period. *J. Mol. Evol.* 47:21–31.

Hoffecker, J. F., W. R. Powers, and T. Goebel. 1993. The colonization of Beringia and the peopling of the New World. *Science* 259:46–53.

Hoffmann, A. A., M. Turelli, and G. M. Simmons. 1986. Unidirectional incompatibility between populations of *Drosophila simulans. Evolution* 40:692–701.

Hogan, K. M., M. C. Hedin, H. S. Koh, S. K. Davis, and I. F. Greenbaum. 1993. Systematic and taxonomic implications of karyotypic, electrophoretic, and mitochondrial-DNA variation in *Peromyscus* from the Pacific Northwest. *J. Mammal.* 74:819–831.

Hongyo, T., G. S. Buzard, R. J. Calvert, and C. M. Weghorst. 1993. 'Cold SSCP': A simple, rapid and non-radioactive method for optimizing single-strand conformational polymorphism analyses. *Nucleic Acids Res.* 21:3637–3642.

Horai, S. and K. Hayasaka. 1990. Intraspecific nucleotide sequence differences in the major noncoding region of the human mitochondrial DNA. *Amer. J. Hum. Genet.* 46:828–842.

Horai, S., K. Hayasaka, K. Hirayama, S. Takenaka, and I. H. Pan. 1987. Evolutionary implications of mitochondrial DNA polymorphism in human populations. In *Human Genetics: Proceedings of the 7th International Congress,* F. Vogel and K. Sperling (eds.). Heidelberg: Springer-Verlag, pp. 177–181.

Horai, S., K. Hayasaka, R. Kondo, K. Tsugane, and N. Takahata. 1995. Recent African origin of modern humans revealed by complete sequences of hominoid mitochondrial DNAs. *Proc. Natl. Acad. Sci. USA* 92:532–536.

Horai, S., R. Kondo, Y. Nakagawa-Hattori, S. Hayashi, S. Sonoda, and K. Tajima. 1993. Peopling of the Americas founded by four major lineages of mitochondrial DNA. *Mol. Biol. Evol.* 10:23–47.

Hosaka, K. and R. E. Hanneman, Jr. 1988. Origin of chloroplast DNA diversity in the Andean potatoes. *Theoret. Appl. Genet.* 76:333–340.

Howard, D. J. and S. H. Berlocher (eds.). 1998. *Endless Forms: Species and Speciation.* New York: Oxford University Press.

Howell, N., S. Halvorson, I. Kubacka, D. A. McCullough, L. A. Bindoff, and D. M. Turnbull. 1992. Mitochondrial gene segregation in mammals: Is the bottleneck always narrow? *Human Genet.* 90:117–120.

Huang, W., Y-X. Fu, B. H.-J. Chang, X. Gu, L. B. Jorde, and W.-H. Li. 1998. Sequence variation in ZFX introns in human populations. *Mol. Biol. Evol.* 15: 138–142.

Hudson, R. R. 1982. Estimating genetic variability with restriction endonucleases. *Genetics* 100:711–719.

———1983. Testing the constant-rate neutral allele model with protein sequence data. *Evolution* 37:203–217.

———1987. Estimating the recombination parameter of a finite population model without selection. *Genet. Res.* 50:245–250.

———1990. Gene genealogies and the coalescent process. *Oxford Surv. Evol. Biol.* 7:1–44.

———1998. Island models and the coalescent process. *Mol. Ecol.* 7:413–418.

Hudson, R. R. and N. L. Kaplan. 1985. Statistical properties of the number of recombination events in the history of a sample of DNA sequences. *Genetics* 111:147–164.

———1996. The coalescent process and background selection. In *New Uses for New Phylogenies,* P. H. Harvey, A. J. Leigh Brown, J. Maynard Smith, and S. Nee (eds.). New York: Oxford University Press, pp. 57–65.

Hudson, R. R., M. Slatkin, and W. P. Maddison. 1992. Estimation of levels of gene flow from DNA sequence data. *Genetics* 132:583–589.

Huelsenbeck, J. P. and D. Hillis. 1993. Success of phylogenetic methods in the four-taxon case. *Syst. Biol.* 42:247–264.

Hull, D. L. 1997. The ideal species concept—and why we can't get it. In *Species: The Units of Biodiversity,* M. F. Claridge, H. A. Dawah, and M. R. Wilson (eds.). New York: Chapman & Hall, pp. 357–380.

Humphries, C. J. and L. R. Parenti. 1986. *Cladistic Biogeography.* Oxford: Clarendon Press.

Humphries, C. J., P. H. Williams, and R. I. Vane-Wright. 1995. Measuring biodiversity value for conservation. *Annu. Rev. Ecol. Syst.* 26:93–111.

Hunt, W. G. and R. K. Selander. 1973. Biochemical genetics of hybridization in European house mice. *Heredity* 31:11–33.

Hurwood, D. A. and J. M. Hughes. 1998. Phylogeography of the freshwater fish, *Mogurnda adspersa,* in streams of northeastern Queensland, Australia: Evidence for altered drainage patterns. *Mol. Ecol.* 7:1507–1517.

Hutchinson, C. A., III, J. E. Newbold, S. S. Potter, and M. H. Edgell. 1974. Maternal inheritance of mammalian mitochondrial DNA. *Nature* 251:536–538.

Ibrahim, K. M., R. A. Nichols, and G. M. Hewitt. 1996. Spatial patterns of genetic variation generated by different forms of dispersal during range expansion. *Heredity* 77:282–291.

Ioerger, T. R., A .G. Clark, and T.-H. Kao. 1990. Polymorphism at the self-incompatibility locus in Solanaceae predates speciation. *Proc. Natl. Acad. Sci. USA* 87:9732–9735.

Irwin, D. M., T. D. Kocher, and A. C. Wilson. 1991. Evolution of the cytochrome *b* gene of mammals. *J. Mol. Evol.* 32:128–144.

Ishibashi, Y., T. Saitoh, S. Abe, and M. C. Yoshida. 1997. Sex-related spatial kin structure in a spring population of grey-sided voles *Clethrionomys rufocanus* as revealed by mitochondrial and microsatellite DNA analyses. *Mol. Ecol.* 6:63–71.

Jaarola, M. and H. Tegelström. 1995. Colonization history of north European field voles (*Microtus agrestis*) revealed by mitochondrial DNA. *Mol. Ecol.* 4:299–310.

————1996. Mitochondrial DNA variation in the field vole (*Microtus agrestis*): Regional population structure and colonization history. *Evolution* 50:2073–2085.

Jabbour-Zahab, R., J. P. Pointier, J. Jourdane, P. Jarne, J. A. Oviedo, M. D. Bargues, S. Mas-Coma, R. Anglés, G. Perera, C. Balzan, K. Khallayoune, and F. Renaud. 1997. Phylogeography and genetic divergence of some lymnaeid snails, intermediate hosts of human and animal fascioliasis with special reference to lymnaeids from the Bolivian Altiplano. *Acta Tropica* 64:191–203.

Jackman, T., J. B. Losos, A. Larson, and K. de Queiroz. 1997. Phylogenetic studies of convergent adaptive radiations in Caribbean *Anolis* lizards. In *Molecular Evolution and Adaptive Radiation*, T. J. Givnish and K. Systma (eds.). Cambridge: Cambridge University Press, pp. 535–557.

James, F. C. 1983. Environmental component of morphological differentiation in birds. *Science* 221:184–186.

Jenuth, J. P., A. C. Peterson, K. Fu, and E. A. Shoubridge. 1996. Random genetic drift in the female germline explains the rapid segregation of mammalian mitochondrial DNA. *Nature Genet.* 14:146–151.

Jerry, D. R. and P. R. Baverstock. 1998. Consequences of a catadromous life-strategy for levels of mitochondrial DNA differentiation among populations of the Australian bass, *Macquaria novemaculeata*. *Mol. Ecol.* 7:1003–1013.

Johns, G. and J. C. Avise. 1998a. A comparative summary of genetic distances in the vertebrates from the mitochondrial cytochrome *b* gene. *Mol. Biol. Evol.* 15:1481–1490.

————1998b. Tests for ancient species flocks based on molecular phylogenetic appraisals of *Sebastes* rockfishes and other marine fishes. *Evolution* 52:1135–1146.

Johnson, M. S., D. C. Wallace, S. D. Ferris, M. C. Rattazzi, and L. L. Cavalli-Sforza. 1983. Radiation of human mitochondrial DNA types analyzed by restriction endonuclease cleavage patterns. *J. Mol. Evol.* 19:255–271.

Johnson, W. E., M. Culver, J. A. Iriarte, E. Eizirik, K. L. Seymour, and S. J. O'Brien. 1998. Tracking the evolution of the elusive Andean mountain cat (*Oreailurus jacobita*) from mitochondrial DNA. *J. Heredity* 89:227–232.

Johnstone, R. A. and G. D. D. Hurst. 1996. Maternally inherited male-killing microorganisms may confound interpretation of mitochondrial DNA variability. *Biol. J. Linn. Soc.* 58:453–470.

Jordan, D. S. 1908. The law of geminate species. *Amer. Nat.* 42:73–80.

Jorde, L. B., M. Bamshad, and A. R. Rogers. 1998. Using mitochondrial and nuclear DNA markers to reconstruct human evolution. *BioEssays* 20:126–136.

Joseph, L. and C. Moritz. 1994. Mitochondrial DNA phylogeography of birds in eastern Australian rainforests: First fragments. *Aust. J. Zool.* 42:385–403.

Joseph, L., C. Moritz, and A. Hugall. 1995. Molecular support for vicariance as a source of diversity in rainforest. *Proc. Roy. Soc. Lond. B* 260:177–182.

Juan, C., K. M. Ibrahim, P. Oromí, and G. M. Hewitt. 1996. Mitochondrial DNA sequence variation and phylogeography of *Pimelia* darkling beetles on the Island of Tenerife (Canary Islands). *Heredity* 77:589–598.

Juan, C., P. Oromí, and G. M. Hewitt. 1997. Molecular phylogeny of darkling beetles from the Canary Islands: Comparison of inter island colonization patterns in two genera. *Biochem. Syst. Ecol.* 25:121–130.

Kambhampati, S., P. Luykx, and C. A. Nalepa. 1996. Evidence for sibling species in *Cryptocercus punctulatus*, the wood roach, from variation in mitochondrial DNA and karyotype. *Heredity* 76:485–496.

Kambysellis, M. P., K.-F. Ho, E. M. Craddock, F. Piano, M. Parisi, and J. Cohen. 1995. Patterns of ecological shifts in the diversification of Hawaiian *Drosophila* inferred from a molecular phylogeny. *Curr. Biol.* 5:1129–1139.

Kaneda, H., J.-I. Hayashi, S. Takahama, C. Taya, K. F. Lindahl, and H. Yonekawa. 1995. Elimination of paternal mitochondrial DNA in interspecific crosses during early mouse embryogenesis. *Proc. Natl. Acad. Sci. USA* 92:4542–4546.

Kann, L. M. and K. Wishner. 1996. Genetic population structure of the copepod *Calanus finmarchicus* in the Gulf of Maine—Allozyme and amplified mitochondrial DNA variation. *Marine Biol.* 125:65–75.

Kaplan, N., R. R. Hudson, and M. Ilzuka. 1991. The coalescent process in models with selection, recombination and geographic subdivision. *Genet. Res.* 57: 83–91.

Kaplan, N. and C. H. Langley. 1979. A new estimate of sequence divergence of mitochondrial DNA using restriction endonuclease mappings. *J. Mol. Evol.* 13:295–304.

Kaplan, N. and K. Risko. 1981. An improved method for estimating sequence divergence of DNA using restriction endonuclease mappings. *J. Mol. Evol.* 17: 156–172.

Karl, S.A. and J. C. Avise. 1992. Balancing selection at allozyme loci in oysters: Implications from nuclear RFLPs. *Science* 256:100–102.

Karl, S. A., B. W. Bowen, and J. C. Avise. 1992. Global population genetic structure and male-mediated gene flow in the green turtle (*Chelonia mydas*): RFLP analyses of anonymous nuclear loci. *Genetics* 131:163–173.

Keohavong, P. and W. G. Thilly. 1989. Fidelity of DNA polymerases in DNA amplification. *Proc. Natl. Acad. Sci. USA* 86:9253–9257.

Kerr, J. T. 1997. Species richness, endemism, and the choice of areas for conservation. *Conserv. Biol.* 11:1094–1100.

Kessler, L. G. and J. C. Avise. 1985. Microgeographic lineage analysis by mitochondrial genotype: Variation in the cotton rat (*Sigmodon hispidus*). *Evolution* 39:831–837.

Kim, I., C. J. Phillips, J. A. Monjeau, E. C. Birney, K. Noack, E. Pumo, R. S. Sikes, and J. A. Dole. 1998. Habitat islands, genetic diversity, and gene flow in a Patagonian rodent. *Mol. Ecol.* 7:667–678.

Kimura, M. 1953. "Stepping-stone" model of population. *Annu. Rep. Natl. Inst. Genet. Japan* 3:62–63.

———1980. A simple method for estimating evolutionary rate of base substitutions through comparative studies of nucleotide sequences. *J. Mol. Evol.* 16:111–120.

Kimura, M. and T. Ohta. 1969. The average number of generations until fixation of a mutant gene in a finite population. *Genetics* 61:763–771.

King, R. A. and C. Ferris. 1998. Chloroplast DNA phylogeography of *Alnus glutinosa* (L.) Gaertn. *Mol. Ecol.* 7:1151–1161.

Kingman, J. F. C. 1982a. The coalescent. *Stoch. Process. Appl.* 13:235–248.

———1982b. On the genealogy of large populations. *J. Appl. Prob.* 19A:27–43.

Kirkpatrick, M. and R. K. Selander. 1979. Genetics of speciation in lake whitefishes in the Allegash basin. *Evolution* 33:478–485.

Klein, J. 1986. *Natural History of the Major Histocompatibility Complex*. New York: John Wiley.

Klein, J., N. Takahata, and F. J. Ayala. 1993. MHC polymorphism and human origins. *Sci. Amer.* 269:78–83.

Klein, N. K. and W. M. Brown. 1995. Intraspecific molecular phylogeny in the yellow warbler (*Dendroica petechia*), and implications for avian biogeography in the West Indies. *Evolution* 48:1914–1932.

Klicka, J. and R. M. Zink. 1997. The importance of recent Ice Ages in speciation: A failed paradigm. *Science* 277:1666–1669.

Klinman, R. M. and J. Hey. 1993. Reduced natural selection associated with low recombination in *Drosophila melanogaster*. *Mol. Biol. Evol.* 10:1239–1258.

Knowlton, N. 1993. Sibling species in the sea. *Annu. Rev. Ecol. Syst.* 24:189–216.

Knowlton, N., L. A. Weight, L. A. Solórzano, D. K. Mills, and E. Bermingham. 1993. Divergence of proteins, mitochondrial DNA, and reproductive compatibility across the Isthmus of Panama. *Science* 260:1629–1632.

Kocher, T. D., J. A. Conroy, K. R. McKaye, and J. R. Stauffer. 1993. Similar morphologies of cichlid fish in Lakes Tanganyika and Malawi are due to convergence. *Mol. Phylogen. Evol.* 2:158–165.

Kocher, T. D., W. K. Thomas, A. Meyer, S. V. Edwards, S. Pääbo, F. X. Villablanca, and A. C. Wilson. 1989. Dynamics of mitochondrial DNA evolution in animals: Amplification and sequencing with conserved primers. *Proc. Natl. Acad. Sci. USA* 86:6196–6200.

Kocher, T. D. and A. C. Wilson. 1991. Sequence evolution of mitochondrial DNA in humans and chimpanzees: control region and a protein-coding region. In *Evolution of Life: Fossils, Molecules, and Culture,* S. Osawa and T. Honjo (eds.). New York: Springer-Verlag, pp. 391–413.

Koehn, R. K. and G. C. Williams. 1978. Genetic differentiation without isolation in the American eel, *Anguilla rostrata.* II. Temporal stability of geographic patterns. *Evolution* 32:624–637.

Kolman, C. J., E. Bermingham, R. Cooke, R. H. Ward, T. D. Arias, and F. Guionneau-Sinclair. 1995. Reduced mtDNA diversity in the Ngöbe Amerinds of Panamá. *Genetics* 140:275–283.

Kolman, C. J., N. Sambuughin, and E. Bermingham. 1996. Mitochondrial DNA analysis of Mongolian populations and implications for the origin of New World founders. *Genetics* 142:1321–1334.

Kondo, R., E. T. Matsuura, H. Ishima, N. Takahata, and S. I. Chigusa. 1990. Incomplete maternal transmission of mitochondrial DNA in *Drosophila. Genetics* 126:657–663.

Kornfield, I. and S. M. Bogdanowicz. 1987. Differentiation of mitochondrial DNA in Atlantic herring, *Clupea harengus. Fish. Bull.* 85:561–568.

Kotoulas, G., A. Magoulas, N. Tsimenides, and E. Zouros. 1995. Marked mitochondrial DNA differences between Mediterranean and Atlantic populations of the swordfish, *Xiphias gladius. Mol. Ecol.* 4:473–481.

Koufopanou, V., A. Burt, and J. W. Taylor. 1997. Concordance of gene genealogies reveals reproductive isolation in the pathogenic fungus *Coccidioides immitis. Proc. Natl. Acad. Sci. USA* 94:5478–5482.

Krajewski, C. 1994. Phylogenetic measures of biodiversity: A comparison and critique. *Biol. Cons.* 69:33–39.

Krings, M., A. Stone, R. W. Schmitz, H. Krainitzki, M. Stoneking, and S. Pääbo. 1997. Neanderthal DNA sequences and the origin of modern humans. *Cell* 90:19–30.

Kristmundsdóttir, A. Y. and J. R. Gold. 1996. Systematics of the blacktail shiner (*Cyprinella venusta*) inferred from analysis of mitochondrial DNA. *Copeia* 1996:773–783.

Kroon, A. M. and C. Saccone (eds.). 1974. *The Biogenesis of Mitochondria.* New York: Academic Press.

Kuhner, M. K. and J. Felsenstein. 1994. A simulation comparison of phylogeny algorithms under equal and unequal evolutionary rates. *Mol. Biol. Evol.* 11:459–468.

Kuhner, M. K., J. Yamato, and J. Felsenstein. 1995. Estimating effective population size and mutation rate from sequence data using Metropolis-Hastings sampling. *Genetics* 140:1421–1430.

———1997. Applications of Metropolis-Hastings genealogy sampling. In *Progress in Population Genetics and Human Evolution,* P. Donnelly and S. Tavaré (eds.). New York: Springer-Verlag, pp. 183–192.

———1998. Maximum likelihood estimation of population growth rates based on the coalescent. *Genetics* 149:429–434.

Kumar, S. and S. B. Hedges. 1998. A molecular timescale for vertebrate evolution. *Nature* 392:917–920.

Laerm, J., J. C. Avise, J. C. Patton, and R. A. Lansman. 1982. Genetic determination of the status of an endangered species of pocket gopher in Georgia. *J. Wildl. Manage.* 46:513–518.

Lahanas, P. N., K. A. Bjorndal, A. B. Bolten, S. E. Encalada, M. M. Miyamoto, R. A. Valverde, and B. W. Bowen. 1998. Genetic composition of a green turtle (*Chelonia mydas*) feeding ground population: Evidence for multiple origins. *Marine Biol.* 130:345–352.

Lamb, T. and J. C. Avise. 1986. Directional introgression of mitochondrial DNA in a hybrid population of tree frogs: the influence of mating behavior. *Proc. Natl. Acad. Sci. USA* 83:2526–2530.

———1992. Molecular and population genetic aspects of mitochondrial DNA variability in the diamondback terrapin, *Malaclemys terrapin. J. Heredity* 83:262–269.

Lamb, T., J. C. Avise, and J. W. Gibbons. 1989. Phylogeographic patterns in mitochondrial DNA of the desert tortoise (*Xerobates agassizi*), and evolutionary relationships among the North American gopher tortoises. *Evolution* 43:76–87.

Lamb, T., T. R. Jones, and J. C. Avise. 1992. Phylogeographic histories of representative herpetofauna of the southwestern U.S.: mitochondrial DNA variation in the desert iguana (*Dipsosaurus dorsalis*) and the chuckwalla (*Sauromalus obesus*). *J. Evol. Biol.* 5:465–480.

Lamb, T., T. R. Jones, and P. J. Wettstein. 1997. Evolutionary genetics and phylogeography of tassel-eared squirrels (*Sciurus aberti*). *J. Mammal.* 78:117–133.

Lamb, T., C. Lydeard, R. B. Walker, and J. W. Gibbons. 1994. Molecular systematics of map turtles (*Graptemys*): A comparison of mitochondrial restriction site versus sequence data. *Syst. Biol.* 43:543–559.

Lansman, R. A., J. C. Avise, C. F. Aquadro, J. F. Shapira, and S. W. Daniel. 1983a. Extensive genetic variation in mitochondrial DNA's among geographic populations of the deer mouse, *Peromyscus maniculatus. Evolution* 37:1–16.

Lansman, R. A., J. C. Avise, and M. D. Huettel. 1983b. Critical experimental test of the possibility of "paternal leakage" of mitochondrial DNA. *Proc. Natl. Acad. Sci. USA* 80:1969–1971.

Lansman, R. A., R. O. Shade, J. F. Shapira, and J. C. Avise. 1981. The use of restriction endonucleases to measure mitochondrial DNA sequence relatedness in natural populations. III. Techniques and potential applications. *J. Mol. Evol.* 17:214–226.

Lansman, R. A., J. F. Shapira, C. Aquadro, S. W. Daniel, and J. C. Avise. 1982. Mitochondrial DNA and evolution in *Peromyscus:* A preliminary report. In *Mitochondrial Genes,* Cold Spring Harbor, N.Y.: Cold Spring Harbor Publ., pp. 133–136.

Larsen, A. H., J. Sigurjónsson, N. Øien, G. Vikingsson, and P. Palsbøll. 1996. Population genetic analysis of nuclear and mitochondrial loci in skin biopsies collected from central and northeastern North Atlantic humpback whales (*Megaptera novaeangliae*): Population identity and migratory destinations. *Proc. Roy. Soc. Lond. B* 263:1611–1618.

Latorre, A., A. Moya, and F. J. Ayala. 1986. Evolution of mitochondrial DNA in *Drosophila subobscura. Proc. Natl. Acad. Sci. USA* 83:8649–8653.

Latta, R. G. and J. B. Mitton. 1997. A comparison of population differentiation across four classes of gene marker in limber pine (*Pinus flexilis* James). *Genetics* 146:1153–1163.

Laurent, L., J. Lescure, L. Excoffier, B. Bowen, M. Domingo, M. Hadjichristophorou, L. Kornaraky, and G. Trabuchet. 1993. Genetic studies of relationships between Mediterranean and Atlantic populations of loggerhead *Caretta caretta* with a mitochondrial marker. *Comptes Rendus de l'Académie des Sciences, Paris* 316:1233–1239.

Laurent, L. and 17 others. 1998. Molecular resolution of marine turtle stock composition in fishery bycatch: A case study in the Mediterranean. *Mol. Ecol.* 7:1529–1542.

Lavery, S., C. Moritz, and D. R. Fielder. 1996. Indo-Pacific population structure and evolutionary history of the coconut crab *Birgus latro. Mol. Ecol.* 5: 557–570.

Lavin, M., S. Mathews, and C. Hughes. 1992. Chloroplast DNA variation in *Gliricidia sepium* (Leguminosae): Intraspecific phylogeny and tokogeny. *Amer. J. Bot.* 78:1576–1585.

Lawlor, D. A., J. Zenmour, P. P. Ennis, and P. Parham. 1988. HLA-A and B polymorphisms predated the divergence of humans and chimpanzees. *Nature* 335:268–271.

Lee, K., J. Feinstein, and J. Cracraft. 1997. The phylogeny of ratite birds: Resolving conflicts between molecular and morphological data sets. In *Avian Molecular Evolution and Systematics*, D. P. Mindell (ed.). New York: Academic Press, pp. 173–195.

Lee, T. E., Jr., B. R. Riddle, and P. L. Lee. 1996. Speciation in the desert pocket mouse (*Chaetodipus penicillatus* Woodhouse). *J. Mammal.* 77:58–68.

Lehman, N., A. Eisenhawer, K. Hansen, L. D. Mech, R. O. Peterson, P. J. P. Gogan, and R. K. Wayne. 1991. Introgression of coyote mitochondrial DNA into sympatric North American gray wolf populations. *Evolution* 45:104–119.

Lehman, N. and R. K. Wayne. 1991. Analysis of coyote mitochondrial DNA genotype frequencies: Estimation of effective number of alleles. *Genetics* 128: 405–416.

Lento, G. M., R. H. Mattlin, G. K. Chambers, and C. S. Baker. 1994. Geographic distribution of mitochondrial cytochrome *b* DNA haplotypes in New Zealand fur seals (*Arctocephalus forsteri*). *Can. J. Zool.* 72:293–299.

Lessa, E. P. 1990. Multidimensional analysis of geographic genetic structure. *Syst. Zool.* 39:242–252.

———1992. Rapid surveying of DNA sequence variation in natural populations. *Mol. Biol. Evol.* 9:323–330.

Lessa, E. P. and G. Applebaum. 1993. Screening techniques for detecting allelic variation in DNA sequences. *Mol. Ecol.* 2:119–129.

Lessios, H. A. 1979. Use of Panamanian sea urchins to test the molecular clock. *Nature* 280:599–601.

———1981. Divergence in allopatry: Molecular and morphological differentiation between sea urchins separated by the Isthmus of Panama. *Evolution* 35:618–634.

Lessios, H. A., B. D. Kessing, and D. R. Robinson. 1998. Massive gene flow across the world's most potent marine biogeographic barrier. *Proc. Roy. Soc. Lond. B* 265:583–588.

Lewin, R. 1993. *Human Evolution: An Illustrated Introduction,* 3d ed. Oxford: Blackwell.

Lewontin, R. C. 1972. The apportionment of human diversity. *Evol. Biol.* 6:381–398.

Lewontin, R. C. and J. L. Hubby. 1966. A molecular approach to the study of genic heterozygosity in natural populations. II. Amount of variation and degree of heterozygosity in natural populations of *Drosophila pseudoobscura. Genetics* 54:595–609.

Lewontin, R. C. and J. Krakauer. 1973. Distribution of gene frequency as a test of the theory of the selective neutrality of polymorphisms. *Genetics* 74:175–195.

Li, C. C. 1955. *Population Genetics.* Chicago: University of Chicago Press.

Li, G. and D. Hedgecock. 1998. Genetic heterogeneity, detected by PCR-SSCP, among samples of larval Pacific oysters (*Crassostrea gigas*) supports the hypothesis of large variance in reproductive success. *Can. J. Fish. Aquat. Sci.* 55:1025–1033.

Li, W.-H. 1981. A simulation study of Nei and Li's model for estimating DNA divergence from restriction enzyme maps. *J. Mol. Evol.* 17:251–255.

———1997. *Molecular Evolution.* Sunderland, Mass.: Sinauer.

Li, W.-H. and L. A. Sadler. 1992. DNA variation in humans and its implications for human evolution. *Oxford Surv. Evol. Biol.* 8:111–134.

Lightowlers, R. N., P. F. Chinnery, D. M. Turnbull, and N. Howell. 1997. Mammalian mitochondrial genetics: heredity, heteroplasmy and disease. *Trends Genet.* 13:450–455.

Linn, S. and W. Arber. 1968. Host specificity of DNA produced by *Escherichia coli.* X. In vitro restriction of phage fd replicative form. *Proc. Natl. Acad. Sci. USA* 59:1300–1306.

Linnaeus, C. 1758. *Systema Naturae.* Stockholm: Laurentius Galvius.

Liu, H.-P., J. B. Mitton, and S.-K. Wu. 1996. Paternal mitochondrial DNA differentiation far exceeds maternal mitochondrial DNA and allozyme differentiation in the freshwater mussel, *Anodonta grandis grandis. Evolution* 50:952–957.

Lodge, D. M. 1993. Biological invasions: Lessons for ecology. *Trends Ecol. Evol.* 8:133–137.

Losos, J. B., T. R. Jackman, A. Larson, K. de Queiroz, and L. Rodriguez-Schettino. 1998. Contingency and determinism in replicated adaptive radiations of island lizards. *Science* 279:2115–2118.

Lotka, A. J. 1931a. Population analysis—the extinction of families—I. *J. Wash. Acad. Sci.* 21:377–380.

———1931b. Population analysis—the extinction of families—II. *J. Wash. Acad. Sci.* 21:453–459.

Lovette, I. J., E. Bermingham, G. Seutin, and R. E. Ricklefs. 1998. Evolutionary differentiation in three endemic West Indian warblers. *Auk* 115:890–903.

Lu, G., S. Li, and L. Bernatchez. 1997. Mitochondrial DNA diversity, population structure, and conservation genetics of four native carps within the Yangtze River, China. *Can. J. Fish. Aquat. Sci.* 54:47–58.

Luikart, G. and F. W. Allendorf. 1996. Mitochondrial-DNA variation and genetic-population structure in Rocky Mountain bighorn sheep (*Ovis canadensis canadensis*). *J. Mammal.* 77:109–123.

Lunt, D. H. and B. C. Hyman. 1997. Animal mitochondrial DNA recombination. *Nature* 387:247.

Lunt, D. H., L. E. Whipple, and B. C. Hyman. 1998. Mitochondrial DNA variable number tandem repeats (VNTRs): Utility and problems in molecular ecology. *Mol. Ecol.* 7: 1441–1455.

Lynch, M. and T. J. Crease. 1990. The analysis of population survey data on DNA sequence variation. *Mol. Biol. Evol.* 7:377–394.

Lyons-Weiler, J. and M. C. Milinkovitch. 1997. A phylogenetic approach to the problem of differential lineage sorting. *Mol. Biol. Evol.* 14:968–975.

Lyrholm, T. and U. Gyllensten. 1998. Global matrilineal population structure in sperm whales as indicated by mitochondrial DNA sequences. *Proc. Roy. Soc. Lond. B* 265:1679–1684.

Lyrholm, T., O. Leimer, and U. Gyllensten. 1996. Low diversity and biased substitution patterns in the mitochondrial DNA control region of sperm whales: Implications for estimates of time since common ancestry. *Mol. Biol. Evol.* 13:1318–1326.

MacNeil, D. and C. Strobeck. 1987. Evolutionary relationships among colonies of Columbian ground squirrels as shown by mitochondrial DNA. *Evolution* 41:873–881.

Maddison, D. R. 1991. African origin of human mitochondrial DNA reexamined. *Syst. Zool.* 40:355–363.

——1994. Phylogenetic methods for inferring the evolutionary history and processes of change in discretely valued characters. *Annu. Rev. Entomol.* 39:267–292.

Maddison, D. R., M. Ruvolo, and D. L. Swofford. 1992. Geographic origins of human mitochondrial DNA: Phylogenetic evidence from control region sequences. *Syst. Biol.* 41:111–124.

Maddison, W. P. 1995. Phylogenetic histories within and among species. In *Experimental and Molecular Approaches to Plant Biosystematics*, P. C. Hoch and A. G. Stephenson (eds.). Monogr. Syst. Missouri Bot. Gard. 53, pp. 273–287.

——1996. Molecular approaches and the growth of phylogenetic biology. In *Molecular Zoology*, J. D. Ferraris and S. R. Palumbi (eds.). New York: Wiley-Liss, pp. 47–63.

——1997. Gene trees in species trees. *Syst. Biol.* 46:523–536.

Maddison, W. P. and D. R. Maddison. 1992. MacClade, version 3.0. Sunderland, Mass.: Sinauer.

Magoulas, A., N. Tsimenides, and E. Zouros. 1996. Mitochondrial DNA phylogeny and the reconstruction of the population history of a species: The case of the European anchovy (*Engraulis encrasicolus*). *Mol. Biol. Evol.* 13: 178–190.

Magoulas, A. and E. Zouros. 1993. Restriction-site heteroplasmy in anchovy (*Engraulis encrasicolus*) indicates incidental biparental inheritance of mitochondrial DNA. *Mol. Biol. Evol.* 10:319–325.

Magurran, A. E. 1998. Population differentiation without speciation. *Phil. Trans. Roy. Soc. Lond. B* 353:275–286.

Maldonado, J. E., F. O. Davila, B. S. Stewart, E. Geffen, and R. K. Wayne. 1995. Intraspecific genetic differentiation in California sea lions (*Zalophus californianus*) from southern California and the Gulf of California. *Marine Mammal Sci.* 11:46–58.

Malecot, G. 1948. *Les mathématiques de l'hérédité.* Paris: Masson et Cie.

Mallet, J. 1995. A species definition for the Modern Synthesis. *Trends Ecol. Evol.* 10:294–299.

Marchant, A. D., M. L. Arnold, and P. Wilkinson. 1988. Gene flow across a chromosomal tension zone. I. Relicts of ancient hybridization. *Heredity* 61:321–328.

Marchington, D. R., G. M. Hartshorne, D. Barlow, and J. Poulton. 1997. Homopolymeric tract heteroplasmy in mtDNA from tissues and single oocytes: Support for a genetic bottleneck. *Amer. J. Human Genet.* 60:408–416.

Margules, C. R., A. O. Nicholls, and R. L. Pressey. 1988. Selecting networks of reserves to maximise biological diversity. *Biol. Cons.* 43:63–76.

Marjoram, P. and P. Donnelly. 1994. Pairwise comparisons of mitochondrial DNA sequences in subdivided populations and implications for early human evolution. *Genetics* 136:673–683.

———1997. Human demography and the time since mitochondrial Eve. In *Progress in Population Genetics and Human Evolution,* P. Donnelly and S. Tavaré (eds.). New York: Springer-Verlag, pp. 107–131.

Marshall, H. D. and A. J. Baker. 1997. Structural conservation and variation in the mitochondrial control region of Fringilline finches (*Fringilla* spp.) and the greenfinch (*Carduelis chloris*). *Mol. Biol. Evol.* 14:173–184.

Martel, R. K. B. and W. Chapco. 1995. Mitochondrial DNA variation in North American Oedipodinae. *Biochem. Genet.* 33:1–11.

Martin, A. P. 1995. Metabolic rate and directional nucleotide substitution in animal mitochondrial DNA. *Mol. Biol. Evol.* 12:1124–1131.

Martin, A. P., R. Humphreys, and S. R. Palumbi. 1992a. Population genetic structure of the armorhead, *Pseudopentaceros wheeleri,* in the North Pacific Ocean: Application of the polymerase chain reaction to fisheries problems. *Can. J. Fish. Aquat. Sci.* 49:2386–2391.

Martin, A. P., G. J. P. Naylor, and S. R. Palumbi. 1992b. Rates of mitochondrial DNA evolution in sharks are slow compared with mammals. *Nature* 357:153–155.

Martin, A. P. and S. R. Palumbi. 1993. Body size, metabolic rate, generation time, and the molecular clock. *Proc. Natl. Acad. Sci. USA* 90:4087–4091.

Martin, G. 1996. Birds in double trouble. *Nature* 380:666–667.

Martins, E. P. (ed.). 1996. *Phylogenies and the Comparative Method in Animal Behavior.* New York: Oxford University Press.

Martins, E. P. and T. F. Hansen. 1997. Phylogenies and the comparative method: A general approach to incorporating phylogenetic information into the analysis of interspecific data. *Amer. Nat.* 149:534–557.

Maruyama, T. and M. Kimura. 1974. Geographical uniformity of selectively neutral polymorphisms. *Nature* 249:30–32.

Matthee, C. A. and T. J. Robinson. 1996. Mitochondrial DNA differentiation among geographical populations of *Pronolagus rupestris,* Smith's red rock rabbit (Mammalia, Lagomorpha). *Heredity* 76:514–523.

———1997. Mitochondrial DNA phylogeography and comparative cytogenetics of the springhare *Pedetes capensis* (Mammalia: Rodentia). *J. Mammal. Evol.* 4:53–73.

Maxam, A. M. and W. Gilbert. 1977. A new method for sequencing DNA. *Proc. Natl. Acad. Sci. USA* 74:560–564.

Mayden, R. L. 1988. Vicariance biogeography, parsimony, and evolution of North American freshwater fishes. *Syst. Zool.* 37:329–355.

———1997. A hierarchy of species concepts: The denouement in the saga of the species problem. In *Species: The Units of Biodiversity,* M. F. Claridge, H. A. Dawah, and M. R. Wilson (eds.). New York: Chapman & Hall, pp. 381–424.

Maynard Smith, J. 1992. Analyzing the mosaic nature of genes. *J. Mol. Evol.* 34:126–129.

Maynard Smith, J. and N. H. Smith. 1998. Detecting recombination from gene trees. *Mol. Biol. Evol.* 15:590–599.

Maynard Smith, J. and E. Szathmáry. 1995. *The Major Transitions in Evolution.* New York: Freeman.

Mayr, E. 1940. Speciation phenomena in birds. *Amer. Nat.* 74:249–278.

———1942. *Systematics and the Origin of Species.* New York: Columbia University Press.

———1963. *Animal Species and Evolution.* Cambridge, Mass.: Harvard University Press.

———1982. Processes of speciation in animals. In *Mechanisms of Speciation,* C. Barigozzi (ed.). New York: Alan R. Liss, pp. 1–19.

McCauley, D. E. 1991. Genetic consequences of local population extinction and re-colonization. *Trends Ecol. Evol.* 6:5–8.

———1994. Contrasting the distribution of chloroplast DNA and allozyme poly-morphism among local populations of *Silene alba:* Implications for studies of gene flow in plants. *Proc. Natl. Acad. Sci. USA* 91:8127–8131.

McCauley, D. E., J. E. Stevens, P. A. Peroni, and J. A. Raveill. 1996. The spatial distri-bution of chloroplast DNA and allozyme polymorphisms within a population of *Silene alba* (Caryophyllaceae). *Am. J. Bot.* 83:727–731.

McCune, A. R. 1997. How fast is speciation? Molecular, geological, and phylo-genetic evidence from adaptive radiations of fishes. In *Molecular Evolution and Adaptive Radiation,* T. J. Givnish and K. J. Sytsma (eds.). Cambridge: Cam-bridge University Press, pp. 585–610.

McCune, A. R. and N. J. Lovejoy. 1998. The relative rate of sympatric and allopatric speciation in fishes: Tests using DNA sequence divergence between sister species and among clades. In *Endless Forms: Species and Speciation,* D. Howard and S. Berlocher (eds.). New York: Oxford University Press, pp. 172–185.

McDonald, J. H., R. Seed, and R. K. Koehn. 1991. Allozymes and morphometric characters of three species of *Mytilus* in the Northern and Southern Hemi-spheres. *Marine Biol.* 111:323–333.

McDonald, J. H., B. C. Verrelli, and L. B. Geyer. 1996. Lack of geographic variation in anonymous nuclear polymorphisms in the American oyster, *Crassostrea vir-ginica. Mol. Biol. Evol.* 13:1114–1118.

McGuigan, K., K. McDonald, K. Parris, and C. Moritz. 1998. Mitochondrial DNA diversity and historical biogeography of a wet forest-restricted frog (*Litoria pearsoniana*) from mid-east Australia. *Mol. Ecol.* 7:175–186.

McGuire, G., F. Wright, and M. J. Prentice. 1997. A graphical method for detecting recombination in phylogenetic data sets. *Mol. Biol. Evol.* 14:1125–1131.

McKitrick, M. C. and R. M. Zink. 1988. Species concepts in ornithology. *Condor* 90:1–14.

McKnight, M. L. 1995. Mitochondrial DNA phylogeography of *Perognathus amplus* and *Perognathus longimembris* (Rodentia: Heteromyidae): A possible mam-malian ring species. *Evolution* 49:816–826.

McKnight, M. L. and H. B. Shaffer. 1997. Large, rapidly evolving intergenic spacers in the mitochondrial DNA of the salamander family Ambystomatidae (Am-phibia: Caudata). *Mol. Biol. Evol.* 14:1167–1176.

McMichael, M. and H. G. Hall. 1996. DNA RFLPs at a highly polymorphic locus distinguish European and African subspecies of the honey bee *Apis mellifera* L. and suggest geographical origins of New World honey bees. *Mol. Ecol.* 5:403–416.

McMillan, W. O. and E. Bermingham. 1996. The phylogeographic pattern of mitochondrial DNA variation in the Dall's porpoise *Phocoenoides dalli*. *Mol. Ecol.*
  5:47–61.

McMillan, W. O. and S. R. Palumbi. 1995. Concordant evolutionary patterns
  among Indo-west Pacific butterflyfishes. *Proc. Roy. Soc. Lond. B* 260:229–236.

———1997. Rapid rate of control-region evolution in Pacific butterflyfishes
  (Chaetodontidae). *J. Mol. Evol.* 45:473–484.

McMillan, W. O., R. A. Raff, and S. R. Palumbi. 1992. Population genetic consequences of developmental evolution in sea urchins (genus *Heliocidaris*). *Evolution* 46:1299–1312.

McNab, B. K. 1971. On the ecological significance of Bergmann's rule. *Ecology*
  52:845–854.

Meehan, B. W. 1985. Genetic comparison of *Macoma balthica* (Bivalvia, Telinidae)
  from the eastern and western North Atlantic Ocean. *Mar. Ecol. Progr. Ser.*
  22:69–76.

Meehan, B. W., J. T. Carlton, and R. Wenne. 1989. Genetic affinities of the bivalve
  Macoma balthica from the Pacific coast of North America: Evidence for recent
  introduction and historical distribution. *Marine Biol.* 102:235–241.

Melnick, D. J. and G. A. Hoelzer. 1992. Differences in male and female macaque dispersal lead to contrasting distributions of nuclear and mitochondrial DNA
  variation. *Int. J. Primatol.* 13:379–393.

Mengel, R. N. 1964. The probable history of species formation in some northern
  wood warblers (Parulidae). *Living Bird* 3:9–43.

Mercure, A., K. Ralls, K. P. Koepfli, and R. K. Wayne. 1993. Genetic subdivisions
  among small canids: Mitochondrial DNA differentiation of swift, kit, and Arctic foxes. *Evolution* 47:1313–1328.

Merilä, J., M. Björkland, and A. J. Baker. 1997. Historical demography and present
  day population structure of the greenfinch, *Carduelis chloris*—an analysis of
  mtDNA control-region sequences. *Evolution* 51:946–956.

Merriweather, D. A., A. G. Clark, S. W. Ballinger, T. G. Schurr, H. Soodyall, T. Jenkins, S. T. Sherry, and D. W. Wallace. 1991. The structure of human mitochondrial DNA variation. *J. Mol. Evol.* 33:543–555.

Merriweather, D. A., F. Rothhammer, and R. E. Ferrell. 1995. Distribution of the
  four founding lineage haplotypes in native Americans suggests a single wave
  of migration for the New World. *Amer. J. Phys. Anthropol.* 98:411–430.

Meselson, M. and R. Yuan. 1968. DNA restriction enzyme from *E. coli*. *Nature*
  217:1110–1114.

Meusel, M. S. and R. F. A. Moritz. 1993. Transfer of paternal mitochondrial DNA during fertilization of honeybee (*Apis mellifera* L.) eggs. *Curr. Genet.* 24: 539–543.

Meyer, A. 1994. Shortcomings of the cytochrome *b* gene as a molecular marker. *Trends Ecol. Evol.* 9:278–280.

Meyer, A., L. L. Knowles, and E. Verheyen. 1996. Widespread geographical distribution of mitochondrial haplotypes in rock-dwelling cichlid fishes from Lake Tanganyika. *Mol. Ecol.* 5:341–350.

Meyer, A., T. D. Kocher, P. Basasibwaki, and A. C. Wilson. 1990. Monophyletic origin of Lake Victoria cichlid fishes suggested by mitochondrial DNA sequences. *Nature* 347:550–553.

Meylan, A. B., B. W. Bowen, and J. C. Avise. 1990. A genetic test of the natal homing versus social facilitation models for green turtle migration. *Science* 248: 724–727.

Michaux, J. R., M.-G. Filippucci, R. M. Libois, R. Fons, and R. F. Matagne. 1996. Biogeography and taxonomy of *Apodemus sylvaticus* (the woodmouse) in the Tyrrhenian region: Enzymatic variations and mitochondrial DNA restriction pattern analysis. *Heredity* 76:267–277.

Milligan, B. G., J. Leebens-Mack, and A. E. Strand. 1994. Conservation genetics: Beyond the maintenance of marker diversity. *Mol. Ecol.* 3:423–435.

Mindell, D. P. and C. E. Thacker. 1996. Rates of molecular evolution: Phylogenetic issues and applications. *Annu. Rev. Ecol. Syst.* 27:279–303.

Mishler, B. D. and R. N. Brandon. 1987. Individuality, pluralism, and the phylogenetic species concept. *Biol. Philos.* 2:397–414.

Miththapala, S., J. Seidensticker, and S. J. O'Brien. 1996. Phylogeographic subspecies recognition in leopards (*Panthera pardus*): Molecular genetic variation. *Conserv. Biol.* 10:1115–1132.

Mittermeier, R. A., N. Myers, J. B. Thomsen, G. A. B. da Fonseca, and S. Olivieri. 1998. Biodiversity hotspots and major tropical wilderness areas: Approaches to setting conservation priorities. *Conserv. Biol.* 12:516–520.

Mitton, J. B. 1997. *Selection in Natural Populations.* Oxford: Oxford University Press.

Miya, M. and M. Nishida. 1997. Speciation in the open ocean. *Nature* 389:803–804.

Miyaki, C. M., S. R. Matioli, T. Burke, and A. Wajntal. 1998. Parrot evolution and paleogeographical events: mitochondrial DNA evidence. *Mol. Biol. Evol.* 15:544–551.

Monehan, T. M. 1994. Molecular genetic analysis of Adélie penguin populations, Ross Island, Antarctica. Master's thesis, University of Auckland, Auckland, New Zealand.

Monnerot, M., J.-C. Mounolou, and M. Solignac. 1984. Intra-individual length heterogeneity of *Rana esculenta* mitochondrial DNA. *Biol. Cell* 52:213–218.

Montagna, W. 1942. The sharp-tailed sparrows of the Atlantic coast. *Wilson. Bull.* 54:107–120.

Moore, W. S. 1995. Inferring phylogenies from mtDNA variation: Mitochondrial-gene trees versus nuclear-gene trees. *Evolution* 49:718–726.

———1997. Mitochondrial-gene trees versus nuclear-gene trees, a reply to Hoelzer. *Evolution* 51:627–629.

Moore, W. S., J. H. Graham, and J. T. Price. 1991. Mitochondrial DNA variation in the northern flicker (*Colaptes auratus,* Aves). *Mol. Biol. Evol.* 8:327–344.

Morales, J. C., P. M. Andau, J. Supriatna, Z.-Z. Zainuddin, and D. J. Melnick. 1997. Mitochondrial DNA variability and conservation genetics of the Sumatran rhinoceros. *Conserv. Biol.* 11:539–543.

Morell, V. 1998. Genes may link ancient Eurasians, Native Americans. *Science* 280:520.

Morin, P. A., J. J. Moore, R. Chakraborty, L. Jin, J. Goodall, and D. S. Woodruff. 1994. Kin selection, social structure, gene flow, and the evolution of chimpanzees. *Science* 265:1193–1201.

Morin, P. A., J. Wallis, J. J. Moore, R. Chakraborty, and D. S. Woodruff. 1993. Non-invasive sampling and DNA amplification for paternity exclusion, community structure, and phylogeography in wild chimpanzees. *Primates* 34:347–356.

Moritz, C. C. 1991. The origin and evolution of parthenogenesis in *Heteronotia binoei* (Gekkonidae): Evidence for recent and localized origins of widespread clones. *Genetics* 129:211–219.

———1994a. Defining "evolutionarily significant units" for conservation. *Trends Ecol. Evol.* 9:373–375.

———1994b. Applications of mitochondrial DNA analysis in conservation: A critical review. *Mol. Ecol.* 3:401–411.

———1995. Uses of molecular phylogenies for conservation. *Phil. Trans. Roy. Soc. Lond. B* 349:113–118.

Moritz, C. C., T. J. Case, D. T. Bolger, and S. Donnellan. 1993a. Genetic diversity and the history of pacific island house geckos (*Hemidactylus* and *Lepidodactylus*). *Biol. J. Linn. Soc.* 48:113–133.

Moritz, C. C., T. E. Dowling, and W. M. Brown. 1987. Evolution of animal mitochondrial DNA: Relevance for population biology and systematics. *Annu. Rev. Ecol. Syst.* 18:269–292.

Moritz, C. C. and D. P. Faith. 1998. Comparative phylogeography and the identification of genetically divergent areas for conservation. *Mol. Ecol.* 7:419–429.

Moritz, C. C. and A. Heideman. 1993. The origin and evolution of parthenogenesis in *Heteronotia binoei* (Gekkonidae): Reciprocal origins and diverse mitochondrial DNA in western populations. *Syst. Biol.* 129:211–219.

Moritz, C. C., A. Heideman, E. Geffen, and P. McRae. 1997. Genetic population structure of the greater bilby *Macrotis lagotis,* a marsupial in decline. *Mol. Ecol.* 6:925–936.

Moritz, C. C., L. Joseph, and M. Adams. 1993b. Cryptic diversity in an endemic rainforest skink *Gnypetoscincus queenslandiae. Biodiv. Conserv.* 2:412–425.

Moritz, C. C., C. J. Schneider, and D. B. Wake. 1992. Evolutionary relationships within the *Ensatina eschscholtzii* complex confirm the ring species interpretation. *Syst. Biol.* 41:273–291.

Moriyama, E. N. and J. R. Powell. 1997. Synonymous substitution rates in *Drosophila:* Mitochondrial versus nuclear genes. *J. Mol. Evol.* 45:378–391.

Morone, J. J. and J. V. Crisci. 1995. Historical biogeography: Introduction to methods. *Annu. Rev. Ecol. Syst.* 26:373–401.

Mountain, J. L. and L. L. Cavalli-Sforza. 1994. Inferences of human evolution through cladistic analysis of nuclear DNA restriction polymorphisms. *Proc. Natl. Acad. Sci. USA* 91:6515–6519.

Muir, C. C., B. M. F. Galdikas, and A. T. Beckenbach. 1998. Is there sufficient evidence to elevate the orangutan of Borneo and Sumatra to separate species? *J. Mol. Evol.* 46:378–381.

Mukai, T., K. Naruse, T. Sato, A. Shima, and M. Morisawa. 1997. Multiregional introgressions inferred from the mitochondrial DNA phylogeny of a hybridizing species complex of gobiid fishes, genus *Tridentiger. Mol. Biol. Evol.* 14:1258–1265.

Mulligan, T. J., R. W. Chapman, and B. L. Brown. 1992. Mitochondrial DNA analysis of walleye pollock, *Theragra chalcogramma,* from the eastern Bering Sea and Shelikof Strait, Gulf of Alaska. *Can. J. Fish. Aquat. Sci.* 49:319–326.

Mullis, K. and F. Faloona. 1987. Specific synthesis of DNA in vitro via a polymerase catalyzed chain reaction. *Meth. Enzymol.* 155:335–350.

Mullis, K., F. Faloona, S. Scharf, R. Saiki, G. Horn, and H. Erlich. 1986. Specific enzymatic amplification of DNA in vitro: The polymerase chain reaction. *Cold Spring Harb. Symp. Quant. Biol.* 51:263–273.

Murphy, W. J. and G. E. Collier. 1997. A molecular phylogeny for Aplocheiloid fishes (Atherinomorpha, Cyprinodontiformes): The role of vicariance and the origins of annualism. *Mol. Biol. Evol.* 14:790–799.

Murray-McIntosh, R. P., B. J. Scrimshaw, P. J. Harfield, and D. Penny. 1998. Testing migration patterns and estimating founding population size in Polynesia by using human mtDNA sequences. *Proc. Natl. Acad. Sci. USA* 95:9047–9052.

Myers, A. A. and P. S. Giller (eds.). 1988. *Analytical Biogeography.* London: Chapman & Hall.

Myers, N. 1988. Threatened biotas: 'Hot-spots' in tropical forests. *Environmentalist* 8:187–208.

———1990. The biodiversity challenge: Expanded hot-spots analysis. *Environmentalist* 10:243–256.

Myers, R. M., T. Maniatis, and L. S. Lerman. 1986. Detection and localization of single base changes by denaturing gradient gel electrophoresis. *Methods Enzymol.* 155:501–527.

Myers, R. M., V. C. Sheffield, and D. R. Cox. 1989a. Mutation detection, GC-clamps, and denaturing gradient gel electrophoresis. In *PCR Technology: Principles and Applications for DNA Amplification,* H. A. Erlich (ed.). New York: Stockton Press, pp. 71–88.

———1989b. Polymerase chain reaction and denaturing gradient gel electrophoresis. In *Polymerase Chain Reaction,* H. A. Erlich, R. Gibbs, and H. H. Kazazian (eds.). Cold Spring Harbor, N.Y.: Cold Spring Harbor Laboratory, pp. 177–181.

Nedbal, M. A. and J. J. Flynn. 1998. Do the combined effects of the asymmetric process of replication and DNA damage from oxygen radicals produce a mutation-rate signature in the mitochondrial genome? *Mol. Biol. Evol.* 15:219–223.

Nedbal, M. A. and D. P. Philipp. 1994. Differentiation of mitochondrial DNA in largemouth bass. *Trans. Amer. Fish. Soc.* 123:460–468.

Nee, S., E. C. Holmes, and P. H. Harvey. 1995. Inferring population history from molecular phylogenies. *Phil. Trans. Roy. Soc. Lond. B* 349:25–31.

Nee, S., E. C. Holmes, A. Rambaut, and P. H. Harvey. 1996a. Inferring population history from molecular phylogenies. In *New Uses for New Phylogenies,* P. H. Harvey, A. J. Leigh Brown, J. Maynard Smith, and S. Nee (eds.). New York: Oxford University Press, pp. 66–80.

Nee, S., A. F. Read, and P. H. Harvey. 1996b. Why phylogenies are necessary for comparative analysis. In *Phylogenies and the Comparative Method in Animal Behavior,* E. P. Martins (ed.). New York: Oxford University Press, pp. 399–411.

Nei, M. 1987. *Molecular Evolutionary Genetics.* New York: Columbia University Press.

———1995. Genetic support for the out-of-Africa theory of human evolution. *Proc. Natl. Acad. Sci. USA* 92:6720–6722.

———1996. Phylogenetic analysis in molecular evolutionary genetics. *Annu. Rev. Genet.* 30:371–403.

Nei, M. and R. K. Chesser. 1983. Estimation of fixation indices and gene diversities. *Ann. Hum. Genet.* 47:253–259.

Nei, M. and D. Graur. 1984. Extent of protein polymorphism and the neutral mutation theory. *Evol. Biol.* 17:73–118.

Nei, M. and A. L. Hughes. 1991. Polymorphism and evolution of the major histocompatibility complex loci in mammals. In *Evolution at the Molecular Level,* R. K. Selander, A. G. Clark, and T. S. Whittam (eds.). Sunderland, Mass.: Sinauer, pp. 222–247.

Nei, M. and W.-H. Li. 1979. Mathematical model for studying genetic variation in terms of restriction endonucleases. *Proc. Natl. Acad. Sci. USA* 76: 5269–5273.

Nei, M., T. Maruyama, and R. Chakraborty. 1975. The bottleneck effect and genetic variability in populations. *Evolution* 29:1–10.

Nei, M. and A. K. Roychoudhury. 1982. Genetic relationship and evolution of human races. *Evol. Biol.* 14:1–59.

Nei, M. and F. Tajima. 1981. DNA polymorphism detectable by restriction endonucleases. *Genetics* 97:145–163.

Nei, M. and N. Takahata. 1993. Effective population size, genetic diversity, and coalescence time in subdivided populations. *J. Mol. Evol.* 37:240–244.

Nei, M. and N. Takezaki. 1996. The root of the phylogenetic tree of human populations. *Mol. Biol. Evol.* 13:170–177.

Neigel, J. E. 1997. A comparison of alternative strategies for estimating gene flow from genetic markers. *Annu. Rev. Ecol. Syst.* 28:105–128.

Neigel, J. E. and J. C. Avise. 1986. Phylogenetic relationships of mitochondrial DNA under various demographic models of speciation. In *Evolutionary Processes and Theory*, E. Nevo and S. Karlin (eds.). New York: Academic Press, pp. 515–534.

———1993. Application of a random-walk model to geographic distributions of animal mitochondrial DNA variation. *Genetics* 135:1209–1220.

Neigel, J. E., R. M. Ball, Jr, and J. C. Avise. 1991. Estimation of single generation migration distances from geographic variation in animal mitochondrial DNA. *Evolution* 45:423–432.

Nelson, G. J. and N. I. Platnick. 1981. *Systematics and Biogeography: Cladistics and Vicariance.* New York: Columbia University Press.

Nelson, G. J. and D. E. Rosen (eds.). 1981. *Vicariance Biogeography: A Critique.* New York: Columbia University Press.

Neuhauser, C., S. M. Krone, and H.-C. Kang. 1997. A note on the stepping stone model with extinction and recolonization. In *Progress in Population Genetics and Human Evolution*, P. Donnelly and S. Tavaré (eds.). New York: Springer-Verlag, pp. 299–307.

Nielsen, J. L., C. A. Gan, J. M. Wright, D. B. Morris, and W. K. Thomas. 1994. Bio-geographic distributions of mitochondrial and nuclear markers for southern steelhead. *Mol. Mar. Biol. Biotech.* 3:281–293.

Nielsen, J. L., M. C. Fountain, and J. M. Wright. 1997. Biogeographic analysis of Pa-cific trout (*Oncorhynchus mykiss*) in California and Mexico based on mitochon-drial DNA and nuclear microsatellites. In *Molecular Systematics of Fishes,* T. D. Kocher and C. A. Stepien (eds.). San Diego: Academic Press, pp. 53–69.

Nixon, K. C. and Q. D. Wheeler. 1990. An amplification of the phylogenetic species concept. *Cladistics* 6:211–223.

Norman, J. A., C. Moritz, and C. J. Limpus. 1994. Mitochondrial DNA control re-gion polymorphisms: Genetic markers for ecological studies of marine turtles. *Mol. Ecol.* 3:363–373.

Norrgard, J. W. and J. E. Graves. 1996. Determination of the natal origin of a juve-nile loggerhead turtle (*Caretta caretta*) population in Chesapeake Bay using mitochondrial DNA analysis. In *Proceedings of the International Symposium on Sea Turtle Conservation Genetics,* B. W. Bowen and W. N. Witzell (eds.). NOAA Tech. Memo. NMFS-SEFSC-396, pp. 129–136.

O'Brien, S. J. and J. F. Evermann. 1988. Interactive influence of infectious disease and genetic diversity in natural populations. *Trends Ecol. Evol.* 3:254–259.

O'Brien, S. J. and E. Mayr. 1991. Bureaucratic mischief: Recognizing endangered species and subspecies. *Science* 251:1187–1188.

O'Brien, S. J., M. E. Roelke, N. Yuhki, K. W. Richards, W. E. Johnson, W. L. Franklin, A. E. Anderson, O. L. Bass, Jr., R. C. Belden, and J. S. Martenson. 1990. Genetic introgression within the Florida panther *Felis concolor coryi. Natl. Geog. Res.* 6:485–494.

O'Corry-Crowe, G. M., R. S. Suydam, A. Rosenberg, K. J. Frost, and A. E. Dizon. 1997. Phylogeography, population structure and dispersal patterns of the be-luga whale *Delphinapterus leucas* in the western Nearctic revealed by mito-chondrial DNA. *Mol. Ecol.* 6:955–970.

O'Foighil, D. and M. J. Smith. 1996. Phylogeography of an asexual marine clam complex, *Lasaea,* in the northeastern Pacific based on cytochrome oxidase III sequence variation. *Mol. Phylogen. Evol.* 6:134–142.

O'Hara, R. J. 1993. Systematic generalization, historical fate, and the species prob-lem. *Syst. Biol.* 42:231–246.

Ohta, T. 1980. Two-locus problems in transmission genetics of mitochondria and chloroplasts. *Genetics* 96:543–555.

Okazaki, T., T. Kobayashi, and Y. Uozumi. 1996. Genetic relationships of pilchards (genus: *Sardinops*) with anti-tropical distributions. *Marine Biol.* 126:585–590.

Okumura, N. and A. Goto. 1996. Genetic variation and differentiation of the two river sculpins, *Cottus nozawae* and *C. amblystomopsis,* deduced from allozyme and restriction enzyme-digested mtDNA fragment length polymorphism analysis. *Ichthyol. Res.* 43:399–416.

Oliveira, R. P., N. E. Broude, A. M. Macedo, C. R. Cantor, C. L. Smith, and S. D. J. Pena. 1998. Probing the genetic population structure of *Trypanosoma cruzi* with polymorphic microsatellites. *Proc. Natl. Acad. Sci. USA* 95:3776–3780.

Olivo, P. D., M. J. Van de Walle, P. J. Laipis, and W. W. Hauswirth. 1983. Nucleotide sequence evidence for rapid genotypic shifts in the bovine mitochondrial D-loop. *Nature* 306:400–402.

Olson, D. M. and E. Dinerstein. 1998. The Global 200: A representation approach to conserving the Earth's most biologically valuable ecoregions. *Conserv. Biol.* 12:502–515.

O'Reilly, P., T. E. Reimchen, R. Beech, and C. Strobeck. 1993. Mitochondrial DNA in *Gasterosteus* and Pleistocene glacial refugium on the Queen Charlotte Islands, British Columbia. *Evolution* 47:678–684.

Orita, M., H. Iwahana, H. Kanazawa, K. Hayashi, and T. Sekiya. 1989a. Detection of polymorphisms of human DNA by gel electrophoresis as single-strand conformation polymorphisms. *Proc. Natl. Acad. Sci. USA* 86:2766–2770.

Orita, M., Y. Suzuki, T. Sekiya, and K. Hayashi. 1989b. Rapid and sensitive detection of point mutations and DNA polymorphisms using the polymerase chain reaction. *Genomics* 5:874–879.

Ortí, G., M. A. Bell, T. E. Reimchen, and A. Meyer. 1994. Global survey of mitochondrial DNA sequences in the threespine stickleback: Evidence for recent migrations. *Evolution* 48:608–622.

Ortí, G, M. P. Hare, and J. C. Avise. 1997. Detection and isolation of nuclear haplotypes by PCR-SSCP. *Mol. Ecol.* 6:575–580.

Ortí, G. and A. Meyer. 1997. The radiation of characiform fishes and the limits of resolution of mitochondrial ribosomal DNA sequences. *Syst. Biol.* 46:75–100.

Osentoski, M. F. and T. Lamb. 1995. Intraspecific phylogeography of the gopher tortoise, *Gopherus polyphemus:* RFLP analysis of amplified mtDNA segments. *Mol. Ecol.* 4:709–718.

Otte, D. and J. A. Endler (eds.). 1989. *Speciation and Its Consequences.* Sunderland, Mass.: Sinauer.

Ovenden, J. R. 1990. Mitochondrial DNA and marine stock assessment: A review. *Aust. J. Marine Freshwater Res.* 41:835–853.

Ovenden, J. R., D. J. Brasher, and R. W. G. White. 1992. Mitochondrial DNA analyses of the red rock lobster *Jasus edwardsii* supports an apparent absence of population subdivision throughout Australasia. *Marine Biol.* 112:319–326.

Ovenden, J. R., A. J. Smolenski, and R. W. G. White. 1989. Mitochondrial DNA restriction site variation in Tasmanian populations of orange roughy (*Hoplostethus atlanticus*), a deep-water marine teleost. *Aust. J. Marine Freshwater Res.* 40:1–9.

Paetkau, D. and C. Strobeck. 1996. Mitochondrial DNA and the phylogeography of Newfoundland black bears. *Can. J. Zool.* 74:192–196.

Page, R. D. M. 1990. Temporal congruence and cladistic analysis of biogeography and cospeciation. *Syst. Zool.* 39:205–226.

———1994. Maps between trees and cladistic analysis of historical associations among genes, organisms, and areas. *Syst. Biol.* 43:58–77.

Page, R. D. M. and E. C. Holmes. 1998. *Molecular Evolution: A Phylogenetic Approach.* Oxford: Blackwell.

Palmer, J. D. 1985. Evolution of chloroplast and mitochondrial DNA in plants and algae. In *Molecular Evolutionary Genetics,* R. J. MacIntyre (ed.). New York: Plenum Press, pp. 131–240.

———1990. Contrasting modes and tempos of genome evolution in land plant organelles. *Trends Genet.* 6:115–120.

———1992. Mitochondrial DNA in plant systematics: Applications and limitations. In *Molecular Systematics of Plants,* P. S. Soltis, J. E. Soltis, and J. J. Doyle (eds.). New York: Chapman & Hall, pp. 36–48.

Palmer, J. D. and L. A. Herbon. 1988. Plant mitochondrial DNA evolves rapidly in structure, but slowly in sequence. *J. Mol. Evol.* 28:87–97.

Palsbøll, P. J., P. J. Chapham, D. K. Mattila, F. Larsen, R. Sears, H. R. Siegismund, J. Sigurjónsson, O. Vasquez, and P. Arctander. 1995. Distribution of mtDNA haplotypes in North Atlantic humpback whales: The influence of behaviour on population structure. *Mar. Ecol. Progr. Ser.* 116:1–10.

Palsbøll, P. J., M. P. Heide-Jørgensen, and R. Dietz. 1997. Population structure and seasonal movements of narwhals, *Monodon monoceros,* determined from mtDNA analysis. *Heredity* 78:284–292.

Palumbi, S. R. 1994. Genetic divergence, reproductive isolation, and marine speciation. *Annu. Rev. Ecol. Syst.* 25:547–572.

———1995. Using genetics as an indirect estimator of larval dispersal. In *Ecology of Marine Invertebrate Larvae,* L. McEdward (ed.). Boca Raton, Fla.: CRC Press, pp. 369–387.

————1996a. Nucleic acids II: The polymerase chain reaction. In *Molecular Systematics*, D. M. Hillis, C. Moritz, and B. K. Mable (eds.). Sunderland, Mass.: Sinauer, pp. 205–247.

————1996b. Macrospatial genetic structure and speciation in marine taxa with high dispersal abilities. In *Molecular Zoology*, J. D. Ferraris and S. R. Palumbi (eds.). New York: Wiley-Liss, pp. 101–117.

Palumbi, S. R. and C. S. Baker. 1994. Contrasting population structure from nuclear intron sequences and mtDNA of humpback whales. *Mol. Biol. Evol.* 11:426–435.

————1996. Nuclear genetic analysis of population structure and genetic variation using intron primers. In *Molecular Genetic Approaches in Conservation*, T. B. Smith and R. K. Wayne (eds.). New York: Oxford University Press, pp. 25–37.

Palumbi, S. R. and F. Cipriano. 1998. Species identification using genetic tools: The value of nuclear and mitochondrial gene sequences in whale conservation. *J. Heredity* 89:459–464.

Palumbi, S. R., G. Grabowsky, T. Duda, L. Geyer, and N. T. Tachino. 1997. Speciation and population genetic structure in tropical Pacific sea urchins. *Evolution* 51:1506–1517.

Palumbi, S. R. and B. D. Kessing. 1991. Population biology of the trans-Arctic exchange: MtDNA sequence similarity between Pacific and Atlantic sea urchins. *Evolution* 45:1790–1805.

Palumbi, S. R. and E. C. Metz. 1991. Strong reproductive isolation between closely related tropical sea urchins (genus *Echinometra*). *Mol. Biol. Evol.* 8:227–239.

Palumbi, S. R. and A. C. Wilson. 1990. Mitochondrial DNA diversity in the sea urchins *Strongylocentrotus purpuratus* and *S. droebachiensis*. *Evolution* 44: 403–415.

Park, L. K., M. A. Brainard, D. A. Dightman, and G. A. Winans. 1993. Low levels of intraspecific variation in the mitochondrial DNA of chum salmon (*Oncorhynchus keta*). *Mol. Mar. Biol. Biotech.* 2:362–370.

Patarnello, T., L. Bargelloni, F. Caldara, and L. Colombo. 1993. Mitochondrial DNA sequence variation in the European sea bass, *Dicentrarchus labrax* L. (Serranidae): Evidence of differential haplotype distribution in natural and farmed populations. *Mol. Marine Biol. Biotech.* 2:333–337.

Patton, J. L. and M. N. F. da Silva. 1997. Definition of species of pouched four-eyed opossums (Didelphidae, *Philander*). *J. Mammal.* 78:90–102.

Patton, J. L., M. N. F. da Silva, M. C. Lara, and M. A. Mustrangi. 1997. Diversity, differentiation, and the historical biogeography of non-volant small mammals of the neotropical forests. In *Tropical Forest Remnants: Ecology, Management, and*

Conservation of Fragmented Communities, W. F. Laurance and R. O. Bierregaard, Jr. (eds.). Chicago: University of Chicago Press, pp. 455–465.

Patton, J. L., M. N. F. da Silva, and J. R. Malcolm. 1994. Gene genealogy and differentiation among arboreal spiny rats (Rodentia: Echimyidae) of the Amazon basin: A test of the riverine barrier hypothesis. *Evolution* 48:1314–1323.

———1996. Hierarchical genetic structure and gene flow in three sympatric species of Amazonian rodents. *Mol. Ecol.* 5:229–238.

Patton, J. L. and M. F. Smith. 1989. Population structure and the genetic and morphological divergence among pocket gophers (genus *Thomomys*). In *Speciation and its Consequences*, D. Otte and J. A. Endler (eds.). Sunderland, Mass.: Sinauer, pp. 284–304.

———1992. MtDNA phylogeny of Andean mice: A test of diversification across ecological gradients. *Evolution* 46:174–183.

———1994. Paraphyly, polyphyly, and the nature of species boundaries in pocket gophers (genus *Thomomys*). *Syst. Biol.* 43:11–26.

Pena, S., F. Santos, N. Bianchi, C. Bravi, F. Carnese, F. Rothhammer, T. Gerelsaikhan B. Munkhtuja, and T. Oyunsuren. 1995. A major founder Y-chromosome haplotype in Amerindians. *Nature Genet.* 11:15–16.

Pennisi, E. 1998. Genome data shake tree of life. *Science* 280:672–674.

Penny, D., M. Steel, P. J. Waddell, and M. D. Hendy. 1995. Improved analyses of human mtDNA sequences support a recent African origin for *Homo sapiens*. *Mol. Biol. Evol.* 12:863–882.

Peres, C. A., J. L. Patton, and M. N. F. da Silva. 1996. Riverine barriers and gene flow in Amazonian saddle-back tamarins. *Folia Primatologica* 67:113–124.

Pesole, G., E. Sbisa, G. Preparata, and C. Saccone. 1992. The evolution of the mitochondrial D-loop region and the origin of modern man. *Mol. Biol. Evol.* 9:587–598.

Petren, K. and T. J. Case. 1997. A phylogenetic analysis of body size evolution and biogeography in chuckwallas (*Sauromalus*) and other iguanines. *Evolution* 51:206–219.

Petri, B., S. Pääbo, A. von Haeseler, and D. Tautz. 1997. Paternity assessment and population subdivision in a natural population of the larger mouse-eared bat *Myotis myotis*. *Mol. Ecol.* 6:235–242.

Petri, B., A. von Haeseler, and S. Pääbo. 1996. Extreme sequence heteroplasmy in bat mitochondrial DNA. *Biol. Chem.* 377:661–667.

Philipp, D. P., W. F. Childers, and G. S. Whitt. 1983. A biochemical genetic evaluation of the northern and Florida subspecies of largemouth bass. *Trans. Amer. Fish. Soc.* 112:1–20.

Phillips, C. A. 1994. Geographic distribution of mitochondrial DNA variants and the historical biogeography of the spotted salamander, *Ambystoma maculatum*. *Evolution* 48:597–607.

Phillips, C. A., W. W. Dimmick, and J. L. Carr. 1996. Conservation genetics of the common snapping turtle (*Chelydra serpentina*). *Conserv. Biol.* 10:397–405.

Pichler, F. B., S. M. Dawson, E. Slooten, and C. S. Baker. 1998. Geographic isolation of Hector's dolphin populations as described by mitochondrial DNA sequences. *Conserv. Biol.* 12:676–682.

Pielou, E. C. 1991. *After the Ice Age: The Return of Life to Glaciated North America*. Chicago: University of Chicago Press.

Pigeon, D., A. Chouinard, and L. Bernatchez. 1997. Multiple modes of speciation involved in the parallel evolution of sympatric morphotypes of lake whitefish (*Coregonus clupeaformis*, Salmonidae). *Evolution* 51:196–205.

Pires, J. M., T. Dobzhansky, and G. A. Black. 1953. An estimate of the number of species of trees in an Amazonian forest community. *Bot. Gazette* 114:467–477.

Pitelka, L. F. and 22 others. 1997. Plant migration and climate change. *Amer. Sci.* 85:464–473.

Plante, Y., P. T. Boag, and B. N. White. 1989. Microgeographic variation in mitochondrial DNA of meadow voles (*Microtus pennsylvanicus*) in relation to population density. *Evolution* 43:1522–1537.

Platnick, N. I. and G. Nelson. 1978. A method for analysis of historical biogeography. *Syst. Zool.* 27:1–16.

Pope, L. C., A. Sharp, and C. Moritz. 1996. Population structure of the yellow-footed rock-wallaby *Petrogale xanthopus* (Gray, 1854) inferred from mtDNA sequences and microsatellite loci. *Mol. Ecol.* 5:629–640.

Potts, W. K. 1996. PCR-based cloning across large taxonomic distances and polymorphism detection: MHC as a case study. In *Molecular Zoology*, J. D. Ferraris and S. R. Palumbi (eds.). New York: Wiley-Liss, pp. 181–194.

Powers, D. A., T. Lauerman, D. Crawford, and L. DiMichele. 1991. Genetic mechanisms for adapting to a changing environment. *Annu. Rev. Genet.* 25:629–659.

Prance, G. T. (ed.). 1982. *Biological Diversification in the Tropics*. New York: Columbia University Press.

Pressey, R. L., C. J. Humphries, C. R. Margules, R. I. Vane-Wright, and P. H. Williams. 1993. Beyond opportunism: Key principles for systematic reserve selection. *Trends Ecol. Evol.* 8:124–128.

Price, T. 1998. Sexual selection and natural selection in bird speciation. *Phil. Trans. Roy. Soc. Lond. B* 353:251–260.

Prinsloo, P. and T. J. Robinson. 1992. Geographic mitochondrial DNA variation in the rock hyrax, *Procavia capensis*. *Mol. Biol. Evol.* 9:447–456.

Pritchard, P. C. H. 1969. Studies of the systematics and reproductive cycles of the genus *Lepidochelys*. Ph.D. diss., University of Florida, Gainesville.

Prychitko, T. M. and W. S. Moore. 1997. The utility of DNA sequences of an intron from the beta-fibrinogen gene in phylogenetic analysis of woodpeckers (Aves: Picidae). *Mol. Phylogen. Evol.* 8:193–204.

Ptacek, M. B., H. C. Gerhardt, and R. D. Sage. 1994. Speciation by polyploidy in treefrogs: Multiple origins of the tetraploid, *Hyla versicolor*. *Evolution* 48: 898–908.

Pullium, H. R. 1988. Sources, sinks, and population regulation. *Amer. Nat.* 132: 652–661.

Pumo, D. E., E. Z. Goldin, B. Elliot, C. J. Phillips, and H. H. Genoways. 1988. Mitochondrial DNA polymorphism in three Antillean island populations of the fruit bat, *Artibeus jamaicensis*. *Mol. Biol. Evol.* 5:79–89.

Quesada, H., C. M. Beynon, and D. O. F. Skibinski. 1995. A mitochondrial DNA discontinuity in the mussel *Mytilus galloprovincialis* Lmk: Pleistocene vicariance biogeography and secondary intergradation. *Mol. Biol. Evol.* 12:521–524.

Quesada, H., C. Gallagher, D. A. G. Skibinski, and D. O. F. Skibinski. 1998. Patterns of polymorphism and gene flow of gender-associated mitochondrial DNA lineages in European mussel populations. *Mol. Ecol.* 7:1041–1051.

Questiau, S., M.-C. Eybert, A. R. Gaginskaya, L. Gielly, and P. Taberlet. 1998. Recent divergence between two morphologically differentiated subspecies of bluethroat (Aves: Muscicapidae: *Luscinia svecica*) inferred from mitochondrial DNA sequence variation. *Mol. Ecol.* 7:239–245.

Quinn, T. W. 1992. The genetic legacy of mother goose—phylogeographic patterns of lesser snow goose *Chen caerulescens caerulescens* maternal lineages. *Mol. Ecol.* 1:105–117.

Quinn, T. W., G. F. Shields, and A. C. Wilson. 1991. Affinities of the Hawaiian goose based on two types of mitochondrial DNA data. *Auk* 108:585–593.

Radtkey, R. R., S. M. Fallon, and T. J. Case. 1997. Character displacement in some *Cnemidophorus* lizards revisited: A phylogenetic analysis. *Proc. Natl. Acad. Sci. USA* 94:9740–9745.

Raff, R. A., C. R. Marshall, and J. M. Turbeville. 1994. Using DNA sequences to unravel the Cambrian radiation of the animal phyla. *Annu. Rev. Ecol. Syst.* 25:351–375.

Rand, A. L. 1948. Glaciation, an isolating factor in speciation. *Evolution* 2:314–321.

Rand, D. M. 1993. Endotherms, ectotherms, and mitochondrial genome-size variation. *J. Mol. Evol.* 37:281–295.

———1994. Thermal habit, metabolic rate and the evolution of mitochondrial DNA. *Trends Ecol. Evol.* 9:125–131.

Rand, D. M. and R. G. Harrison. 1986. Mitochondrial DNA transmission genetics in crickets. *Genetics* 114:955–970.

Randazzo, A. F. and D. S. Jones (eds.). 1997. *The Geology of Florida.* Gainesville: University Press of Florida.

Randi, E. 1993. Effects of fragmentation and isolation on genetic variability of the Italian populations of the wolf *Canis lupus* and brown bear *Ursus arctos. Acta Theor.* 38:113–120.

Randi, E., L. Gentile, G. Boscagli, D. Huber, and H. U. Roth. 1994. Mitochondrial DNA sequence divergence among some west European brown bear (*Ursus arctos* L.) populations: Lessons for conservation. *Heredity* 73:480–489.

Rapacz, J., L. Chen, E. Butler-Brunner, M.-J. Wu, J. O. Hasler-Rapacz, R. Butler, and V. N. Schumaker. 1991. Identification of the ancestral haplotype for apolipoprotein B suggests an African origin of *Homo sapiens sapiens* and traces their subsequent migration to Europe and the Pacific. *Proc. Natl. Acad. Sci. USA* 88:1403–1406.

Rassmann, K., D. Tautz, F. Trillmich, and C. Gliddon. 1997. The microevolution of the Galápagos marine iguana *Amblyrhynchus cristatus* assessed by nuclear and mitochondrial genetic analysis. *Mol. Ecol.* 6:437–452.

Rawson, P. D. and T. J. Hilbish. 1995. Evolutionary relationships among the male and female mitochondrial DNA lineages in the *Mytilus edulis* species complex. *Mol. Biol. Evol.* 12:893–901.

———1998. Asymmetric introgression of mitochondrial DNA among European populations of blue mussels (*Mytilus* spp.). *Evolution* 52:100–108.

Redd, A. J., N. Takezaki, S. T. Sherry, S. T. McGarvey, A. S. M. Sofro, and M. Stoneking. 1995. Evolutionary history of the COII/tRNA^{Lys} intergenic 9 base pair deletion in human mitochondrial DNAs from the Pacific. *Mol. Biol. Evol.* 12:604–615.

Reeb, C. A. and J. C. Avise. 1990. A genetic discontinuity in a continuously distributed species: Mitochondrial DNA in the American oyster, *Crassostrea virginica. Genetics* 124:397–406.

Reich, D. E. and D. B. Goldstein. 1998. Genetic evidence for a Paleolithic human population expansion in Africa. *Proc. Natl. Acad. Sci. USA* 95:8119–8123.

Reid, D. G., E. Rumbak, and R. H. Thomas. 1996. DNA, morphology and fossils: Phylogeny and evolutionary rates of the gastropod genus *Littorina. Phil. Trans. Roy. Soc. Lond. B* 351:877–895.

Remington, C. L. 1968. Suture-zones of hybrid interaction between recently joined biotas. *Evol. Biol.* 2:321–428.

Ribeiro, S. and G. B. Golding. 1998. The mosaic nature of the eukaryotic nucleus. *Mol. Biol. Evol.* 15:779–788.

Rich, S. M., D. A. Caporale, S. R. Telford III, T. D. Kocher, D. L. Hartl, and A. Spielman. 1995. Distribution of *Ixodes ricinus*-like ticks of eastern North America. *Proc. Natl. Acad. Sci. USA* 92:6284–6288.

Rich, S. M., M. C. Licht, R. R. Hudson, and F. J. Ayala. 1998. Malaria's Eve: Evidence of a recent population bottleneck throughout the world populations of *Plasmodium falciparum. Proc. Natl. Acad. Sci. USA* 95:4425–4430.

Richardson, L. R. and J. R. Gold. 1993. Mitochondrial DNA variation in red grouper (*Epinephelus morio*) and greater amberjack (*Seriola dumerili*) from the Gulf of Mexico. *ICES J. Marine Sci.* 50:53–62.

———1995. Evolution of the *Cyprinella lutrensis* species group. III. Geographic variation in the mitochondrial DNA of *Cyprinella lutrensis*—the influence of Pleistocene glaciation on population dispersal and vicariance. *Mol. Ecol.* 4:163–171.

Richter, C. 1992. Reactive oxygen and DNA damage in mitochondria. *Mutat. Res.* 275:249–255.

Ricklefs, R. E. (ed.). 1993. *Species Diversity in Ecological Communities: Historical and Geographical Perspectives.* Chicago: University of Chicago Press.

Riddle, B. R. 1995. Molecular biogeography in the pocket mice (*Perognathus* and *Chaetodipus*) and grasshopper mice (*Onychomys*): The late Cenozoic development of a North American aridlands rodent guild. *J. Mammal.* 76:283–301.

———1996. The molecular phylogeographic bridge between deep and shallow history in continental biotas. *Trends Ecol. Evol.* 11:207–211.

Riddle, B. R. and R. L. Honeycutt. 1990. Historical biogeography in North American arid regions: An approach using mitochondrial-DNA phylogeny in grasshopper mice (genus *Onychomys*). *Evolution* 44:1–15.

Riddle, B. R., R. L. Honeycutt, and P. L. Lee. 1993. Mitochondrial DNA phylogeography in northern grasshopper mice (*Onychomys leucogaster*)—the influence of Quaternary climatic oscillations on population dispersion and divergence. *Mol. Ecol.* 2:183–193.

Rieseberg, L. H. 1991. Homoploid reticulate evolution in *Helianthus* (Asteraceae): Evidence from ribosomal genes. *Amer. J. Bot.* 78:1218–1237.

———1997. Hybrid origins of plant species. *Annu. Rev. Ecol. Syst.* 28:359–389.

Rieseberg, L. H. and L. Brouillet. 1994. Are many plant species paraphyletic? *Taxon* 43:21–32.

Rieseberg, L. H., R. Carter, and S. Zona. 1990. Molecular tests of the hypothesized hybrid origin of two diploid *Helianthus* species (Asteraceae). *Evolution* 44:1498–1511.

Rieseberg, L. H., J. Whitton, and C. R. Linder. 1996. Molecular marker incongruence in plant hybrid zones and phylogenetic trees. *Acta Bot. Neerl.* 45:243–262.

Rising, J. D. and J. C. Avise. 1993. An application of genealogical concordance principles to the taxonomy and evolutionary history of the sharp-tailed sparrow (*Ammodramus caudacutus*). *Auk* 110:844–856.

Ritchie, M. G., R. K. Butlin, and G. M. Hewitt. 1989. Assortative mating across a hybrid zone in *Chorthippus parallelus* (Orthoptera: Acrididae). *J. Evol. Biol.* 2: 339–352.

Robinson, N. A. 1995. Implications from mitochondrial DNA for management to conserve the eastern barred bandicoot (*Perameles gunnii*). *Conserv. Biol.* 9: 114–125.

Roderick, G. K. 1996. Geographic structure of insect populations: Gene flow, phylogeography, and their uses. *Annu. Rev. Entomol.* 41:325–362.

Roderick, G. K. and R. G. Gillespie. 1998. Speciation and phylogeography of Hawaiian terrestrial arthropods. *Mol. Ecol.* 7:519–531.

Roehrdanz, R. L. and D. A. Johnson. 1988. Mitochondrial DNA variation among geographical populations of the screwworm fly, *Cochliomyia hominivorax*. *J. Med. Entomol.* 25:136–141.

Rogers, A. R. 1997. Population structure and modern human origins. In *Progress in Population Genetics and Human Evolution,* P. Donnelly and S. Tavaré (eds.). New York: Springer-Verlag, pp. 55–79.

Rogers, A. R. and H. Harpending. 1992. Population growth makes waves in the distribution of pairwise genetic differences. *Mol. Biol. Evol.* 9:552–569.

Rogers, A. R. and L. B. Jorde. 1995. Genetic evidence on the origin of modern humans. *Hum. Biol.* 67:1–36.

Rogers, J., P. B. Samallow, and A. G. Comuzzie. 1996. Estimating the age of the common ancestor of men from the ZFY intron. *Science* 272:1360–1361.

Roman, J., S. Santhuff, P. Moler, and B. W. Bowen. 1999. Cryptic evolution and population structure in the alligator snapping turtle (*Macroclemys temminckii*). *Conserv. Biol.* 13:1–9.

Ronquist, R. 1997. Dispersal-vicariance analysis: A new approach to the quantification of historical biogeography. *Syst. Biol.* 46:195–203.

Rosel, P. E. 1992. Genetic population structure and systematics of some small cetaceans inferred from mitochondrial DNA sequence variation. Ph.D. diss., University of California, San Diego.

Rosel, P. E. and B. A. Block. 1996. Mitochondrial control region variability and global population structure in the swordfish, *Xiphias gladius*. *Marine Biol.* 125:11–22.

Rosel, P. E., A. E. Dizon, and M. G. Haygood. 1995. Variability of the mitochondrial control region in populations of the harbour porpoise, *Phocoena phocoena*, on interoceanic and regional scales. *Can. J. Fish. Aquat. Sci.* 52:1210–1219.

Rosel, P. E., A. E. Dizon, and J. E. Heyning. 1994. Genetic analysis of sympatric morphotypes of common dolphins (genus *Delphinus*). *Marine Biol.* 119:159–167.

Rosen, D. E. 1975. A vicariance model for Caribbean biogeography. *Syst. Zool.* 24:431–464.

———1979. Fishes from the uplands and intermontane basins of Guatemala: Revisionary studies and comparative geography. *Bull. Amer. Mus. Natur. Hist.* 162:267–376.

Rosenblum, L. L., J. Supriatna, and D. J. Melnick. 1997. Phylogeographic analysis of pigtail macaque populations (*Macaca nemestrina*) inferred from mitochondrial DNA. *Amer. J. Phys. Anthropol.* 104:35–45.

Rosenzweig, M. L. 1995. *Species Diversity in Space and Time.* Cambridge: Cambridge University Press.

Ross, K. G., M. J. B. Krieger, D. D. Shoemaker, E. L. Vargo, and L. Keller. 1997. Hierarchical analysis of genetic structure in native fire ant populations: Results from three classes of molecular markers. *Genetics* 147:643–655.

Rossi, M., E. Barrio, A. Latorre, J. E. Quezada-Díaz, E. Hasson, A. Moya, and A. Fontdevilla. 1996. The evolutionary history of *Drosophila buzzatti*. XXX. Mitochondrial DNA polymorphism in original and colonizing populations. *Mol. Biol. Evol.* 13:314–323.

Routman, E. 1993. Mitochondrial DNA variation in *Cryptobranchus alleganiensis*, a salamander with extremely low allozyme diversity. *Copeia* 1993:407–416.

Routman, E., R. Wu, and A. R. Templeton. 1994. Parsimony, molecular evolution, and biogeography: The case of the North American giant salamander. *Evolution* 48:1799–1809.

Rowan, R. G. and J. A. Hunt. 1991. Rates of DNA change and phylogeny from the DNA sequences of the alcohol dehydrogenase gene for five closely related species of Hawaiian *Drosophila*. *Mol. Biol. Evol.* 8:49–70.

Roy, M. S., E. Geffen, D. Smith, E. Ostrander, and R. K. Wayne. 1994b. Patterns of differentiation and hybridization in North American wolf-like canids revealed by analysis of microsatellite loci. *Mol. Biol. Evol.* 11:553–570.

Roy, M. S., D. J. Girman, and R. K. Wayne. 1994a. The use of museum specimens to reconstruct the genetic variability and relationships of extinct populations. *Experientia* 50:551–557.

Rozas, A., J. M. Hernandez, V. M. Cabrera, and A. Prerosti. 1990. Colonization in America by *Drosophila subobscura:* Effect of the founder event on mitochondrial DNA polymorphism. *Mol. Biol. Evol.* 7:103–109.

Rubinoff, I. and E. G. Leigh. 1990. Dealing with diversity: The Smithsonian Tropical Research Institute and tropical biology. *Trends Ecol. Evol.* 5:115–118.

Ruedi, M., M. F. Smith, and J. L. Patton. 1997. Phylogenetic evidence of mitochondrial DNA introgression among pocket gophers in New Mexico (family Geomyidae). *Mol. Ecol.* 6:453–462.

Ruttner, F. 1988. *Biogeography and Taxonomy of Honeybees.* New York: Springer-Verlag.

Ruvolo, M., D. Pan, S. Zehr, T. Goldberg, T. R. Disotell, and M. von Dornum. 1994. Gene trees and hominoid phylogeny. *Proc. Natl. Acad. Sci. USA* 91:8900–8904.

Ruvolo, M., S. Zehr, M. von Dornum, D. Pan, B. Chang, and J. Lin. 1993. Mitochondrial COII sequences and modern human origins. *Mol. Biol. Evol.* 10: 1115–1135.

Ryan, M. J. 1996. Phylogenetics in behavior: some cautions and expectations. In *Phylogenies and the Comparative Method in Animal Behavior,* E. P. Martins (ed.). New York: Oxford University Press, pp. 1–21.

Ryder, O. A. 1986. Species conservation and the dilemma of subspecies. *Trends Ecol. Evol.* 1:9–10.

Ryder, O. A. and L. G. Chemnick. 1993. Chromosomal and mitochondrial-DNA variation in orangutans. *J. Heredity* 84:405–409.

Ryman, N. and F. Utter (eds.). 1987. *Population Genetics and Fishery Management.* Seattle: University of Washington Press.

Saccone, C. and A. M. Kroon (eds.). 1976. *The Genetic Function of Mitochondrial DNA.* Amsterdam: North-Holland.

Salem, A.-H., F. M. Badr, M. F. Gaballah, and S. Pääbo. 1996. The genetics of traditional living: Y-chromosomal and mitochondrial lineages in the Sinai Peninsula. *Amer. J. Hum. Genet.* 59:741–743.

Saltonstall, K., G. Amato, and J. Powell. 1998. Mitochondrial DNA variability in Grauer's gorillas of Kahuzi-Biega National Park. *J. Heredity* 8:129–135.

Sambuughin, N., Y. G. Rychkov, and V. N. Petrishchev. 1992. Genetic differentiation of Mongolian population: The geographical distribution of mtDNA RFLPs, mitotypes and population estimation of mutation rate for mitochondrial genome. *Genetika* 28:136–153.

Sanger, F., S. Nicklen, and A. R. Coulson. 1977. DNA sequencing with chain-terminating inhibitors. *Proc. Natl. Acad. Sci. USA* 74:5463–5467.

Santos, M., R. H. Ward, and R. Barrantes. 1994. MtDNA variation in the Chibcha Amerindian Heutar from Costa Rica. *Hum. Biol.* 66:963–977.

Santucci, F., B. C. Emerson, and G. M. Hewitt. 1998. Mitochondrial DNA phylogeography of European hedgehogs. *Mol. Ecol.* 7:1163–1172.

Sarver, S. K., M. C. Landrum, and D. W. Foltz. 1992. Genetics and taxonomy of ribbed mussels (*Geukensia* spp.). *Marine Biol.* 113:385–390.

Satta, Y. and N. Takahata. 1990. Evolution of *Drosophila* mitochondrial DNA and the history of the *melanogaster* subgroup. *Proc. Natl. Acad. Sci. USA* 87:9558–9562.

Satta, Y., N. Toyohara, C. Ohtaka, Y. Tatsuno, T. K. Watanabe, E. T. Matsura, S. I. Chigusa, and N. Takahata. 1988. Dubious maternal inheritance of mitochondrial DNA in *D. simulans* and evolution of *D. mauritiana. Genet. Res.* 52:1–6.

Saunders, N. C., L. G. Kessler, and J. C. Avise. 1986. Genetic variation and geographic differentiation in mitochondrial DNA of the horseshoe crab, *Limulus polyphemus. Genetics* 112:613–627.

Saville, B. J., Y. Kohli, and J. B. Anderson. 1998. mtDNA recombination in a natural population. *Proc. Natl. Acad. Sci. USA* 95:1331–1335.

Sawyer, S. 1989. Statistical tests for detecting gene conversion. *Mol. Biol. Evol.* 6:526–538.

Schaal, B. A., D. A. Hayworth, K. M. Olsen, J. T. Rauscher, and W. A. Smith. 1998. Phylogeographic studies in plants: problems and prospects. *Mol. Ecol.* 7:465–474.

Schaffer, H. 1970. The fate of neutral mutants as a branching process. In *Mathematical Topics in Population Genetics*, K. Kojima (ed.). New York: Springer-Verlag, pp. 317–336.

Scharf, S. J., G. T. Horn, and H. A. Erlich. 1986. Direct cloning and sequence analysis of enzymatically amplified genomic sequences. *Science* 233:1076–1078.

Schliewen, U. K., D. Tautz, and S. Pääbo. 1994. Sympatric speciation suggested by monophyly of crater lake cichlids. *Nature* 368:629–632.

Schluter, D. 1995. Uncertainty in ancient phylogenies. *Nature* 377:108–109.

Schluter, D., T. Price, A. O. Mooers, and D. Ludwig. 1997. Likelihood of ancestor states in adaptive radiation. *Evolution* 51:1699–1711.

Schneider, C. J., M. Cunningham, and C. Moritz. 1998. Comparative phylogeography and the history of endemic vertebrates in the Wet Tropics rainforests of Australia. *Mol. Ecol.* 7:487–498.

Schubart, C. D., R. Diesel, and S. B. Hedges. 1998. Rapid evolution to terrestrial life in Jamaican crabs. *Nature* 393:363–364.

Schurr, T. G., S. W. Ballinger, Y. Y. Gan, J. A. Hodge, D. A. Merriwether, D. N. Lawrence, W. C. Knowler, K. M. Weiss, and D. C. Wallace. 1990. Amerindian mitochondrial DNAs have rare Asian mutations at high frequencies, suggesting they derived from four primary maternal lineages. *Amer. J. Hum. Genet.* 46:613–623.

Schweigert, J. F. and R. E. Withler. 1990. Genetic differentiation of Pacific herring based on enzyme electrophoresis and mitochondrial DNA analysis. *Amer. Fish. Soc. Symp.* 7:459–469.

Scoles, D. R. and J. E. Graves. 1993. Genetic analysis of the population structure of yellowfin tuna *Thunnus albacares* in the Pacific Ocean. *Fish. Bull.* 91:690–698.

Scott, J. M. and B. Csuti. 1997. Gap analysis for biodiversity surveys and maintenance. In *Biodiversity II: Understanding and Protecting Our Biological Resources,* M. L. Reaka-Kudla et al. (eds.). Washington, D.C.: Joseph Henry Press, pp. 321–340.

Scott, J. M., B. Csuti, J. D. Jacobi, and S. Caicco. 1990. Gap analysis: Assessing protection needs. In *Landscape Linkages and Biodiversity,* W. E. Hudson (ed.). Washington, D.C.: Island Press, pp. 15–26.

Scott, J. M., B. Csuti, J. D. Jacobi, and J. E. Estes. 1987. Species richness: A geographic approach to protecting future biological diversity. *BioScience* 37:782–788.

Scribner, K. T. and J. C. Avise. 1993. Cytonuclear genetic architecture in mosquitofish populations and the possible roles of introgressive hybridization. *Mol. Ecol.* 2:139–149.

Scribner, K. T., J. Bodkin, B. Ballachey, S. R. Fain, M. A. Cronin, and M. Sanchez. 1997. Population genetic studies of the sea otter (*Enhydra lutris*): A review and interpretation of available data. In *Molecular Genetics of Marine Mammals,* A. E. Dizon, S. J. Chivers, and W. F. Perrin (eds.). Special Publ. 3, Society for Marine Mammalogy, pp. 197–208.

Sears, C. J., B. W. Bowen, R. W. Chapman, S. B. Galloway, S. R. Hopkins-Murphy, and C. M. Woodley. 1995. Demographic composition of the feeding population of juvenile loggerhead sea turtles (*Caretta caretta*) off Charleston, South Carolina: Evidence from mitochondrial DNA markers. *Marine Biol.* 123:869–874.

Sedberry, G.R., J. L. Carlin, R. W. Chapman, and B. Eleby. 1996. Population structure in the pan-oceanic wreckfish, *Polyprion americanus* (Teleostei: Polyprionidae), as indicated by mtDNA variation. *J. Fish. Biol.* 49 (supplement A): 318–329.

Seddon, J. M., P. R. Baverstock, and A. Georges. 1998. The rate of mitochondrial 12S rRNA evolution is similar in freshwater turtles and marsupials. *J. Mol. Evol.* 46:460–464.

Seehausen, O., J. J. M. van Alphen, and F. Witte. 1997. Cichlid fish diversity threatened by eutrophication that curbs sexual selection. *Science* 277:1808–1811.

Seielstad, M. T., E. Minch, and L. L. Cavalli-Sforza. 1998. Genetic evidence for a higher female migration rate in humans. *Nature Genet.* 20:278–280.

Selander, R. K. 1971. Systematics and speciation in birds. In *Avian Biology,* vol. 1, D. S. Farmer and J. R. King (eds.). New York: Academic Press, pp. 57–147.

Seutin, G., J. Brawn, R. E. Ricklefs, and E. Bermingham. 1993. Genetic divergence among populations of a tropical passerine, the streaked saltator (*Saltator albicollis*). *Auk* 110:117–126.

Seutin, G., N. K. Klein, R. E. Ricklefs, and E. Bermingham. 1994. Historical biogeography of the bananaquit (*Coereba flaveola*) in the Caribbean region: A mitochondrial DNA assessment. *Evolution* 48:1041–1061.

Seutin, G., L. M. Ratcliffe, and P. T. Boag. 1995. Mitochondrial DNA homogeneity in the phenotypically diverse redpoll finch complex (Aves: Carduelinae: *Carduelis flammea-hornemanni*). *Evolution* 49:962–973.

Shaffer, H. B. and M. L. McKnight. 1996. The polytypic species revisited: genetic differentiation and molecular phylogenetics of the tiger salamander *Ambystoma tigrinum* (Amphibia: Caudata) complex. *Evolution* 50:417–433.

Shaw, D. M. and C. H. Langley. 1979. Inter- and intraspecific variation in restriction maps of *Drosophila* mitochondrial DNAs. *Nature* 281:696–699.

Shedlock, A. M., J. D. Parker, D. A. Crispin, T. W. Pietsch, and G. C. Burmer. 1992. Evolution of the salmonid mitochondrial control region. *Mol. Phylogen. Evol.* 1:179–192.

Sheldon, F. H. and L. A. Whittingham. 1997. Phylogeny in studies of bird ecology, behavior, and morphology. In *Avian Molecular Evolution and Systematics,* D. P. Mindell (ed.). New York: Academic Press, pp. 279–299.

Sherry, S. T., A. R. Rogers, H. Harpending, H. Soodyall, T. Jenkins, and M. Stoneking. 1994. Mismatch distributions of mtDNA reveal recent human population expansions. *Hum. Biol.* 66:761–775.

Shields, G. F. and J. R. Gust. 1995. Lack of geographic structure in mitochondrial DNA sequences of Bering Sea walleye pollock, *Theragra chalcogramma. Mol. Mar. Biol. Biotech.* 4:69–82.

Shields, G. F. and T. D. Kocher. 1991. Phylogenetic relationships of North American ursids based on analysis of mitochondrial DNA. *Evolution* 45:218–221.

Shields, G. F., A. M. Schmiechen, B. L. Frazier, A. Redd., M. I. Voevoda, J. K. Reed, and R. H. Ward. 1993. Mitochondrial DNA sequences suggest a recent evolutionary divergence for Beringian and northern North American populations. *Amer. J. Hum. Genet.* 53:549–562.

Shields, G. F. and A .C. Wilson. 1987. Calibration of mitochondrial DNA evolution in geese. *J. Mol. Evol.* 24:212–217.

Shitara, H., J-I. Hayashi, S. Takahama, H. Kaneda, and H. Yonekawa. 1998. Maternal inheritance of mouse mtDNA in interspecific hybrids: Segregation of the leaked paternal mtDNA followed by the prevention of subsequent paternal leakage. *Genetics* 148:851–857.

Shubin, N. 1998. Evolutionary cut and paste. *Nature* 394:12–13.

Shulman, M. J. and E. Bermingham. 1995. Early life histories, ocean currents, and the population genetics of Caribbean reef fishes. *Evolution* 49:897–910.

Sibley, C. G. 1991. Phylogeny and classification of birds from DNA comparisons. *Acta XX Congressus Internationalis Ornithologici* I:111–126.

Sibley, C. G. and J. E. Ahlquist. 1986. Reconstructing bird phylogeny by comparing DNAs. *Sci. Amer.* 254(2):82–93.

———1990. *Phylogeny and Classification of Birds—A Study in Molecular Evolution.* New Haven, Conn.: Yale University Press.

Silberman, J. D., S. K. Sarver, and P. J. Walsh. 1994. Mitochondrial DNA variation and population structure in the spiny lobster *Panulirus argus. Marine Biol.* 120:601–608.

Simon, C., F. Frati, A. Beckenbach, B. Crespi, H. Liu, and P. Flook. 1994. Evolution, weighting, and phylogenetic utility of mitochondrial gene sequences and a compilation of conserved polymerase chain reaction primers. *Ann. Ent. Soc. Amer.* 87:651–701.

Simonsen, B. T., H. R. Siegismund, and P. Arctander. 1998. Population structure of African buffalo inferred from mtDNA sequences and microsatellite loci: High variation but low differentiation. *Mol. Ecol.* 7:225–237.

Simpson, B. B. and J. Haffer. 1978. Speciation patterns in the Amazonian forest biota. *Annu. Rev. Ecol. Syst.* 9:497–518.

Simpson, G. G. 1945. The principles of classification and a classification of mammals. *Bull. Amer. Mus. Natur. Hist.* 85:1–350.

Skibinski, D. O. F., C. Gallagher, and C. M. Beynon. 1994. Mitochondrial DNA inheritance. *Nature* 368:817–818.

Slade, R. W. and C. Moritz. 1998. Phylogeography of *Bufo marinus* from its natural and introduced range. *Proc. Roy. Soc. Lond. B* 265:769–777.

Slade, R. W., C. Moritz, and A. Heideman. 1994. Multiple nuclear-gene phylogenies: Application to pinnipeds and comparison with a mitochondrial DNA gene phylogeny. *Mol. Biol. Evol.* 11:341–356.

Slade, R. W., C. Moritz, A. Heideman, and P. T. Hale. 1993. Rapid assessment of single-copy nuclear DNA variation in diverse species. *Mol. Ecol.* 2:359–373.

Slatkin, M. 1977. Gene flow and genetic drift in a species subject to frequent local extinctions. *Theor. Pop. Biol.* 12:253–262.

———1985a. Gene flow in natural populations. *Annu. Rev. Ecol. Syst.* 16:393–430.

———1985b. Rare alleles as indicators of gene flow. *Evolution* 39:53–65.

———1987. Gene flow and the geographic structure of natural populations. *Science* 236:787–792.

———1989. Detecting small amounts of gene flow from phylogenies of alleles. *Genetics* 121:609–612.

———1991. Inbreeding coefficients and coalescence times. *Genet. Res.* 58:167–175.

Slatkin, M. and H. E. Arter. 1991. Spatial autocorrelation methods in population genetics. *Amer. Nat.* 138:499–517.

Slatkin, M. and N. H. Barton. 1989. A comparison of three indirect methods for estimating average levels of gene flow. *Evolution* 43:1349–1368.

Slatkin, M. and R. R. Hudson. 1991. Pairwise comparisons of mitochondrial DNA sequences in stable and exponentially growing populations. *Genetics* 129:555–562.

Slatkin, M. and W. P. Maddison. 1989. A cladistic measure of gene flow inferred from the phylogenies of alleles. *Genetics* 123:603–613.

Smith, D. R. 1991. African bees in the Americas: insights from biogeography and genetics. *Trends Ecol. Evol.* 6:17–21.

Smith, D. R. and W. M. Brown. 1990. Restriction endonuclease cleavage site and length polymorphisms in mitochondrial DNA of *Apis mellifera mellifera* and *A.m. carnica* (Hymenoptera: Apidae). *Ann. Entomol. Soc. Amer.* 83:81–88.

Smith, D. R., O. R. Taylor, and W. M. Brown. 1989. Neotropical Africanized honey bees have African mitochondrial DNA. *Nature* 339:213–215.

Smith, M. F. 1998. Phylogenetic relationships and geographic structure in pocket gophers in the genus *Thomomys*. *Mol. Phylogen. Evol.* 9:1–14.

Smith, M. F. and J. L. Patton. 1993. The diversification of South American murid rodents: Evidence from mitochondrial DNA sequence data for the akodontine tribe. *Biol. J. Linn. Soc.* 50:149–177.

Smith, M. J., A. Arndt, S. Gorski, and E. Fajber. 1993. The phylogeny of echinoderm classes based on mitochondrial gene arrangements. *J. Mol. Evol.* 36:545–554.

Smith, M. W., R. W. Chapman, and D. A. Powers. 1998. Mitochondrial DNA analysis of Atlantic Coast, Chesapeake Bay, and Delaware Bay populations of the teleost *Fundulus heteroclitus* indicates temporally unstable distributions over geologic time. *Mol. Marine Biol. Biotech.* 7:79–87.

Smith, T. B. and R. K. Wayne (eds.). 1996. *Molecular Genetic Approaches in Conservation.* New York: Oxford University Press.

Smolenski, A. J., J. R. Ovenden, and R. W. G. White. 1993. Evidence of stock separation in southern hemisphere orange roughy (*Hoplostethus atlanticus,* Trachichthyidae) from restriction-enzyme analysis of mitochondrial DNA. *Marine Biol.* 116:219–230.

Smouse, P. E. 1998. To tree or not to tree. *Mol. Ecol.* 7:399–412.

Smouse, P. E., T. E. Dowling, J. A. Tworek, W. R. Hoeh, and W. M. Brown. 1991. Effects of intraspecific variation on phylogenetic inference: A likelihood analysis of mtDNA restriction site data in cyprinid fishes. *Syst. Zool.* 40: 393–409.

Sneath, P. H. A. and R. R. Sokal. 1973. *Numerical Taxonomy.* San Francisco: Freeman.

Sokal, R. R., R. M. Harding, and N. L. Oden. 1989a. Spatial patterns of human gene frequencies in Europe. *Amer. J. Phys. Anthropol.* 80:267–294.

Sokal, R. R., G. M. Jacquez, and M. C. Wooten. 1989b. Spatial autocorrelation analysis of migration and selection. *Genetics* 121:845–855.

Solignac, M., J. Genermont, M. Monnerot, and J.-C. Mounolou. 1984. Genetics of mitochondria in *Drosophila:* Inheritance in heteroplasmic strains of *D. mauritiana. Mol. Gen. Genet.* 197:183–188.

Solignac, M., M. Monnerot, and J.-C. Mounolou. 1983. Mitochondrial DNA heteroplasmy in *Drosophila mauritiana. Proc. Natl. Acad. Sci. USA* 80:6942–6946.

Soltis, D. E., M. A. Gitzendanner, D. D. Strenge, and P. S. Soltis. 1997. Chloroplast DNA intraspecific phylogeography of plants from the Pacific Northwest of North America. *Plant Syst. Evol.* 206:353–373.

Soltis, D. E., M. S. Mayer, P. S. Soltis, and M. Edgerton. 1991. Chloroplast DNA variation in *Tellima grandiflora* (Saxifragaceae). *Amer. J. Bot.* 78:1379–1390.

Soltis, D. E., P. S. Soltis, R. K. Kuzoff, and T. L. Tucker. 1992b. Geographic structuring of chloroplast DNA genotypes in *Tiarella trifoliata* (Saxifragaceae). *Plant Syst. Evol.* 181:203–216.

Soltis, D. E., P. S. Soltis, and B. G. Milligan. 1992a. Intraspecific chloroplast DNA variation: Systematic and phylogenetic implications. In *Molecular Systematics of Plants,* P. S. Soltis, D. E. Soltis, and J. J. Doyle (eds.). New York: Chapman & Hall, pp. 117–150.

Soltis, D. E., P. S. Soltis, T. A. Ranker, and B. D. Ness. 1989. Chloroplast DNA variation in a wild plant *Tolmiea menziesii*. *Genetics* 121:819–826.

Sperling, F. A. H. and R. G. Harrison. 1994. Mitochondrial DNA variation within and between species of the *Papilio machaon* group of swallowtail butterflies. *Evolution* 48:408–422.

Sperling, F. A. H. and D. A. Hickey. 1994. Mitochondrial DNA sequence variation in the spruce budworm species complex (*Choristoneura*: Lepidoptera). *Mol. Biol. Evol.* 11:656–665.

Spiess, E. B. 1977. *Genes in Populations*. New York: John Wiley & Sons.

Stanhope, M. J., B. Hartwick, and D. Baillie. 1993. Molecular phylogeographic evidence for multiple shifts in habitat preference in the diversification of an amphipod species. *Mol. Ecol.* 2:99–112.

Stanley, H. F., S. Casey, J. M. Carnahan, S. Goodman, J. Harwood, and R. K. Wayne. 1996. Worldwide patterns of mitochondrial DNA differentiation in the harbor seal (*Phoca vitulina*). *Mol. Biol. Evol.* 13:368–382.

Staton, J. L., L. L. Daehler, and W. M. Brown. 1997. Mitochondrial gene arrangement of the horseshoe crab *Limulus polyphemus* L.: Conservation of major features among arthropod classes. *Mol. Biol. Evol.* 14:867–874.

Stenico, M., L. Nigro, and G. Barbujani. 1998. Mitochondrial lineages in Ladin-speaking communities of the eastern Alps. *Proc. Roy. Soc. Lond. B* 265:555–561.

Stephens, J. C. 1985. Statistical methods of DNA sequence analysis: Detection of intragenic recombination or gene conversion. *Mol. Biol. Evol.* 2:539–556.

Stepien, C. A. 1995. Population genetic divergence and geographic patterns from DNA sequences: Examples from marine and freshwater fishes. *Amer. Fish. Soc. Symp.* 17:263–287.

Stewart, D. T. and A. J. Baker. 1994. Patterns of sequence variation in the mitochondrial D-loop region of shrews. *Mol. Biol. Evol.* 11:9–21.

Stoneking, M. 1997. Recent African origin of human mitochondrial DNA: Review of the evidence and current status of the hypothesis. In *Progress in Population Genetics and Human Evolution*, P. Donnelly and S. Tavaré (eds.). New York: Springer-Verlag, pp. 1–13.

———1998. Women on the move. *Nature Genet.* 20:219–220.

Stoneking, M., K. Bhatia, and A. C. Wilson. 1986. Mitochondrial DNA variation in eastern highlanders of Papua New Guinea. In *Genetic Variation and Its Maintenance*, D. F. Roberts and G. F. DeStefano (eds.). Cambridge: Cambridge University Press, pp. 87–100.

Stoneking, M., L. B. Jorde, K. Bhatia, and A. C. Wilson. 1990. Geographic variation in human mitochondrial DNA from Papua New Guinea. *Genetics* 124: 717–733.

Stoneking, M. and A. C. Wilson. 1989. Mitochondrial DNA. In *The Colonization of the Pacific: A Genetic Trail,* A. V. S. Hill and S. W. Serjeantson (eds.). New York: Oxford University Press, pp. 215–245.

Strange, R. M. and B. M. Burr. 1997. Intraspecific phylogeography of North American highland fishes: A test of the Pleistocene vicariance hypothesis. *Evolution* 51:885–897.

Streelman, J. T., R. Zardoya, A. Meyer, and S. A. Karl. 1998. Multilocus phylogeny of cichlid fishes (Pisces: Perciformes): evolutionary comparison of microsatellite and single-copy nuclear loci. *Mol. Biol. Evol.* 15:798–808.

Strenge, D. 1994. The intraspecific phylogeography of *Polystichum munitum* and *Alnus rubra.* Master's thesis, Washington State University, Pullman.

Stringer, C. B. and P. Andrews. 1988. Genetic and fossil evidence for the origin of modern humans. *Science* 239:1263–1268.

Sturmbauer, C., E. Verheyen, L. Rüber, and A. Meyer. 1997. Phylogeographic patterns in populations of cichlid fishes from rocky habitats in Lake Tanganyika. In *Molecular Systematics of Fishes,* T. D. Kocher and C. A. Stepien (eds.), San Diego: Academic Press, pp. 97–111.

Sullivan, J., J. A. Markert, and C. W. Kilpatrick. 1997. Phylogeography and molecular systematics of the *Peromyscus aztecus* species group (Rodentia: Muridae) inferred using parsimony and likelihood. *Syst. Biol.* 46:426–440.

Suzuki, H., S. Minato, S. Sakurai, K. Tsuchiya, and I. M. Fokin. 1997. Phylogenetic position and geographic differentiation of the Japanese dormouse, *Glirulus japonicus,* revealed by variations among rDNA, mtDNA and the SRY gene. *Zool. Sci.* 14:167–173.

Suzuki, H., S. Wakana, H. Yonekawa, K. Moriwaki, S. Sakurai, and E. Nevo. 1996. Variations in ribosomal DNA and mitochondrial DNA among chromosomal species of subterranean mole rats. *Mol. Biol. Evol.* 13:85–92.

Swift, C. C., C. R. Gilbert, S. A. Bortone, G. H. Burgess, and R. W. Yerger. 1985. Zoogeography of the southeastern United States: Savannah River to Lake Ponchartrain. In *Zoogeography of North American Freshwater Fishes,* C. H. Hocutt and E. O. Wiley (eds.). New York: Wiley, pp. 213–265.

Swofford, D. L. 1996. PAUP*: Phylogenetic Analysis Using Parsimony (and Other Methods), version 4.0. Sunderland, Mass.: Sinauer.

Swofford, D. L., G. J. Olsen, P. J. Waddell, and D. M. Hillis. 1996. Phylogenetic inference. In *Molecular Systematics* 2d. ed., D. M. Hillis, C. Moritz, and B. K. Mable (eds.). Sunderland, Mass.: Sinauer, pp. 407–514.

Taberlet, P. 1996. The use of mitochondrial DNA control region sequencing in conservation genetics. In *Molecular Genetic Approaches in Conservation*, T. B. Smith and R. K. Wayne (eds.). New York: Oxford University Press, pp. 125–142.

Taberlet, P. and J. Bouvet. 1994. Mitochondrial DNA polymorphism, phylogeography, and conservation genetics of the brown bear *Ursus arctos* in Europe. *Proc. Roy. Soc. Lond. B* 255:195–200.

Taberlet, P., L. Fumagalli, and J. Hausser. 1994. Chromosomal versus mitochondrial DNA evolution: Tracking the evolutionary history of the southwestern European populations of the *Sorex araneus* group (Mammalia, Insectivora). *Evolution* 48:623–636.

Taberlet, P., L. Fumagalli, A.-G. Wust-Saucy, and J.-F. Cosson. 1998. Comparative phylogeography and postglacial colonization routes in Europe. *Mol. Ecol.* 7:453–464.

Taberlet, P., A. Meyer, and J. Bouvet. 1992. Unusual mitochondrial DNA polymorphism in two local populations of blue tit *Parus caeruleus. Mol. Ecol.* 1:27–36.

Taberlet, P., J. E. Swenson, F. Sandegren, and A. Bjarvall. 1995. Localization of a contact zone between two highly divergent mitochondrial DNA lineages of the brown bear *Ursus arctos* in Scandinavia. *Conserv. Biol.* 9:1255–1261.

Taib, Z. 1997. Branching processes and evolution. In *Progress in Population Genetics and Human Evolution*, P. Donnelly and S. Tavaré (eds.). New York: Springer-Verlag, pp. 321–329.

Tajima, F. 1983. Evolutionary relationship of DNA sequences in finite populations. *Genetics* 105:437–460.

———1989. The effect of change in population size on DNA polymorphism. *Genetics* 123:597–601.

Takahata, N. 1988. The coalescent in two partially isolated diffusion populations. *Genet. Res.* 52:213–222.

———1989. Gene genealogy in three related populations: Consistency probability between gene and population trees. *Genetics* 122:957–966.

———1990. A simple genealogical structure of strongly balanced allelic lines and trans-species evolution of a polymorphism. *Proc. Natl. Acad. Sci. USA* 87:2419–2423.

———1991. Genealogy of neutral genes and spreading of selected mutations in a geographically structured population. *Genetics* 129:585–595.

———1993. Allelic genealogy and human evolution. *Mol. Biol. Evol.* 10:2–22.

———1995. A genetic perspective on the origin and history of humans. *Annu. Rev. Ecol. Syst.* 26:343–372.

Takahata, N. and T. Maruyama. 1981. A mathematical model of extranuclear genes and the genetic variability maintained in a finite population. *Genet. Res.* 37:291–302.

Takahata, N. and M. Nei. 1990. Allelic genealogy under overdominant and frequency-dependent selection and polymorphism of major histocompatibility complex loci. *Genetics* 124:967–978.

Takahata, N. and S. R. Palumbi. 1985. Extranuclear differentiation and gene flow in the finite island model. *Genetics* 109:441–457.

Takahata, N., Y. Satta, and J. Klein. 1995. Divergence time and population size in the lineage leading to modern humans. *Theor. Pop. Biol.* 48:198–221.

Takahata, N. and M. Slatkin. 1990. Genealogy of neutral genes in two partially isolated populations. *Theor. Pop. Biol.* 38:331–350.

Talbot, S. L. and G. F. Shields. 1996. Phylogeography of brown bears (*Ursos arctos*) of Alaska and paraphyly within the Ursidae. *Mol. Phylogen. Evol.* 5:477–494.

Tan, A.-M. and D. B. Wake. 1995. MtDNA phylogeography of the California newt, *Taricha torosa* (Caudata, Salamandridae). *Mol. Phylogen. Evol.* 4:383–394.

Tarr, C. L. and R. C. Fleischer. 1993. Mitochondrial DNA variation and evolutionary relationships in the amakihi complex. *Auk* 110:825–831.

Tashian, R. and G. Lasker (eds.). 1996. Molecular anthropology: Toward a new evolutionary paradigm. *Mol. Phylogen. Evol.* 5:1–285.

Tateno, Y., M. Nei, and F. Tajima. 1989. Accuracy of estimated phylogenetic trees from molecular data. I. Distantly related species. *J. Mol. Evol.* 18:387–404.

Tavaré, S. 1984. Line-of-descent and genealogical processes, and their applications in population genetic models. *Theor. Pop. Biol.* 26:119–164.

Taylor, D. J., P. D. N. Hebert, and J. K. Colbourne. 1996. Phylogenetics and evolution of the *Daphnia longispina* group (Crustacea) based on 12S rDNA sequence and allozyme variation. *Mol. Phylogen. Evol.* 5:495–510.

Taylor, E. B. and J. J. Dodson. 1994. A molecular analysis of relationships and biogeography within a species complex of Holarctic fish (genus *Osmerus*). *Mol. Ecol.* 3:235–248.

Taylor, J. A. (ed.). 1984. *Themes in Biogeography.* London: Croom Helm.

Tegelström, H. 1987. Transfer of mitochondrial DNA from the northern red-backed vole (*Clethrionomys rutilus*) to the bank vole (*C. glareolus*). *J. Mol. Evol.* 24: 218–227.

Templeton, A. R. 1992. Human origins and analysis of mitochondrial DNA sequences. *Science* 255:737.

——1993. The "Eve" hypothesis: a genetic critique and reanalysis. *Amer. Anthropol.* 95:51–72.

———1994. The role of molecular genetics in speciation studies. In *Molecular Approaches to Ecology and Evolution*, B. Schierwater, B. Streit, G. P. Wagner, and R. DeSalle (eds.). Basel: Birkhäuser Verlag, pp. 455–477.

———1996. Gene lineages and human evolution. *Science* 272:1363.

———1998. Nested clade analyses of phylogeographic data: Testing hypotheses about gene flow and population history. *Mol. Ecol.* 7:381–397.

Templeton, A. R., E. Boerwinkle, and C. F. Sing. 1987. A cladistic analysis of phenotypic associations with haplotypes inferred from restriction endonuclease mapping. I. Basic theory and an analysis of alcohol dehydrogenase activity in *Drosophila. Genetics* 117:343–351.

Templeton, A. R., K. A. Crandall, and C. F. Sing. 1992. A cladistic analysis of phenotypic associations with haplotypes inferred from restriction endonuclease mapping and DNA sequence data. III. Cladogram estimation. *Genetics* 132:619–633.

Templeton, A. R. and N. J. Georgiadis. 1996. A landscape approach to conservation genetics: Conserving evolutionary processes in the African Bovidae. In *Conservation Genetics: Case Histories from Nature*, J. C. Avise and J. L. Hamrick (eds.). New York: Chapman & Hall, pp. 398–430.

Templeton, A. R., E. Routman, and C. A. Phillips. 1995. Separating population structure from population history: A cladistic analysis of the geographical distribution of mitochondrial DNA haplotypes in the tiger salamander, *Ambystoma tigrinum. Genetics* 140:767–782.

Templeton, A. R. and C. F. Sing. 1993. A cladistic analysis of phenotypic associations with haplotypes inferred from restriction endonuclease mapping. IV. Nested analyses with cladogram uncertainty and recombination. *Genetics* 134:659–669.

Theimer, T. C. and P. Keim. 1994. Geographic patterns of mitochondrial-DNA variation in collared peccaries. *J. Mammal.* 75:121–128.

Thomas, W. K., S. Pääbo, F. X. Villablanca, and A. C. Wilson. 1990. Spatial and temporal continuity of kangaroo rat populations shown by sequencing mitochondrial DNA from museum specimens. *J. Mol. Evol.* 31:101–112.

Thomas, W. K., R. E. Withler, and A. T. Beckenbach. 1986. Mitochondrial DNA analysis of Pacific salmonid evolution. *Can. J. Zool.* 64:1059–1064.

Thomaz, D., A. Guillar, and B. Clarke. 1996. Extreme divergence of mitochondrial DNA within species of pulmonate land snails. *Proc. Roy. Soc. Lond. B* 263: 363–368.

Thompson, C. E., E. B. Taylor, and J. D. McPhail. 1997. Parallel evolution of lake-stream pairs of threespine sticklebacks (*Gasterosteus*) inferred from mitochondrial DNA variation. *Evolution* 51:1955–1965.

Thorpe, R. S. 1996. The use of DNA divergence to help determine the correlates of evolution of morphological characters. *Evolution* 50:524–531.

Thorpe, R. S., H. Black, and A. Malhotra. 1996. Matrix correspondence tests on the DNA phylogeny of the Tenerife lacertid elucidate both historical causes and morphological adaptation. *Syst. Biol.* 45:335–343.

Thorpe, R. S., A. Malhotra, H. Black, J. C. Daltry, and W. Wüster. 1995. Relating geographic pattern to phylogenetic process. *Phil. Trans. Roy. Soc. Lond. B* 349:61–68.

Thorpe, R. S., D. P. McGregor, and A. M. Cumming. 1993. Population evolution of western Canary Island lizards (*Gallotia galloti*): 4-base endonuclease restriction fragment length polymorphisms of mitochondrial DNA. *Biol. J. Linn. Soc.* 49:219–227.

Thorpe, R. S., D. P. McGregor, A. M. Cumming, and W. C. Jordan. 1994. DNA evolution and colonization sequence of island lizards in relation to geological history: mtDNA RFLP, cytochrome *b*, cytochrome oxidase, 12S rRNA sequence, and nuclear RAPD analysis. *Evolution* 48:230–240.

Thrailkill, K. M., C. W. Birky, G. Luckermann, and K. Wolf. 1980. Intracellular population genetics: Evidence for random drift of mitochondrial allele frequencies in *Saccharomyces cerevisiae* and *Saccharomyces pombe*. *Genetics* 96: 237–262.

Tibayrenc, M., F. Kjellberg, and F. J. Ayala. 1990. A clonal theory of parasitic protozoa: The population structures of *Entamoeba, Giardia, Leishmania, Naegleria, Plasmodium, Trichomonas,* and *Trypanosoma* and their medical and taxonomical consequences. *Proc. Natl. Acad. Sci. USA* 87:2414–2418.

———1991. The clonal theory of parasitic protozoa. *BioScience* 41:767–774.

Tibbets, C. A. and T. E. Dowling. 1996. Effects of intrinsic and extrinsic factors on population fragmentation in three species of North American minnows (Teleostei: Cyprinidae). *Evolution* 50:1280–1292.

Tishkoff, S. A., and 14 others. 1996. Global patterns of linkage disequilibrium at the CD4 locus and modern human origins. *Science* 271:1380–1387.

Tivy, J. 1993. *Biogeography: A Study of Plants in the Ecosphere,* 3d. ed. Essex, England: Longman.

Todaro, M. A., J. W. Fleeger, Y. P. Hu, A. W. Hrincevich, and D. W. Foltz. 1996. Are meiofaunal species cosmopolitan? Morphological and molecular analysis of *Xenotrichula intermedia* (Gastrotricha: Chaetonotida). *Marine Biol.* 125:735–742.

Toline, C. A. and A. J. Baker. 1995. Mitochondrial DNA variation and population genetic structure of the northern redbelly dace (*Phoxinus eos*). *Mol. Ecol.* 4:745–753.

Torroni, A., Y.-S. Chen, O. Semino, A. S. Santachiara-Beneceretti, C. R. Scott, M. T. Lott, M. Winter, and D. C. Wallace. 1994a. mtDNA and Y-chromosome polymorphisms in four native American populations from southern Mexico. *Amer. J. Hum. Genet.* 54:303–318.

Torroni, A., J. A. Miller, L. G. Moore, S. Zamudio, J. Zhuang, T. Droma, and D. C. Wallace. 1994b. Mitochondrial DNA analysis in Tibet: Implications for the origin of the Tibetan population and its adaptation to high altitude. *Amer. J. Phys. Anthropol.* 93:189–199.

Torroni, A., T. G. Schurr, M. F. Cabell, M. D. Brown, J. V. Neel, M. Larsen, D. G. Smith, C. M. Vullo, and D. C. Wallace. 1993a. Asian affinities and continental radiation of the four founding native American mtDNAs. *Amer. J. Hum. Genet.* 53:563–590.

Torroni, A., T. G. Schurr, C.-C. Yang, E. J. E. Szathmary, R. C. Williams, M. S. Schanfield, G. A. Troup, W. C. Knowler, D. N. Lawrence, K. M. Weiss, and D. C. Wallace. 1992. Native American mitochondrial DNA analysis indicates that the Amerind and the Nadene populations were founded by two independent migrations. *Genetics* 130:153–162.

Torroni, A., R. I. Sukernik, T. G. Schurr, Y. B. Starikovshaya, M. F. Cabell, M. H. Crawford, A. G. Comuzzie, and D. C. Wallace. 1993b. mtDNA variation of aboriginal Siberians reveals distinct genetic affinities with Native Americans. *Amer. J. Hum. Genet.* 53:591–608.

Tuomisto, H., K. Ruokolainen, R. Kalliola, A. Linna, W. Danjoy, and Z. Rodriguea. 1995. Dissecting Amazonian biodiversity. *Science* 269:63–66.

Turelli, M. and A.A. Hoffmann. 1991. Rapid spread of an inherited incompatibility factor in California *Drosophila. Nature* 353:440–442.

———1995. Cytoplasmic incompatibility in *Drosophila simulans:* Dynamics and parameter estimates from natural populations. *Genetics* 140:1319–1338.

Turnbull, D. M. and R. N. Lightowlers. 1998. An essential guide to mtDNA maintenance. *Nature Genet.* 18:199–200.

Turner, T. F., J. C. Trexler, D. N. Kuhn, and H. W. Robison. 1996. Life-history variation and comparative phylogeography of darters (Pisces: Percidae) from the North American central highlands. *Evolution* 50:2023–2036.

Upholt, W. B. 1977. Estimation of DNA sequence divergence from comparison of restriction endonuclease digests. *Nucleic Acids Res.* 4:1257–1265.

Upholt, W. B. and I. B. Dawid. 1977. Mapping of mitochondrial DNA of individual sheep and goats: Rapid evolution in the D loop region. *Cell* 11: 571–583.

Vandijk, P. and T. Bakzschotman. 1997. Chloroplast DNA phylogeography and cytotype geography in autopolyploid *Plantago media*. *Mol. Ecol.* 6:345–352.

Vane-Wright, R. I., C. J. Humphries, and P. H. Williams. 1991. What to protect—systematics and the agony of choice. *Biol. Cons.* 55:235–254.

Vanlerberghe, F., P. Boursot, J. T. Nielsen, and F. Bonhomme. 1988. A steep cline for mitochondrial DNA in Danish mice. *Genet. Res.* 52:185–193.

van Oppen, M. J. H., O. E. Diekmann, C. Wiencke, W. T. Stam, and J. L. Olsen. 1994. Tracking dispersal routes: Phylogeography of the Arctic-Antarctic disjunct seaweed *Acrosiphonia arcta* (Chlorophyta). *J. Phycol.* 30:67–80.

van Oppen, M. J. H., S. G. A. Draisma, J. L. Olsen, and W. T. Stam. 1995. Multiple trans-Arctic passages in the red alga *Phycodrys rubens:* Evidence from nuclear rDNA ITS sequences. *Marine Biol.* 123:179–188.

Van Syoc, R. J. 1994. Genetic divergence between subpopulations of the eastern Pacific goose barnacle *Pollicipes elegans:* Mitochondrial cytochrome *c* subunit 1 nucleotide sequences. *Mol. Mar. Biol. Biotech.* 3:338–346.

Van Vuuren, B. J. and T. J. Robinson. 1997. Genetic population structure in the yellow mongoose, *Cynictis penicillata*. *Mol. Ecol.* 6:1147–1153.

Van Wagner, C. E. and A. J. Baker. 1990. Association between mitochondrial DNA and morphological evolution in Canada geese. *J. Mol. Evol.* 31:373–382.

Varvio, S.-L., R. K. Koehn, and R. Väinölä. 1988. Evolutionary genetics of the *Mytilus edulis* complex in the North Atlantic region. *Marine Biol.* 98:51–60.

Vasquez, P., S. J. B. Cooper, J. Gosalvez, and G. M. Hewitt. 1994. Nuclear DNA introgression across a Pyrenean hybrid zone between parapatric sub-species of the grasshopper *Chorthippus parallelus*. *Heredity* 73:436–443.

Vawter, A. T., R. Rosenblatt, and G. C. Gorman. 1980. Genetic divergence among fishes of the eastern Pacific and the Caribbean: Support for the molecular clock. *Evolution* 34:705–711.

Verheyen, E., L. Ruber, J. Snoeks, and A. Meyer. 1996. Mitochondrial phylogeography of rock-dwelling cichlid fishes reveals evolutionary influence of historical lake level fluctuations of Lake Tanganyika, Africa. *Proc. Roy. Soc. Lond. B* 351:797–805.

Vermeij, G. J. 1978. *Biogeography and Adaptation*. Cambridge, Mass.: Harvard University Press.

———1991a. Anatomy of an invasion: The trans-Arctic interchange. *Paleobiology* 17:281–307.

———1991b. When biotas meet: Understanding biotic interchange. *Science* 253: 1099–1104.

Vigilant, L., R. Pennington, H. Harpending, T. D. Kocher, and A. C. Wilson. 1989. Mitochondrial DNA sequences in single hairs from a southern African population. *Proc. Natl. Acad. Sci. USA* 86:9350–9354.

Vigilant, L., M. Stoneking, H. Harpending, K. Hawkes, and A. C. Wilson. 1991. African populations and the evolution of human mitochondrial DNA. *Science* 253:1503–1507.

Vila, C., P. Savolainen, J. E. Maldonado, I. R. Amorim, J. E. Rice, R. L. Honeycutt, K. A. Crandall, J. Lundeberg, and R. K. Wayne. 1997. Multiple and ancient origins of the domestic dog. *Science* 276:1687–1689.

Villablanca, F. X., G. K. Roderick, and S. R. Palumbi. 1998. Invasion genetics of the Mediterranean fruit fly: Variation in multiple nuclear introns. *Mol. Ecol.* 7:547–560.

Vogler, A. P. and R. DeSalle. 1993. Phylogeographic patterns in coastal North American tiger beetles (*Cicindela dorsalis* Say) inferred from mitochondrial DNA sequences. *Evolution* 47:1192–1202.

———1994a. Evolution and phylogenetic information content of the ITS-1 region in the tiger beetle *Cicindela dorsalis*. *Mol. Biol. Evol.* 11:393–405.

———1994b. Diagnosing units of conservation management. *Conserv. Biol.* 8: 354–363.

Vogler, A. P., C. B. Knisley, S. B. Glueck, J. M. Hill, and R. DeSalle. 1993. Using molecular and ecological data to diagnose endangered populations of the puritan tiger beetle *Cicindela puritana*. *Mol. Ecol.* 2:375–383.

Wade, M. J. and D. E. McCauley. 1984. Group selection: The interaction of local deme size and migration in the differentiation of small populations. *Evolution* 38:1047–1058.

———1988. Extinction and recolonization: Their effects on the genetic differentiation of local populations. *Evolution* 42:995–1005.

Wagner, W. L. and V. A. Funk (eds.). 1995. *Hawaiian Biogeography: Evolution on a Hot Spot Archipelago*. Washington, D.C.: Smithsonian Institution Press.

Waits, L. P., S. L. Talbot, R. H. Ward, and G. F. Shields. 1998. Mitochondrial DNA phylogeography of the North American brown bear and implications for conservation. *Conserv. Biol.* 12:408–417.

Wake, D. B. 1997. Incipient species formation in salamanders of the *Ensatina* complex. *Proc. Natl. Acad. Sci. USA* 94:7761–7767.

Walker, D. and J. C. Avise. 1998. Principles of phylogeography as illustrated by freshwater and terrestrial turtles in the southeastern United States. *Annu. Rev. Ecol. Syst.* 29:23–58.

Walker, D., V. J. Burke, I. Barák, and J. C. Avise. 1995. A comparison of mtDNA restriction sites vs. control region sequences in phylogeographic assessment of the musk turtle (*Sternotherus minor*). *Mol. Ecol.* 4:365–373.

Walker, D., P. E. Moler, K. A. Buhlmann, and J. C. Avise. 1998a. Phylogeographic patterns in *Kinosternon subrubrum* and *K. baurii* based on mitochondrial DNA restriction analyses. *Herpetologica* 54:174–184.

———1998b. Phylogeographic uniformity in mitochondrial DNA of the snapping turtle (*Chelydra serpentina*). *Anim. Conserv.* 1:55–60.

Walker, D., W. S. Nelson, K. A. Buhlmann, and J. C. Avise. 1997. Mitochondrial DNA phylogeography and subspecies issues in the monotypic freshwater turtle *Sternotherus odoratus*. *Copeia* 1997:16–21.

Walker, D., G. Ortí, and J. C. Avise. 1998c. Phylogenetic distinctiveness of a threatened aquatic turtle (*Sternotherus depressus*). *Conserv. Biol.* 12:639–645.

Wallace, A. R. 1849. On the monkeys of the Amazon. *Proc. Roy. Soc. Lond.* 20:107–110.

———1865. On the phenomenon of variation and geographic distribution as illustrated by the Papilionidae of the Mayalan region. *Trans. Linn. Soc. Lond.* 25:1–71.

Wallace, D. C. 1986. Mitochondrial genes and disease. *Hospital Practice* 21:77–92.

———1992. Mitochondrial genetics: A paradigm for aging and degenerative diseases? *Science* 256:628–632.

Wallis, G. P. and J. W. Arntzen. 1989. Mitochondrial-DNA variation in the crested newt superspecies: Limited cytoplasmic gene flow among species. *Evolution* 43:88–104.

Walpole, D. K., S. K. Davis, and I. F. Greenbaum. 1997. Variation of mitochondrial DNA in populations of *Peromyscus eremicus* from the Chihuahuan and Sonoran deserts. *J. Mammal.* 78:397–404.

Wang, J.Y.-C. 1993. Mitochondrial DNA analysis of the harbour porpoise. Master's thesis, University of Quelph, Guelph, Ontario.

Wang, R. L., J. Wakeley, and J. Hey. 1997. Gene flow and natural selection in the origin of *Drosophila pseudoobscura* and close relatives. *Genetics* 147:1091–1106.

Waples, R. S. 1991. Pacific salmon, *Oncorhynchus* spp., and the definition of a "species" under the Endangered Species Act. *Mar. Fish. Rev.* 53:11–22.

———1998. Separating the wheat from the chaff: Patterns of genetic differentiation in high gene flow species. *J. Heredity* 89:438–450.

Ward, R. D., N. G. Elliott, P. M. Grewe, and A. J. Smolenski. 1994b. Allozyme and mitochondrial DNA variation in yellowfin tuna (*Thunnus albacares*) from the Pacific Ocean. *Marine Biol.* 118:531–539.

Ward, R. D., D. O. F. Skibinski, and M. Woodwark. 1992. Protein heterozygosity, protein structure, and taxonomic differentiation. *Evol. Biol.* 26:73–159.

Ward, R. D., M. Woodward, and D. O. F. Skibinski. 1994a. A comparison of genetic diversity levels in marine, freshwater, and anadromous fishes. *J. Fish Biol.* 44:213–227.

Ward, R. H. 1997. Phylogeography of human mtDNA: An Amerindian perspective. In *Progress in Population Genetics and Human Evolution*, P. Donnelly and S. Tavaré (eds.). New York: Springer-Verlag, pp. 33–53.

Ward, R. H., B. L. Frazier, K. Dew-Jaeger, and S. Pääbo. 1991. Extensive mitochondrial diversity within a single Amerindian tribe. *Proc. Natl. Acad. Sci. USA* 88:8720–8724.

Ward, R. H., A. Redd, D. Valencia, B. Frazier, and S. Pääbo. 1993. Genetic and linguistic differentiation in the Americas. *Proc. Natl. Acad. Sci. USA* 90:10663–10667.

Ward, R. H. and C. Stringer. 1997. A molecular handle on the Neanderthals. *Nature* 388:225–226.

Waters, J. M. and J. A. Cambray. 1997. Intraspecific phylogeography of the Cape galaxias from South Africa: Evidence from mitochondrial DNA sequences. *J. Fish Biol.* 50:1329–1338.

Watterson, G. A. 1975. On the number of segregating sites in genetical models without recombination. *Theor. Pop. Biol.* 7:256–276.

————1984. Lines of descent and the coalescent. *Theor. Pop. Biol.* 26:77–92.

Wayne, R. K. 1996. Conservation genetics in the Canidae. In *Conservation Genetics: Case Histories from Nature*, J. C. Avise and J. L. Hamrick (eds.). New York: Chapman & Hall, pp. 75–118.

Wayne, R. K. and S. M. Jenks. 1991. Mitochondrial DNA analysis supports extensive hybridization of the endangered red wolf (*Canis rufus*). *Nature* 351: 565–568.

Wayne, R. K., N. Lehman, M. W. Allard, and R. L. Honeycutt. 1992. Mitochondrial DNA variability in the gray wolf—genetic consequences of population decline and habitat fragmentation. *Conserv. Biol.* 6:559–569.

Wayne, R. K., A. Meyer, N. Lehman, B. Van Valkenburgh, P. W. Kat, T. K. Fuller, D. Girman, and S. J. O'Brien. 1990. Large sequence divergence among mitochondrial DNA genotypes within populations of eastern African black-backed jackals. *Proc. Natl. Acad. Sci. USA* 87:1772–1776.

Webb, S.D. 1990. Historical biogeography. In *Ecosystems of Florida*, R. I. Myers and J. J. Ewel (eds.). Orlando: University of Central Florida Press, pp. 70–100.

Webb, T., III and P. J. Bartlein. 1992. Global changes during the last 3 million years: Climatic controls and biotic responses. *Annu. Rev. Ecol. Syst.* 23:141–173.

Weider, L. J. and A. Hobaek. 1997. Postglacial dispersal, glacial refugia, and clonal structure in Russian/Siberian populations of the Arctic *Daphnia pulex* complex. *Heredity* 78:363–372.

Weider, L. J., A. Hobaek, T. J. Crease, and H. Stibor. 1996. Molecular characterization of clonal population structure and biogeography of arctic apomictic *Daphnia* from Greenland and Iceland. *Mol. Ecol.* 5:107–118.

Weiller, G. 1998. Phylogenetic profiles: A graphical method for detecting genetic recombinations in homologous sequences. *Mol. Biol. Evol.* 15:326–335.

Weir, B. S. 1996. *Genetic Data Analysis II.* Sunderland, Mass.: Sinauer.

Weir, B. S. and C. C. Cockerham. 1984. Estimating *F*-statistics for the analysis of population structure. *Evolution* 38:1358–1370.

Weiss, G. and A. von Haeseler. 1996. Estimating the age of the common ancestor of men from the ZFY intron. *Science* 272:1359–1360.

Wenink, P. W. and A. J. Baker. 1996. Mitochondrial DNA lineages in composite flocks of migratory and wintering dunlins (*Calidris alpina*). *Auk* 113:744–756.

Wenink, P. W., A. J. Baker, H-U. Rösner, and M. G. J. Tilanus. 1996. Global mitochondrial DNA phylogeography of holarctic breeding dunlins (*Calidris alpina*). *Evolution* 50:318–330.

Wenink, P. W., A. J. Baker, and M. G. J. Tilanus. 1993. Hypervariable-control-region sequences reveal global population structuring in a long-distance migrant shorebird, the dunlin (*Calidris alpina*). *Proc. Natl. Acad. Sci. USA* 90:94–98.

Wheeler, Q. D. and K. C. Nixon. 1990. Another way of looking at the species problem: A reply to de Queiroz and Donoghue. *Cladistics* 6:77–81.

Whitfield, L. S., J. E. Suiston, and P. N. Goodfellow. 1995. Sequence variation of the human Y chromosome. *Nature* 378:379–380.

Whitlock, M. C. and N. H. Barton. 1997. The effective size of a subdivided population. *Genetics* 146:427–441.

Whitlock, M. C. and D. E. McCauley. 1990. Some population genetic consequences of colony formation and extinction: genetic correlations within founding groups. *Evolution* 44:1717–1724.

Whitmore, T. C. and G. T. Prance (eds.). 1987. *Biogeography and Quaternary History of Tropical America.* Oxford: Clarendon Press.

Whittam, T. S., A. G. Clark, M. Stoneking, R. L. Cann, and A. C. Wilson. 1986. Allelic variation in human mitochondrial genes based on patterns of restriction site polymorphism. *Proc. Natl. Acad. Sci. USA* 83:9611–9615.

Wilcox, T. P., L. Hugg, J. A. Zeh, and D. W. Zeh. 1997. Mitochondrial DNA sequencing reveals extreme genetic differentiation in a cryptic species complex of neotropical pseudoscorpions. *Mol. Phylogen. Evol.* 7:208–216.

Wiley, E. O. 1988. Parsimony analysis and vicariance biogeography. *Syst. Zool.* 37:271–290.

Wiley, E. O. and R. H. Hagen. 1997. Mitochondrial DNA sequence variation among the sand darters (Percidae: Teleostei). In *Molecular Systematics of Fishes*, T. D. Kocher and C. A. Stepien (eds.). San Diego: Academic Press, pp. 75–96.

Williams, G. C. and R. K. Koehn. 1984. Population genetics of North Atlantic catadromous eels (*Anguilla*). In *Evolutionary Genetics of Fishes*, B. J. Turner (ed.). New York: Plenum Press, pp. 529–560.

Williams, G. C., R. K. Koehn, and J. B. Mitton. 1973. Genetic differentiation without isolation in the American eel, *Anguilla rostrata. Evolution* 27:192–204.

Williams, S. T. and J. A. H. Benzie. 1998. Evidence of a biogeographic break between populations of a high dispersal starfish: Congruent regions within the Indo-West Pacific defined by color morphs, mtDNA, and allozyme data. *Evolution* 52:87–99.

Wills, C. 1995. When did Eve live? An evolutionary detective story. *Evolution* 49:593–607.

Wilmer, J. W., C. Moritz, L. Hall, and J. Toop. 1994. Extreme population structuring in the threatened ghost bat, *Macroderma gigas:* Evidence from mitochondrial DNA. *Proc. Roy. Soc. Lond. B* 257:193–198.

Wilson, A. C., R. L. Cann, S. M. Carr, M. George, Jr., U. B. Gyllensten, K. M. Helm-Bychowski, R. G. Higuchi, S. R. Palumbi, E. M. Prager, R. D. Sage, and M. Stoneking. 1985. Mitochondrial DNA and two perspectives on evolutionary genetics. *Biol. J. Linn. Soc.* 26:375–400.

Wilson, C. C. and L. Bernatchez. 1998. The ghost of hybrids past: Fixation of arctic charr (*Salvelinus alpinus*) mitochondrial DNA in an introgressed population of lake trout (*S. namaycush*). *Mol. Ecol.* 7:127–132.

Wilson, C. C. and P. D. N. Hebert. 1996. Phylogeographic origins of lake trout (*Salvelinus namaycush*) in eastern North America. *Can. J. Fish. Aquat. Sci.* 53:2764–2775.

Wilson, C. C., P. D. N. Hebert, J. D. Reist, and J. B. Dempson. 1996. Phylogeography and postglacial dispersal of arctic charr *Salvelinus alpinus* in North America. *Mol. Ecol.* 5:187–197.

Wilson, E. O. (ed.). 1988. *Biodiversity.* Washington, D.C.: National Academy Press.

Wilson, E. O. and W. L. Brown, Jr. 1953. The subspecies concept and its taxonomic application. *Syst. Zool.* 2:97–111.

Wilson, G. M., W. K. Thomas, and A. T. Beckenbach. 1985. Intra- and inter-specific mitochondrial DNA sequence divergence in *Salmo:* Rainbow, steelhead, and cutthroat trouts. *Can. J. Zool.* 63:2088–2094.

Wittmann, U., P. Heidrich, M. Wink, and E. Gwinner. 1995. Speciation in the stonechat (*Saxicola torquata*) inferred from nucleotide sequences of the cytochrome B gene. *J. Zool. Syst. Evol. Res.* 33:116–122.

Woese, C.R. and G.E. Fox. 1977. Phylogenetic structure of the prokaryotic domain: The primary kingdoms. *Proc. Natl. Acad. Sci. USA* 74:5088–5090.

Woese, C. R., O. Kandler, and M. L. Wheelis. 1990. Towards a natural system of organisms: Proposal for the domains of Archaea, Bacteria, and Eukarya. *Proc. Natl. Acad. Sci. USA* 87:4576–4579.

Wolfe, K. H., W.-H. Li, and P. M. Sharp. 1987. Rates of nucleotide substitution vary greatly among plant mitochondrial, chloroplast, and nuclear DNAs. *Proc. Natl. Acad. Sci. USA* 84:9054–9058.

Wollenberg, K. and J. C. Avise. 1998. Sampling properties of genealogical pathways underlying population pedigrees. *Evolution* 52:957–966.

Wolpoff, M. H. 1989. Multiregional evolution: The fossil alternative to Eden. In *The Human Revolution: Behavioural and Biological Perspectives on the Origins of Modern Humans,* P. Mellars and C. Stringer (eds.). Edinburgh: Edinburgh University Press, pp. 62–108.

———1992. Theories of modern human origins. In *Continuity or Replacement: Controversies in* Homo sapiens *Evolution,* G. Bräuer and F. H. Smith (eds.). Rotterdam: Balkema, pp. 25–63.

Wooding, S. and R. Ward. 1997. Phylogeography and Pleistocene evolution in the North American black bear. *Mol. Biol. Evol.* 14:1096–1105.

Wooten, M. C. and C. Lydeard. 1990. Allozyme variation in a natural contact zone between *Gambusia affinis* and *Gambusia holbrooki. Biochem. Syst. Ecol.* 18: 169–173.

Wright, H. E. (ed.). 1965. *The Quaternary of the United States.* Princeton, N.J.: Princeton University Press.

Wright, J. W., C. Spolsky, and W. M. Brown. 1983. The origin of the parthenogenetic lizard *Cnemidophorus laredoensis* inferred from mitochondrial DNA analysis. *Herpetologica* 39:410–416.

Wright, S. 1931. Evolution in Mendelian populations. *Genetics* 16:97–159.

———1943. Isolation by distance. *Genetics* 28:114–138.

———1946. Isolation by distance under diverse systems of mating. *Genetics* 31:39–59.

————1951. The genetical structure of populations. *Ann. Eugen.* 15:323–354.

Wrischnik, L. A., R. G. Higuchi, M. Stoneking, H. A. Erlich, N. Arnheim, and A. C. Wilson. 1987. Length mutations in human mitochondrial DNA; direct sequencing of enzymatically amplified DNA. *Nucleic Acids Res.* 15:529–542.

Wu, C.-I. 1991. Inferences of species phylogeny in relation to segregation of ancient polymorphisms. *Genetics* 127:429–435.

Wu, C.-I. and W.-H. Li. 1985. Evidence for higher rates of nucleotide substitution in rodents than in man. *Proc. Natl. Acad. Sci. USA* 82:1741–1745.

Xiong, W., W.-H. Li, I. Posner, T. Yamamura, A. Yamamoto, A. M. Gotto, Jr., and L. Chan. 1991. No severe bottleneck during human evolution: Evidence from two apolipoprotein C-II deficiency alleles. *Amer. J. Hum. Genet.* 48:383–389.

Xu, J., R. W. Kerrigan, A. S. Sonnenberg, P. Callac, P. A. Horgen, and J. B. Anderson. 1998. Mitochondrial DNA variation in natural populations of the mushroom *Agaricus bisporus. Mol. Ecol.* 7:19–33.

Xu, X. and U. Arnason. 1996. The mitochondrial DNA molecule of Sumatran orangutan and a molecular proposal for two (Bornean and Sumatran) species of orangutan. *J. Mol. Evol.* 43:431–437.

Yamagata, T., K. Ohishi, M. O. Faruque, J. S. Masangkay, C. Baloc, D. Vubinh, S. S. Mansjoer, H. Ikeda, and T. Namikawa. 1995. Genetic variation and geographic distribution on the mitochondrial DNA in local populations of the musk shrew, *Suncus murinus. Japan. J. Genet.* 70:321–337.

Yang, Y.-J., R.-S. Lin, J.-L. Wu, and C.-F. Hui. 1994. Variation in mitochondrial DNA and population structure of the Taipei treefrog *Rhacophorus taipeianus* in Taiwan. *Mol. Ecol.* 3:219–228.

Zamudio, K. R. and H. W. Greene. 1997. Phylogeography of the bushmaster (*Lachesis muta:* Viperidae): Implications for neotropical biogeography, systematics, and conservation. *Biol. J. Linn. Soc.* 62:421–442.

Zamudio, K. R., K. B. Jones, and R. H. Ward. 1997. Molecular systematics of short-horned lizards: Biogeography and taxonomy of a widespread species complex. *Syst. Biol.* 46:284–305.

Zardoya, R., D. M. Vollmer, C. Craddock, J. T. Streelman, S. A. Karl, and A. Meyer. 1996. Evolutionary conservation of microsatellite flanking regions and their use in resolving the phylogeny of cichlid fishes (Pisces: Perciformes). *Proc. Roy. Soc. Lond. B* 263:1589–1598.

Zaslavskaya, N. I., S. O. Sergievsky, and A. N. Tatarenkov. 1992. Allozyme similarity of Atlantic and Pacific species of *Littorina* (Gastropoda: Littorinidae). *J. Mollusc. Stud.* 58:377–384.

Zehnder, G. W., L. Sandall, A. M. Tisler, and T. O. Powers. 1992. Mitochondrial DNA diversity among 17 geographic populations of *Leptinotarsa decemlineata* (Coleoptera: Chrysomelidae). *Ann. Ent. Soc. Amer.* 85:234–240.

Zhang, D.-X. and G. M. Hewitt. 1996. An effective method for allele-specific sequencing using restriction enzyme and biotinylation (ASSURE B). *Mol. Ecol.* 5:591–594.

———1997. Insect mitochondrial control region: A review of its structure, evolution, and usefulness in evolutionary studies. *Biochem. Syst. Ecol.* 25:99–120.

Zhang, D.-X., J. M. Szymura, and G. M. Hewitt. 1995. Evolution and structural conservation of the control region of insect mitochondrial DNA. *J. Mol. Evol.* 40:382–391.

Zhang, Y. and O. A. Ryder. 1995. Different rates of mitochondrial DNA sequence evolution in Kirk's dik-dik (*Madoqua kirkii*) populations. *Mol. Phylogen. Evol.* 4:291–297.

Zhi, L., W. B. Karesh, D. N. Janczewski, H. Frazier-Taylor, D. Sajuthi, F. Gombek, M. Andau, J. S. Martenson, and S. J. O'Brien. 1996. Genomic differentiation among natural populations of orang-utan (*Pongo pygmaeus*). *Curr. Biol.* 6:1326–1336.

Zhu, D., S. Degnan, and C. Moritz. 1998. Evolutionary distinctiveness and status of the endangered Lake Eacham rainbowfish (*Melanotaenia eachamensis*). *Conserv. Biol.* 12:80–93.

Zhu, T., B. T. Korber, A. J. Nahmias, E. Hooper, P. M. Sharp, and D. D. Ho. 1998. An African HIV-1 sequence from 1959 and implications for the origin of the epidemic. *Nature* 391:594–597.

Zink, R. M. 1994. The geography of mitochondrial DNA variation, population structure, hybridization, and species limits in the fox sparrow (*Passerella iliaca*). *Evolution* 48:96–111.

———1996. Comparative phylogeography in North American birds. *Evolution* 50:308–317.

———1997. Phylogeographic studies of North American birds. In *Avian Molecular Evolution and Systematics*, D. P. Mindell (ed.). New York: Academic Press, pp. 301–324.

Zink, R. M. and J. C. Avise. 1990. Patterns of mitochondrial DNA and allozyme evolution in the avian genus *Ammodramus. Syst. Zool.* 39:148–161.

Zink, R.M. and R.C. Blackwell. 1998. Molecular systematics and biogeography of aridland gnatcatchers (genus *Polioptila*) and evidence supporting species status of the California gnatcatcher (*Polioptila californica*). *Mol. Phylogen. Evol.* 9:26–32.

Zink, R. M., R. C. Blackwell, and O. Rojassoto. 1997. Species limits in the Le Conte's thrasher. *Condor* 99:132–138.

Zink, R. M. and D. L. Dittmann. 1993a. Gene flow, refugia, and evolution of geographic variation in the song sparrow (*Melospiza melodia*). *Evolution* 47:717–729.

———1993b. Population structure and gene flow in the chipping sparrow and a hypothesis for evolution in the genus *Spizella*. *Wilson Bull.* 105:399–413.

Zink, R. M., J. M. Fitzsimons, D. L. Dittmann, D. R. Reynolds, and R. T. Nishimoto. 1996. Evolutionary genetics of Hawaiian freshwater fish. *Copeia* 1996:330–335.

Zink, R. M. and M. C. McKitrick. 1995. The debate over species concepts and its implications for ornithology. *Auk* 112:701–719.

Zink, R. M. and J. V. Remsen, Jr. 1986. Evolutionary processes and patterns of geographic variation in birds. In *Current Ornithology*, vol. 4, R. F. Johnston (ed.). New York: Plenum Press, pp. 1–69.

Zink, R. M., S. Rohwer, A. V. Andreev, and D. L. Dittmann. 1995. Trans-Beringia comparisons of mitochondrial DNA differentiation in birds. *Condor* 97: 639–649.

Zink, R. M., W. L. Rootes, and D. L. Dittmann. 1991. Mitochondrial DNA variation, population structure, and evolution of the common grackle (*Quiscalus quiscula*). *Condor* 93:318–329.

Zischler, M., H. Geisert, A. von Haeseler, and S. Pääbo. 1995. A nuclear 'fossil' of the mitochondrial D-loop and the origin of modern humans. *Nature* 378: 489–492.

Zouros, E., A. O. Ball, C. Saavedra, and K. R. Freeman. 1994a. Mitochondrial DNA inheritance. *Nature* 368:818.

———1994b. An unusual type of mitochondrial DNA inheritance in the blue mussel *Mytilus*. *Proc. Natl. Acad. Sci. USA* 91:7463–7467.

Zouros, E., R. K. Freeman, A. O. Ball, and G. H. Pogson. 1992. Direct evidence for extensive paternal mitochondrial DNA inheritance in the marine mussel *Mytilus. Nature* 359:412–414.

Zouros, E. and D. M. Rand. 1999. Population genetics and evolution of animal mitochondrial DNA. In *Evolutionary Genetics from Molecules to Morphology*, R. Singh and C. Krimbas (eds). Cambridge, Cambridge University Press, in press.

Zwanenburg, K. C. T., P. Bentzen, and J. M. Wright. 1992. Mitochondrial DNA differentiation in western North Atlantic populations of haddock (*Melanogrammus aeglefinus*). *Can. J. Fish. Aquat. Sci.* 49:2527–2537.

INDEX